Pressure Vessels

Pressure Vessels

THE ASME CODE SIMPLIFIED

Robert Chuse

Stephen M. Eber, P.E.

SIXTH EDITION

McGRAW-HILL BOOK COMPANY

New York St. Louis San Francisco Auckland Bogotá
Hamburg Johannesburg London Madrid Mexico
Montreal New Delhi Panama Paris São Paulo
Singapore Sydney Tokyo Toronto

Library of Congress Cataloging in Publication Data

Chuse, Robert.
  Pressure vessels.

  Bibliography: p.
  Includes index.
  1. Pressure vessels — Standards — United States.
I. Eber, Stephen M.   II. Title.
TS283.C53 1984   681'.76041   83-11311
ISBN 0-07-010874-9

1234567890 DOC/DOC 89876543

ISBN 0-07-010874-9

*The editors for this book were Diane Heiberg, Harold B. Crawford, and Chet
Gottfried, the designer was Naomi Auerbach, and the production supervisor was
Sally Fliess. It was set in Baskerville by Progressive Typographers.
Printed and bound by R. R. Donnelley & Sons Company.*

For Eve and Linda

Contents

Preface

This sixth edition, which has been brought up to date and expanded, contains new sections that outline recent Code changes. New additions include a chapter on ASME Code Section I, "Power Boilers," and additional information about nuclear vessels and the rigorous requirements of ASME Code Section III, Division 1, "Nuclear Power Plant Components." Other major additions that estimators, engineers, and inspectors will find very useful are the thickness requirement tables for internal pressure of cylindrical shells and torispherical and ellipsoidal heads, as well as thickness charts for external pressure of cylindrical shells.

The *ASME Boiler and Pressure Vessel Code* is progressive and viable. Important changes and additions are made when required. These changes are pinpointed to make our readers aware of them. For example, the Code mandatory appendixes have been renumbered. The former Code Par. UA-4 in Appendix I is now Par. 1-4 and the former Code Fig. UA-60 in Appendix II is now Code Fig. 2-4 in Appendix 2.

The mandatory appendixes are presently numbered 1 to 15.

The nonmandatory appendixes are now lettered from Appendix A to Appendix Z. The former Code Par. UA-265 in Appendix L, is now L-1, and the former Table UA-1020 in Appendix W is now Table W-3.

Of all pressure vessels manufactured under the American Society of Mechanical Engineers Boiler and Pressure Vessel Code, more have been produced to the ASME Section VIII, Division 1, "Pressure Vessel" Code than with all other sections of the Code combined. Since its first edition in 1925, this Code has been an important reference for designers and fabricators of pressure vessels to be used in most states, all Canadian provinces, and internationally.

Pressure Vessels—The ASME Code Simplified was first written with the aim of presenting in practical, interesting, and helpful form the information that will enable any person working with the ASME Section VIII, Division 1, "Pressure Vessel," Code to get the most out of it.

The American Society of Mechanical Engineers Code is the result of years of development by individuals who have experience in its application in industry, and it contains a tremendous amount of information. Those who use the Code regularly may have little difficulty in finding the information they want, but the designer, fabricator, or inspector who consults it infrequently may require the assistance offered by this book.

This book illustrates specific problems that are commonly encountered by those persons and directs them to the required Code information. It facilitates understanding of the application of Code rules as it indicates designs and materials for which precautions are necessary. Experience has proved that many important factors are overlooked in the design, fabrication, and inspection of Code pressure vessels.

This book brings many of these overlooked items to the reader's attention and suggests methods for better design, fabrication, and inspection control that will lead to more efficient and economical operations and better quality control systems.

The various charts and tables on design, fabrication, and inspection of Code vessels are created to facilitate the reader's application of Code requirements. For example, estimators, engineers, and inspectors will find the new thickness tables for cylindrical shells and dished heads for internal pressure useful because they provide information on required thicknesses at a glance. All references to specific parts of the Code—subsections, parts, paragraphs, figures, tables, and appendixes—have been prefixed with the word "Code," (for example, Code Fig. 1-4) in order to distinguish them from references to the various figures, tables, and appendixes in this text.

It is the authors' hope that this book will help clarify the reader's understanding of the *ASME Boiler and Pressure Vessel Code* sections and that it will encourage people in the industry to take all the necessary precautions in ordering, designing, fabricating, and inspecting Code vessels.

The data have been selected to answer questions most frequently asked by pressure vessel manufacturers. This information is advisory only. There is no obligation on the part of anyone to adhere to the recommendations made.

It must be remembered that there is no alternative to reading and understanding the ASME Code.

Many individuals, companies, and professional societies have provided information, pictures, and illustrations that help to clarify many items. We are also grateful to our friends in the pressure vessel manufacturing industry for their suggestions and help in checking many of the items in this book.

Robert Chuse
Stephen M. Eber

Pressure Vessels

ONE

Origin, Development, and Jurisdiction of the ASME Code

HISTORY OF THE ASME CODE

On March 20, 1905, a disastrous boiler explosion occurred in a shoe factory in Brockton, Massachusetts, killing 58 persons, injuring 117 others, and leaving a quarter of a million dollars in property damage. For years prior to 1905, boiler explosions had been regarded as either an inevitable evil or "an act of God." But this catastrophic accident had the effect of making the people of Massachusetts see the necessity and desirability of legislating rules and regulations for the construction of steam boilers in order to secure their maximum safety. After much debate and discussion, the state enacted the first legal code of rules for the construction of steam boilers in 1907. In 1908, the state of Ohio passed similar legislation, the Ohio Board of Boiler Rules adopting, with a few changes, the rules of the Massachusetts Board.

Therefore, other states and cities in which explosions had taken place began to realize that accidents could be prevented by the proper design, construction, and inspection of boilers and pressure vessels and began to formulate rules and regulations for this purpose. As regulations differed from state to state and often conflicted with one another, manufacturers began to find it difficult to construct vessels for use in one state that would be accepted in another. Because of this lack of uniformity, both manufacturers and users made an appeal in 1911 to the Council of the American Society of Mechanical Engineers to correct the situation. The Council answered the appeal by appointing a committee "to formulate standard specifications for the construction of steam boilers and other pressure vessels and for their care in service."

Fig. 1.1 The Brockton, Massachusetts shoe factory.

The first committee consisted of seven members, all experts in their respective fields: one boiler insurance engineer, one material manufacturer, two boiler manufacturers, two professors of engineering, and one consulting engineer. The committee was assisted by an advisory committee of eighteen engineers representing various phases of design, construction, installation, and operation of boilers.

Following a thorough study of the Massachusetts and Ohio rules and other useful data, the committee made its preliminary report in 1913 and sent 2000 copies of it to professors of mechanical engineering, engineering departments of boiler insurance companies, chief inspectors of boiler inspection departments of states and cities, manufacturers of steam boilers, editors of engineering journals, and others interested in the construction and operation of steam boilers, with a request for suggestions of changes or additions to the proposed regulations.

After 3 years of countless meetings and public hearings, a final draft of the first *ASME Rules for Construction of Stationary Boilers and for Allowable Working Pressures,* known as the 1914 edition, was adopted in the spring of 1915.

ADDITIONS TO THE CODE

Since 1914, many changes have been made and new sections added to the Code as the need arose.   The present sections are listed in the following order:

Section I.   "Power Boilers"

Section II.   "Material Specifications"
                      "Ferrous Materials, Part A"
                      "Nonferrous Materials, Part B"
                      "Welding Rods, Electrodes, and Filler Metals, Part C"

Secton III, Division 1.   "Nuclear Power Plant Components"

   Subsection NA: "General Requirements"
   Subsection NB: "Class 1 Components"
   Subsection NC: "Class 2 Components"
   Subsection ND: "Class 3 Components"
   Subsection NE: "Class MC Components"
   Subsection NF: "Component Supports"
   Subsection NG: "Core Support Structures"

Section III, Division 2.   "Concrete Reactor Vessel Containments"

Section IV.   "Heating Boilers"

Secton V.   "Nondestructive Examinations"

Secton VI.   "Recommended Rules for Care and Operation of Heating Boilers"

Section VII.   "Recommended Rules for Care of Power Boilers"

Section VIII, Division 1.   "Pressure Vessels"

Fig. 1.2   Shoe factory after the boiler explosion of March 20, 1905 which led to the adoption of many state boiler codes and the *ASME Boiler and Pressure Vessel Code*. (Courtesy of Hartford Steam Boiler Inspection & Insurance Company)

Section VIII, Division 2. "Pressure Vessels — Alternative Rules"
Section IX. "Welding and Brazing Qualifications"
Section X. "Fiberglass-Reinforced Plastic Pressure Vessels"
Section XI. "Rules for Inservice Inspection of Nuclear Power Plant Components"

ASME BOILER AND PRESSURE VESSEL COMMITTEE

The increase in the size of the Code reflects the progress of industry in this country. To keep up with this spontaneous growth, constant revisions have been required.

The ASME Code has been kept up to date by the Boiler and Pressure Vessel Committee (presently consisting of more than 800 volunteer engineers and other technical professionals) which considers the needs of the users, manufacturers, and inspectors of boilers and pressure vessels. In the formulation of its rules for the establishment of design and operating pressures, the Committee considers materials, construction, methods of fabrication, inspection, certification, and safety devices. The ASME works closely with American National Standards Institute (ANSI) to assure that the resulting documents meet the ANSI's criteria for publication as American National Standards.

The members of the Committee do not represent particular organizations or companies but have recognized background and experience by which they are placed in categories which include manufacturers, users of the products for which the codes are written, insurance inspection, regulatory, and general. The Committee meets on a regular basis to consider requests for interpretations and revisions and additions to Code rules as dictated by advances in technology.

Approved revisions and additions are published semiannually in addenda to the Code.

To illustrate, boilers were operating in 1914 at a maximum pressure of 275 psi and temperature of 600°F. Today, boilers are designed for pressures as high as 5000 psi and temperatures of 1100°F, and pressure vessels for pressures of 3000 psi and over and for temperatures ranging from −350°F to more than 1000°F.

Each new material, design, fabrication method, and protective device brought new problems to the Boiler Code Committee, requiring the expert technical advice of many subcommittees in order to expedite proper additions to and revisons of the Code. As a result of the splendid work done by these committees, the *ASME Boiler and Pressure Vessel Code* has been developed; it is a set of standards that assures every state of the safe design and construction of all boiler and pressure vessels used within its borders and is used around the world as a basis for enhancing public health, safety, and welfare. Many foreign manufacturers are accredited under the provisions of the *ASME Boiler and Pressure Vessel Code.*

PROCEDURE FOR OBTAINING THE CODE SYMBOL AND CERTIFICATE

Users of pressure vessels prefer to order ASME Code vessels because they know that such vessels will be designed, fabricated, and inspected to an approved quality control system in compliance with a safe standard.

Pressure vessel manufacturers want the Code Symbol and Certificate of Authorization so that they will be able to bid for Code work, thereby broadening their business opportunities. They also believe that authorization to build Code vessels will enhance the reputation of their shop.

If a company is interested in building Code vessels according to the ASME Section VIII, Division 1, "Pressure Vessels" Code, they should acquaint themselves with Code Par. U-2, which requires the manufacturer to have a contract or agreement with a qualified inspection agency employing Authorized Inspectors. This third party in the manufacturer's plant, by virtue of being authorized by the state to do Code inspection, is the legal representation which permits the manufacturer to fabricate under state laws (the ASME Code).

Manufacturers who want to construct Code vessels covered by Section VIII, Division 1, obligate themselves with respect to quality and documentation. (See Code Appendix 10, "Quality Control Systems.") A survey will be required for the initial issuance of an ASME Certificate of Authorization and for each renewal. The evaluation is performed jointly by the authorized inspection agency and the jurisdictional authority concerned which has adopted, and also administers, the applicable boiler and pressure vessel legislation. Where the jurisdictional authority does not make the survey, or the jurisdiction is the inspection agency, the National Board of Boiler and Pressure Vessel Inspectors will be asked to participate in the survey.

After the survey has been jointly made to establish that a quality control system is actually in practice, the National Board representative, if involved, and the jurisdiction authority will discuss their findings with the manufacturer. If the manufacturer's system does not meet Code requirements, the company will be asked to make the necessary corrections. The survey team will then forward their report to the ASME with the recommendation that the manufacturer receive the Certificate of Authorization or, if failure to meet standards exists, that it not be issued.

All Code shops must follow the above procedures in order to obtain the Code Symbol. If your shop wants the Code Certificate of Authorization, you should write to the Secretary of the Boiler and Pressure Vessel Code Committee, 345 E. 47th St., New York, NY 10017, stating your desire to build such vessels. To help the secretary and the Subcommittee on Code Symbol Stamps evaluate your shop, the application must describe the type and size of vessel the shop is capable of building and the type and size of your equipment, especially fabricating equipment. If possible, you should arrange with some authorized inspection agency, such as that of your state, city, or insurance company, to undertake the inspection service after your shop has received ASME certification. You can then inform the ASME that specific inspection arrangements have already been made.

The secretary will send a statement of your request to the Subcommittee on Code Symbol Stamps. A request will also be made to the chief inspector of the particular state, city, or other inspection agency governing your company and/or the National Board of Boiler and Pressure Vessel Inspectors, to make a survey that will determine whether or not your company has a quality control system that is capable of designing and fabricating pressure vessels by ASME Code rules.

The secretary receives a report of this survey and will then forward the report to the Subcommittee on Code Symbol Stamps. Should they judge favorably, your company will be issued a Code Symbol and Certificate of Authorization. A manufacturer with the ability, integrity, and quality control system to design and fabricate good pressure vessels will have no difficulty in obtaining Code authorization.

THE NATIONAL BOARD OF BOILER AND PRESSURE VESSEL INSPECTORS

The National Board of Boiler and Pressure Vessel Inspectors, an independent, nonprofit organization whose members are the jurisdictional officials responsible for enforcing and administrating the *ASME Boiler and Pressure Vessel Code,* was first organized in 1919. Since then, it has served for the uniform administration and enforcement of the rules of the *ASME Boiler and Pressure Vessel Code.* In drafting the first Code, the Boiler Code Committee realized that it had no authority to write rules to govern administration, and that compliance could be made mandatory only by the legislative bodies of states and cities.

Although some authorities thought that uniformity could be achieved by a code of uniform rules, they proved to be mistaken. Various states and cities adopted the ASME Code with the proviso that boilers be inspected during construction by an inspector qualified under their own regulations; the boiler was then to be stamped with the individual local stamping to indicate its conformity with these regulations. This requirement created an unwieldy situation, for boilers constructed in strict accordance with the ASME Code still had to be stamped with the local stamping, thereby causing needless delay and expense in delivery of the vessel.

It was evident that some arrangement had to be made to overcome such difficulties. Therefore, boiler manufacturers met with the chief inspectors of the states and cities that had adopted the ASME Code and formed the National Board of Boiler and Pressure Vessel Inspectors for the purpose of presenting the ASME Code to governing bodies of all states and cities. Their aim was not only to promote safety and uniformity in the construction, installation, and inspection of boilers and pressure vessels but also to establish reciprocity between political subdivisions of the United States. Such ideals could best be carried out by a central organization under whose auspices chief inspectors, or other officials charged with the enforcement of inspection regulations, could meet and discuss their problems. The efforts of this first group of administrators succeeded in extending National Board membership to all Canadian provinces and most of the states and cities of the United States. It is now possible for an authorized shop to build a boiler or pressure vessel that will be accepted anywhere in the United States or Canada after it has been inspected by an Authorized Inspector holding a National Board Commission.

The ASME Code requires that Inspectors must meet certain minimum requirements of education and experience and pass a written examination before they can be commissioned to perform Code inspections. One of the many functions of the National Board is the commissioning of Authorized Inspectors.

SHOP REVIEWS BY THE NATIONAL BOARD

Before the issuance of an ASME Certificate of Authorization to build Code vessels, the manufacturer must have and demonstrate a quality control system. This system has to include a written description explaining in detail the quality-controlled manufacturing process.

Before the issuance of renewal of a Certificate of Authorization, the manufacturer's facilities and organization are subject to a joint review by the inspection agency and the legal jurisdiction concerned. For those areas where there is no jurisdiction, or where a jurisdiction does not review a manufacturer's facility, and/or the jurisdiction is the inspection agency, the function may be carried out by a representative of the National Board of Boiler and Pressure Vessel Inspectors. When the review of the manufacturer's facilities and organization have been jointly made by the inspection agency, and/or the legal jurisdiction, or a representative of the National Board of Boiler and Pressure Vessel Inspectors, a written report will be made to the American Society of Mechanical Engineers. The National Board of Boiler and Pressure Vessel Inspectors also participates in all nuclear surveys.

The National Board also issues Certificates of Authorization for the use of the National Board R stamp, to be applied to ASME Code and National Board stamped boilers and/or pressure vessels where repairs are to be made. (See the section "Welded Repair or Alteration Procedures" in this chapter.)

Another important service the National Board carries out is the certification of safety valves and safety relief valves. The *ASME Boiler and Pressure Vessel Code* contains specific requirements governing the design and capacity certification of safety valves installed on Code stamped vessels. The certification tests are conducted at a testing laboratory approved by the ASME Boiler and Pressure Vessel Committee.

In addition to safety valve certification, the National Board has its own testing laboratory for its experimental work and for certifying ASME safety and relief valves. The facilities of the laboratory are also available to manufacturers and other organizations for research, development, or other test work.

NATIONAL BOARD REQUIREMENTS

All Canadian provinces and most states require boilers and pressure vessels to be inspected during fabrication by an Inspector holding a National Board Commission and then to be stamped with a National Board standard number. Qualified and authorized boiler and pressure vessel manufacturers must be registered with the National Board of Boiler and Pressure Vessel Inspectors, 1055 Crupper Ave., Columbus, OH 43229. In addition, two data sheets on each vessel must be filed with the National Board, one copy of which is retained by the Board and the other sent to the administrative authority of the state, city, or province in which the vessel is to be used. (See Code Par. UG-120.)

The National Board of Boiler and Pressure Vessel Inspectors is constantly working to assure greater safety of life and property by promoting and securing uniform enforcement of boiler and pressure vessel laws and uniform approval of designs and structural details of these vessels, including the accessories that affect their safe operation; by furthering the establishment of a uniform Code; by espousing one standard of qualifications and examinations for inspectors who are to enforce the requirements of this Code; and by seeing that all relevant data of the Code are made available to members.

CODE CASE INTERPRETATIONS

As the Code does not cover all details of design, construction, and materials, pressure vessel manufacturers sometimes have difficulty in interpreting it when trying to meet specific customer requirements. In such cases the Authorized Inspector should be consulted. If the Inspector is unable to give or has any doubts or questions about the proper interpretation of the intent of the Code, the question can be referred to the Inspector's office. If they are not able to provide a ruling, the manufacturer may then request the assistance of the Boiler and Pressure Vessel Committee, which meets regularly to consider inquiries of this nature. In referring questions to the Boiler and Pressure Vessel Committee, it is necessary to submit complete details, sketches of the construction involved, and references to the applicable paragraphs of the Code; it is also useful to include opinions expressed by others.

Inquiries should be submitted by a letter to the Secretary of the ASME Boiler and Pressure Vessel Committee, 345 E. 47th St., New York, NY 10017. The secretary will distribute copies of the inquiry to the committee members for study. At the next committee meeting, interpretations will be formulated for submission to the ASME Board on Codes and Standards, which has been authorized by the Council of the Society to pass judgment. After a decision has been reached, it is forwarded to the inquiring party and also published in *Mechanical Engineering* magazine. If no further criticism is received, the decision may be formally adopted by the Council of the Society.

This interpretation may then become an addendum to the Code. If so, it will first be printed in the "addenda" supplement of the Code for subsequent inclusion in the latest edition to the Code (printed every 3 years).

The addenda incorporate as many case interpretations as possible in order to keep the open file of cases to a minimum. For their own good, all Code manufactures should subscribe to the "Code Case Interpretation Service" offered by the American Society of Mechanical Engineers. They will then automatically receive all interpretations issued by the Boiler and Pressure Vessel Committee, some of which may prove very useful.

THE UNIFORM BOILER AND PRESSURE VESSEL LAWS SOCIETY

The Uniform Boiler and Pressure Vessel Laws Society is a nonpolitical, non-commercial, and nonprofit technical society. Its objective is to secure uniformity in the laws, rules and regulations, and administration thereof, which affect the boiler and pressure vessel industry, inspection agencies, and users.

The Society cooperates with all organizations and officials charged with the enforcement of boiler and pressure vessel inspection laws and regulations. Though all states should require ASME construction of boilers and pressure vessels, some states and cities do not. The Society publishes a "Synopsis of Boiler and Pressure Vessel Laws, Rules, and Regulations," which gives a summary of applicable laws of states, cities, countries, and provinces in the United States and Canada and has proved valuable to users and manufacturers of boilers and pressure vessels. Members of the Society are kept informed by bulletins about the actions of governmental bodies adopting or revising codes and regulations for the construction, installation, and inspection of boilers and pressure vessels.

Figures 1.3 and 1.4 show political subdivisions of the United States and Canada that require pressure vessels to comply with the *ASME Boiler and Pressure Vessel Code* based on the latest available information. Each year additional states and cities adopt the Code for their minimum standards. This list is thus for guidance only. If there is any question about requirements, communicate with the inspection department of the relevant state, territory, or city, or the Uniform Boiler and Pressure Vessel Laws Society, 2838 Long Beach Rd., P. O. Box 512, Oceanside, NY 11572.

CANADIAN PRESSURE VESSEL REQUIREMENTS

United States manufacturers of ASME Code vessels often receive orders for vessels to be installed in Canada. All provinces of Canada have adopted the ASME Code and require vessels to be shop-inspected by an Inspector holding a National Board Commission and to be stamped with a Provincial Registration Number in addition to the ASME Symbol and National Board stamping.

But before construction begins, one important requirement has to be met in all the provinces of Canada. Each manufacturer must submit blueprints and

UNIFORM BOILER AND PRESSURE VESSEL LAWS SOCIETY

Incorporated •

2838 Long Beach Road
P.O. Box 512
Oceanside, N.Y. 11572
Telephone 516-536-5485

PRINTED IN U.S.A.

Boiler & Pressure
Vessel Law

Boiler Law Only

Cities & Counties
with Boiler & Pressure
Vessel Ordinances

Fig. 1.3   States, cities, and counties in United States and provinces in Canada with boiler and vessel ordinances. *(Uniform Boiler and Pressure Vessel Laws Society)*

TABULATION OF THE BOILER AND PRESSURE VESSEL LAWS OF THE UNITED STATES AND CANADA

Key:
```
    I   - Power Boilers                        XI  - In service inspection - Nuclear
 III(1) - Nuclear Components                   A   - Law Requires ASME Construction
 III(2) - Concrete Reactor Vessels             O   - Require Own Construction Code or ASME
    IV  - Heating Boilers                       N   - Law Does Not Cover
VIII(1) - Pressure Vessels                     *   - Operator's License Required
VIII(2) - Pressure Vessels - alternative rules **  - Limited to specific vessels
    X   - Fiber-glass Reinforced Plastic       *** - Pending Rules and Regulations
          Pressure Vessels
```

STATES & TERRITORIES	I	III(1)	III(2)	IV	VIII(1)	VIII(2)	X	XI
ALABAMA	A	N	N	A	N	N	N	N
ALASKA	A	A	A	A	A	A	A	A
ARIZONA	A*	A	A	A	N	N	N	A
ARKANSAS	A*	A	A	A*	A*	A	A	A
CALIFORNIA	A	A	A	A	A	A	A	A
COLORADO	A	A	A	A	A	N	A	A
CONNECTICUT	A	A	N	A	N	N	N	N
DELAWARE	A	A	A	A	A	A	A	A
DIST. OF COLUMBIA	A*	A	N	A*	A*	A	N	N
FLORIDA	N	N	N	N	N	N	N	N
GEORGIA	A	N	N	A	A	N	N	N
GUAM	A	N	N	A	N	N	N	N
HAWAII	A	A	A	A	A	A	A	A
IDAHO	A	A	N	A	A	A	N	N
ILLINOIS	A	A	A	A	A	A	A	A
INDIANA	A	A	A	A	A	A	N	A
IOWA	A	A	A	A	A	A	A	A
KANSAS	A	A	A	A	N	N	N	A
KENTUCKY	A	A	A	A	A	A	A	A
LOUISIANA	A	N	N	A	A	A	N	N
MAINE	A	A	A	A	A	A	A	A
MARYLAND	A	A	A	A	A	A	A	A
MASSACHUSETTS	A*	A	N	A	A**	N	A	A
MICHIGAN	A	A	A	A	A**	N	A	A
MINNESOTA	A*	A	A	A*	A*	A	A	A
MISSISSIPPI	A	N	N	A	A	A	A	N
MISSOURI	N	N	N	N	N	N	N	N
MONTANA	A*	N	N	A*	N	N	N	N
NEBRASKA	A	A	N	A**	A**	A**	A	N
NEVADA	A	A	A	A	A	A	A	A
NEW HAMPSHIRE	A	A	N	A	A	N	N	N
NEW JERSEY	A*	A	A	A*	A	A	A	A
NEW MEXICO	N	N	N	N	N	N	N	N
NEW YORK	O	A	N	A	N	N	N	N
NORTH CAROLINA	A	A	A	A	A	A	A	A
NORTH DAKOTA	A	A	A	A	A	N	A	A
OHIO	A*	A	A	A*	A	O	A	A
OKLAHOMA	A	N	N	A	A	N	N	N
OREGON	A	A	A	A	A	A	A	A
PANAMA CANAL ZONE	A*	A	N	A*	A	A	N	N
PENNSYLVANIA	A*	A	A	A*	A*	A	A	A
PUERTO RICO	A	A	N	A	A	N	N	N
RHODE ISLAND	A	N	N	A	A	A	A	N
SOUTH CAROLINA	N	N	N	N	N	N	N	N
SOUTH DAKOTA	A	A	A	A	N	N	N	A
TENNESSEE	A	A	A	A	A	A	A	A
TEXAS	A	A	A	A	N	N	N	A
UTAH	A	A	A	A	A	A	A	A
VERMONT	A	A	N	A	A	A	N	N
VIRGINIA	A	A	A	A	A	A	A	A
WASHINGTON	A	A	A	A	A	A	A	A
WEST VIRGINIA	A	N	N	N	N	N	N	N
WISCONSIN	A	A	A	A	A	A	A	A
WYOMING***	A	N	N	A	A	N	N	N

PROVINCES IN CANADA	I	III(1)	III(2)	IV	VIII(1)	VIII(2)	X	XI
ALBERTA	A*	A	A	A*	A	A	A	A
BRITISH COLUMBIA	A*	A	N	A*	A	A	A	N
MANITOBA	A*	A	A	A*	A*	A	A	A
NEW BRUNSWICK	A*	A	A	A*	A*	A	A	A
NEWFOUNDLAND & LABRADOR	A*	A	A	A*	A*	A	A	A
NORTHWEST TERR.	A*	N	N	A*	A*	N	A	N
NOVA SCOTIA	A*	A	A	N	A*	A	A	A
ONTARIO	A*	A	N	A*	A*	A	A	N
PRINCE EDWARD IS.	A	A	N	A	A	N	N	N
QUEBEC	A*	A	N	A*	A*	A	A	N
SASKATCHEWAN	A*	A	A	A*	A*	N	A	A
YUKON TERRITORY	A*	A	N	A*	A*	N	N	N

CITIES AND COUNTIES

	I	III(1)	III(2)	IV	VIII(1)	VIII(2)	X	XI
ALBUQUERQUE, N.M.	A	N	N	A	N	N	N	N
BUFFALO, N.Y.	O*	A	N	A	A	N	N	N
CHICAGO, IL.	A*	A	A	A	A	A	A	A
DEARBORN, MI.	A*	N	N	A*	A*	N	N	N
DENVER, CO.	A*	A	N	A*	A*	A	N	N
DES MOINES, IA.	A*	N	N	A*	N	N	N	N
DETROIT, MI.	A*	A	A*	A	A	A	A	A
E. ST. LOUIS, IL.	A*	N	N	A*	A*	N	N	N
GREENSBORO, N.C.	A	N	N	A	A	N	N	N
KANSAS CITY, MO.	A*	A	N	A*	A*	A	A	N
LOS ANGELES, CA.	A*	A	N	A*	A	A	A	N
MEMPHIS, TN.	A*	A	N	A*	A*	A	N	A
MIAMI, FL.	A*	A	N	A*	A*	A	N	N
MILWAUKEE, WI.	A*	A	N	A*	A*	A	A	A
NEW ORLEANS, LA.	A*	N	N	A*	A*	A	N	N
NEW YORK CITY, N.Y.	O*	N	N	A	N	N	N	N
OKLAHOMA CITY, OK.	A*	N	N	A*	N	N	N	N
OMAHA, NB.	A*	N	N	A*	A*	N	N	N
PHOENIX, AZ.	A	N	N	A	A	A	N	N
ST. JOSEPH, MO.	A*	A	N	A*	A*	A	N	N
ST. LOUIS, MO.	A*	A	N	A*	A*	N	N	N
SAN FRANCISCO, CA.	A	N	N	A	A	N	N	N
SAN JOSE, CA.	A*	N	N	A	A	N	N	N
SEATTLE, WA.	A	A	A	A	A	A	N	N
SPOKANE, WA.	A*	N	N	A*	A	N	N	N
TACOMA, WA.	A*	N	N	A*	A*	N	N	N
TAMPA, FL.	A*	N	N	A*	A*	A	N	N
TUCSON, AZ.	A	N	N	A	A	A	N	N
TULSA, OK.	A*	A	N	A*	A*	A	N	N
UNIVERSITY CITY, MO.	A*	N	N	A*	A*	N	N	N
WHITE PLAINS, N.Y.	O*	N	N	A	N	N	N	N
ARLINGTON CO.,VA.	A	A	N	A	A	A	N	N
DADE CO., FL.	A	N	N	A	A	N	N	N
FAIRFAX CO., VA.	A	A	N	A	A	A	N	N
JEFFERSON PARISH	A	A	N	A*	A*	A	N	N
ST. LOUIS CO.,MO.	A*	A	N	A*	A	A	N	N

1. This condensed Data Sheet does not list all the exemptions and variances in the many laws and regulations. More detailed information is available in the Society's "Synopsis of Boiler and Pressure Vessel Laws, Rules and Regulations." Further information may be obtained from the jurisdictional authority or the Society.

2. Some states not having a boiler or pressure vessel law do require boiler or pressure vessel construction to be in accordance with the ASME Code under their laws for Workmen's Compensation, Liquefied Petroleum Gas, etc. The "Synopsis" gives more detail on special requirements.

3. A Pennsylvania commission is required for shop inspectors of boilers and pressure vessels to be shipped into that state.

4. An Ohio commission is required for shop inspectors of boilers and pressure vessels to be shipped into that state.

5. The New York State Construction Code is identical to the ASME Code.

6. Canadian provincial Boiler Inspection Departments accept the ASME Boiler and Pressure Vessel Code as their minimum standard. Manufacturers must register with the Department and submit designs and specifications for approval before fabricating boilers or pressure vessels for installation in Canada. Also see Canadian Standards Association B51 Code.

Fig. 1.4   Tabulations of the boiler and pressure vessel laws of the United States and Canada. *(Uniform Boiler and Pressure Vessel Laws Society)*

specification sheets in triplicate of all designs for approval and registration by the chief inspector of the province in which the vessel is to be used.

After the chief inspector receives the drawings, they are checked by an engineer to determine their compliance with the Code and also with provincial regulations. If the design does not meet the requirements, a report is sent to the manufacturer explaining why it cannot be approved and requesting that the necessary changes be made and the corrected drawings returned. If the design complies in full, one copy of the drawing is returned to the manufacturer with the stamped approval and a registration number that must be stamped on the vessel in addition to the ASME Symbol and National Board stamping.

Once a design has been approved and registered, any number of vessels of that design can be built and used in the province where it is approved. If a vessel of the same design is to be sent to another Canadian province, the manufacturer's letter to the chief inspector of that province should state the registration number of the original approval. The second province will then use the same registration number with the addition of its own province number. For example, a design first registered in Ontario might have been given the number 764. This number would be followed by a decimal point and then the number 5, which signifies the Province of Ontario. The registration would thus be 764.5. If this same design were used in the Province of Manitoba, it would be given the registration number 764.54, the figure 4 denoting Manitoba.

The Canadian provinces also require their own manufacturer's affidavit form, with the registration number and the shop inspector's signature on the data sheets. Finally, when a vessel is delivered to a purchaser in Canada, an affidavit of manufacture bearing the registration number and the signature of the authorized shop inspector must be sent to the chief inspector of the province for which it is intended.

WELDED REPAIR OR ALTERATION PROCEDURE

Repairs or alterations of ASME Code vessels may be made in any shop that manufactures Code vessels or in the field by any welding contractor qualified to make repairs on such vessels. Recognizing the need for rules for repair and alterations to boilers and pressure vessels, the National Board of Boiler and Pressure Vessel Inspectors, in their inspection code, Chapter III, gives rules that many states have incorporated into their own boiler and pressure vessel laws. Many companies have been using these rules as a standard for repairs to pressure vessels.

The National Board also has Certificates of Authorization for use of the National Board R stamp. Any repair organization may obtain Certificates of Authorization for use of the National Board R stamp to be applied to ASME Code and National Board stamped boilers and/or pressure vessels where repairs have been made.

Before an R stamp is issued, the firm making repairs must submit to a review of its repair methods.   It must also have an acceptable written quality control system covering repair design, materials to be used, welding procedure specifications, welders' qualifications, repair methods, and examination.   The review will be made by a National Board member jurisdiction where the repair company is located.   If there is no National Board member jurisdiction, or, at the request of such jurisdiction, the review will be made by a representative of the National Board.

Manufacturers and assemblers who hold ASME Certificates of Authorization and Code stamps, with the exception of H (cast iron), V, UV, and UM stamps, may obtain the National Board R stamp without review provided their quality control system covers repairs.

Where the R stamp is to be applied to National Board stamped boilers and pressure vessels all repairs are subject to acceptance by an Inspector who holds a valid National Board Commission.

The Authorized Inspector will be guided by the National Board Inspection Code as well as by any existing local rules governing repair of vessels.*   When the work is to be done on location, repairs to a vessel require greater skill than that required when the vessel was first shop-constructed.   Utmost consideration must be given to field repairs, which may necessitate cutting, shaping, fitting, and welding with portable equipment.   The practicability of moving the vessel from the site to a properly equipped repair shop should be considered. Before repairs are made, the method of repair must be approved by an Authorized Inspector.   The Inspector will examine the vessel, identify the material to be welded, and compare it with the material to be used in repair.   A check is then made to ensure that the welding contractor or shop has a qualified welding procedure for the material being welded and that the welder who does the work is properly qualified to weld that material.   If the repair or alteration requires design calculations, the Inspector will review the calculations to assure that the original design requirements are met.

When the repairs or alterations and the necessary tests have been completed, a report called "Record of Welded Repairs" must be signed by the Inspector who authorized the repairs and by the contractor or manufacturer doing the repair work.   Copies of this report must be sent to the proper state authorities, and a copy must be retained for review by the vessel owner and the Authorized Inspector.

A sample form of a "Report of Welded Repair or Alteration" is shown in Fig. 1.5.   Some states and insurance companies have their own forms which must be signed by the contractor or manufacturer making the repair and by the Authorized Inspector.

When an alteration is made to a vessel, it shall comply with the section of the ASME Code to which the original vessel was constructed.

The required inspection shall be made by an Authorized Inspector holding a

---

* The Code is published by the National Board of Boiler and Pressure Vessel Inspectors, 1055 Crupper Ave., Columbus, OH 43229.

FORM R-1, REPORT OF WELDED ☐ REPAIR OR ☐ ALTERATION
As Required by the Provisions of The National Board Inspection Code

1. Work done by _____   _____
   (Name and address of repair or alteration organization)   (Serial No.)

2. Owner _____
   (Name and address of owner)

3. Location of Installation_____
   (Name and address)

4. Unit Identification _____ Name of Manufacturer_____
   (Boiler. Pressure Vessel)

5. Identifying Nos. _____   _____   _____   _____   _____
   (Mfgr. Serial No.)   (National Board No.)   (Jurisdiction)   (Other)   (Year Built)

6. Description of Work: _____
   (Use back. separate sheet, or sketch if necessary)

   _____

   _____

   _____

   _____

   _____ Pressure Test, if Applied_____psi

7. Remarks: Attached are Manufacturer's Partial Data Reports properly identified and signed by Commissioned Inspectors for the following
   items of this report: _____

   _____

   _____

   _____

   _____

   _____

   _____

   _____

   (Name of part. item number. mfgr's name. and identifying stamp)

CERTIFICATE OF COMPLIANCE

We certify that the statements made in this report are correct and that all ____(design)_____
material, construction, and workmanship on this _____conform to
                                                              (repair. alteration)
The National Board Inspection Code.
Date_____Signed _____ by _____
                         (Repair. Alteration Organization)   (Authorized Representative)

Our Certificate of Authorization No. _____ to use the _____Symbol expires_____, 19_____

CERTIFICATE OF INSPECTION

I, the undersigned, holding a valid commission issued by The National Board of Boiler and Pressure Vessel Inspectors or the State or
Province of_____and employed by_____
_____of_____have inspected
the work described in this Data Report on _____, 19_____and state that to the best of my knowledge and belief,
this work has been done in accordance with The National Board Inspection Code.
By signing this certificate, neither the Inspector nor his employer makes any warranty, expressed or implied, concerning the work
described in this Report. Furthermore, neither the Inspector nor his employer shall be liable in any manner for any personal injury or
property damage or a loss of any kind arising from or connected with this inspection, except such liability as may be provided in a policy
of insurance which the Inspector's insurance company may issue upon said object and then only in accordance with the terms of said
policy.
Date_____   _____ Commissions _____
                                      Inspector   Nat'l Board. State. Province and No.

   This form may be obtained from the National Board of Boiler and Pressure Vessel Inspectors, 1055 Crupper Ave., Col's., O. 43229.   NB-66 Rev. 0

Fig. 1.5   Report of welded repair or alteration. *(Courtesy of National Board of Boiler and Pressure Vessel Inspectors)*

National Board Commission. No alteration shall be initiated without the
approval of the Authorized Inspector.

When alterations are completed, a data report for the alterations on the
appropriate ASME Code data report form (U-1 or U-1A, P-1 or P-3) shall be
filled out and the words "Manufactured by" shall be changed to read "Altered

ALTERED BY _____

_____ PSI AT_____F
    (MAWP)   (Temp)

_____
(Manufacturer's alteration number, if used)

_____
(Date altered)

Fig. 1.6   Stamping or nameplate of an altered boiler or pressure vessel.

by." Such data shall clearly indicate what changes or alterations have been made in the original construction. On the data report, under remarks, the name of the original manufacturer, manufacturer's serial number, and the National Board number assigned to the vessel by the orignal manufacturer shall be stated. The original and one copy of the certified data report shall be filled with the National Board of Boiler and Pressure Vessel Inspectors. A copy shall go to the jurisdictional authority, inspection agency, and the customer.

When the stamping plate is made for an altered vessel, it shall contain all the information required by the National Board Inspection Code (see Fig. 1.6). Note the difference in the stamping of the nameplate of an altered vessel from the stamping of the nameplate of a repaired vessel (see Fig. 1.7).

The word "Altered" in letters $5/16$ inch high shall be stamped on the plate in lieu of the Code symbol, National Board R symbol, and National Board number. The date shown on the nameplate shall be the date of alteration. This nameplate shall be permanently attached to the vessel adjacent to the

R

_____
(Name of repair firm)

_____ psi at_____F
(Maximum allowable   (temp)
working pressure)

No._____
(National Board   _____
Repair symbol
stamp no.)

_____
[Date of repair(s)]

Fig. 1.7   Stamping or nameplate of a boiler or pressure vessel repaired by welding.

original manufacturer's nameplate or stamping, which shall remain on the vessel.

If the alteration requires removal of the original manufacturer's nameplate or stamping, it shall be retained and reattached to the altered vessel. Such relocation shall be noted on the data report.

MANAGERIAL FACTORS IN HANDLING CODE WORK

The producer of welded pressure vessels is faced today with many serious problems, not the least of which is managing to stay in business and earn a fair profit.* Severe competition exists throughout the plate fabricating industry, not only because of the growth in the number of fabricating shops but also because of the tremendous technological advances made in recent years. Many of the older, more heavily capitalized plants have been hard hit. Some of these have successfully turned to specialized fields; others produce specialty products that return a good profit when in demand. When demand slackens, these shops often turn to general fabrication, thereby increasing the competitive situation for the general plate fabricator.

For this type of fabricator, who essentially runs a job shop, the profit margin is restricted from the outset because jobs are usually awarded to the lowest bidder. Although there is nothing unusual in this method of competing, it does increase the need to think ahead clearly and to avoid unnecessary costs. For the job shop handling ASME Code vessels as well as non-Code work, the importance of careful planning is greatly increased because of the relatively high costs of Code material and also the variety in techniques and procedures demanded.

A second important factor in the award of a contract is delivery time, and here again planning assumes utmost importance. The buyer of a Code vessel is not necessarily concerned if a producer incurs extra expense through carelessness but is very often acutely aware of delivery time. Good customer relations and even future business depend upon on-time deliveries. Obviously, cost and delivery time are related.

The job shop, then, is faced with three fundamental problems in the production of Code work:

1. The ability to fabricate in accordance with the Code
2. The necessity for keeping costs at a minimum
3. The importance of making deliveries on time

The question naturally arises, "What concrete steps can be taken to assure the best fulfillment of these requirements?" A few general suggestions bear mention. First of all, there is no substitute for reading and understanding the ASME Code. Many protests are heard on this score, even from inspectors and engineers, but a knowledge of those sections of the Code that pertain to the type

---

* The material in the section "Managerial Factors in Handling Code Work" was contributed by C. H. Willer, former editor of *Welding Journal*.

of work in which one is engaged is a necessity. But even this is not enough. Some vessel requirements are extremely simple; others are not. The vast industrial complex of today has created such a wide variety of pressure vessels that they constitute a perpetual challenge to the ingenuity of the fabricator. For the job shop, close cooperation and understanding are required right down the line. Sales, estimating, design, and purchasing must all be backed up by sound shop experience and a reasonable time schedule.

Particular methods for handling an order vary from one company to another. But within the limits set by customer specification and the rules of the Code, the job shop can seek a competitive advantage and avoid unnecessary costs and time losses by determining the following:

1. An overall production schedule
2. Acceptable alternatives in plate material and components
3. Acceptable alternatives in design geometry
4. Acceptable alternatives in fabricating methods
5. Acceptable sources of supply

Some freedom of choice is usually left to the fabricator, and finding the best means of satisfying the requirements at the least cost is the essence of this phase of the manufacturing process.

WORK-FLOW SCHEDULES

By the time an order for a Code job is accepted, it has been estimated, bid, and sold. For many reasons, a job-shop estimator does not completely and finally specify all details of material design, and fabrication at the time of bidding. After an order is received, a work-flow schedule must be drawn up that allots time for shop fabrication, procurement of materials and components, design, drafting, and takeoff. The shop fabrication schedule is based on an estimate of the total quantity of work-hours required, the necessary work-hour sequence (allowing for nonadditive, parallel operations), and finally the stipulated delivery date and the total shop load for the calendar period. This information determines the required delivery date for plate material, purchased components, and shop subassemblies. The lead time for these items sets the date for purchasing and thus for the completion of the design, drafting, and takeoff details initially required. An estimate of the time needed for the latter operations indicates the provisional starting date.

Although this simple computation can be made in many different ways, it is most important that it be made in a specific way, for it forms the basis on which all scheduling and expediting are founded. If the provisional starting date indicates on-time delivery, there is no problem. The sequence of determined dates becomes the production schedule, and expediting is managed accordingly. If the estimated delivery falls considerably ahead of time, the entire schedule can be appropriately delayed or more time can be allotted to any or all of the production steps. In this situation, the cost of maintaining the inventory must be balanced against the future availability of material in a changing market. It must also be remembered that although filler jobs are desirable, a reduction in work rate is not. If the estimated delivery falls after the required date, on the other hand, the operations must be reviewed to determine where and how time can be saved. Sometimes a simple solution is quickly found; more often, pressure has to be applied on all concerned, including the ASME Code Inspector. However, this pressure must be applied in a rational way.

It should be realized that the two hard necessities governing job-shop work—cost saving and time saving—are not always compatible. Earlier deliveries of purchased material usually cost more; heavier pressure on a work force usually results in overtime. Code restrictions on design, fabrication, and materials narrow the field of action. Nevertheless, certain routine procedures exist that will result in considerable savings.

MILL ORDERS

From a cost and delivery angle, the design of a pressure vessel and the purchasing of the plate and heads are often related. Thus the steel mill system of extras, based on the weight of items ordered, varies inversely with the weight, a 10,000-lb item having no extra and lesser quantities taking gradually increased extras according to the published table. The maximum cost difference could amount to many dollars a ton. As an item consists of any number of plates of one width and one thickness of a given quality, it pays to design a vessel with equal-width plates. (See Appendix A.) Various mills also publish extras based on width, and these tables must be consulted. A buyer might be able to save considerably by arbitrarily increasing a plate width a fraction of an inch.

Tonnage mounts up fast in steel plate. The question is often asked, "Is it less expensive to design a given vessel for a higher test steel with less thickness even though the quality extra incurred by the decreased thickness may be higher?" It sometimes works out this way, by using, say, an SA-515 Grade 70 steel instead of an SA-285 Grade C steel. Fabrication cost differences must also be considered.

Sometimes the buyer rather than the designer will note that a plate useful for stock inventory can be added to a special item to bring the total weight into a lower class of extras, thus saving money. This is also true of heads. At some mills, discounts for quantity begin with four to six heads, and very sizable discounts are offered for larger quantities. It is also highly desirable to attempt to keep head diameters, radii, and thicknesses to the standards carried in mill stocks.

Mill conditions often affect the length of time required to fill an order. As mills have different rolling practices, schedules, minimum quantity stipulations, base prices, size limits, and the like, all these details should be discussed with a sales representative before attempting to place an order, especially if the buyer is not experienced. The lead time for most mills in normal periods runs about 45 days, but it is not safe to guess.

WAREHOUSE ORDERS

If delivery requirements make it impossible to obtain steel from a mill, Code plate can be obtained in various sizes from warehouses. Quick delivery can be obtained of certain size plates, usually the lighter gages up to ½ in thick, 96 in wide, and 10 to 20 ft long.   In addition to the higher base price, the buyer must consider the cost of obtaining the size plate desired from the sizes offered. Heads of certain sizes and gages are carried by some warehouses and also are stocked for immediate delivery by some mills.   Standard thickness of ³/₁₆, ¼, ⁵/₁₆, ⅜, and ½ in are typical of flanged and dished heads with diameters up to 96 in. Special heads are also listed in published catalogs.

SHOP ORGANIZATION

By far the biggest opportunities for saving time and money lie in the handling and organizing of Code work in the shop.   Unless shop procedure is properly set up and the work scheduled in advance for inspection, extra cost and delay are likely.   Each visit by a Code Inspector represents a direct cash outlay both for the time used by the Inspector and for the time taken by personnel to show and review the work.   Many shows adopt the practice of arranging for periodic visits so that as much work as possible can be grouped for one visit and the availability of the Inspector can be relied upon.   Lack of planning often results in a last-minute rush to get ready with possible overtime expense or in an incomplete presentation that necessitates an additional inspection for some overlooked detail.   If a job is not carefully watched, it may progress beyond the point at which an inspector can make a satisfactory investigation of a completed operation.   Similarly, the vessel may be found to contain components for which the Code quality standard cannot be established.   Either situation may result in costly rework or even complete loss.

Sometimes it seems that in the complexity of a busy job shop it is almost impossible to plan very far ahead.   Lack of planning, however, only means that more plans must be made, discarded, revised, and started all over again. Before work is started, therefore, the complete specifications should be reviewed, the sequence of work operations determined, and a rough shop schedule including the expected dates of inspection formulated.   Without this schedule a final delivery date cannot safely be determined.   The job must follow this schedule closely under responsible supervision.   When Code material is received, it should be carefully marked and checked.   At least by the time of layout, the reference numbers should be checked aganst material certification and the required physical and chemical properties proved.   At this time also, material should be checked for thickness and surface defects.   This simple routine should help to prevent unnecessary and costly rework and loss of time.

Descriptive Guide to the ASME Code Section VIII, Division 1, Pressure Vessels

An attempt is made in this chapter to assist those wishing to find specific information quickly without having to wade through the large fund of specifications and recommendations encompassed by the Code. Figures and tables outline designs, materials, or performance requirements, indicate conditons for which precautions are necessary, and list Code paragraphs to be read for the proper information. Experience has proved that many of these items are often overlooked. Also included is a handy reference chart (Fig. 2.1) that graphically illustrates the various parts of a pressure vessel and the Code paragraphs that apply to each. It should be remembered, however, that there is no substitute for reading the Code itself.

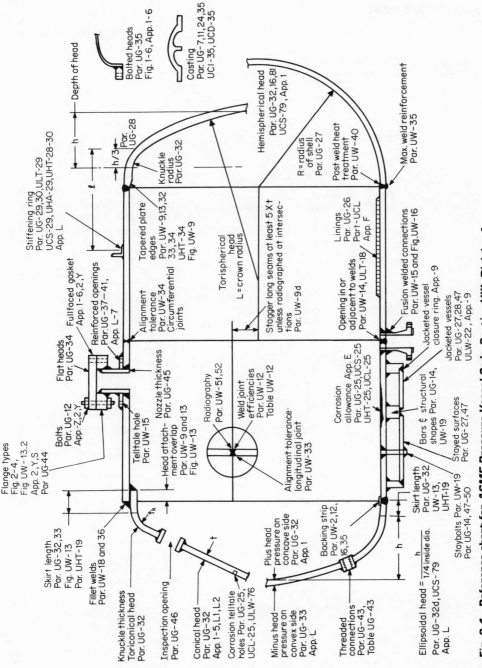

Flange types
Fig. 2-4, UW-13.2
App. 2, Y, S
Par. UG-44

Bolted heads
Par. UG-35
Fig. 1-6, App. 1-6

Casting
Par. UG-7,11,24,35
UCI-35, UCD-35

Depth of head

Stiffening ring
Par. UG-29,30,ULT-29
UCS-29, UHA-29,UHT-28-30
App. L

Par. UG-28

Hemispherical head
Par. UG-32,16,81
UCS-79, App. 1

Knuckle
radius
Par.UG-32

Max. weld reinforcement
Par. UW-35

R = radius
of shell
Par. UG-27

Post weld heat
treatment
Par. UW-40

Fullfaced gasket
App. 1-6,2,Y

Reinforced openings
Par.UG-37-41,
App. L-7

Tapered plate
edges
Par. UW-9,13,32
33, 34
UHT-34
Fig. UW-9

Alignment
tolerance
Par. UW-34
Circumferential
joints

Torispherical
head
L = crown radius

Stagger long seams of least 5 X t
unless radiographed at intersec-
tions
Par. UW-9d

Linings
Par. UG-26
Part-UCL
App. F

Opening in or
adjacent to welds
Par. UW-14, ULT-18

Fusion welded connections
Par. UW-15 and Fig.UW-16

Jacketed vessel
closure ring. App. 9

Jacketed vessels
Par. UG-27,28,47
ULW-22, App. 9

Flat heads
Par. UG-34

Telltale hole
Par. UW-15

Nozzle thickness
Par. UG-45

Head attach-
ment overlap
Par. UW-9 and 13
Fig. UW-13

Radiography
Par. UW-51,52

Weld joint
efficiencies
Par. UW-12
Table UW-12

Corrosion
allowance App. E
Par. UG-25,UCS-25
UHT-25, UCL-25

Bars & structural
shapes Par. UG-14,
UW-19

Stayed surfaces
Par. UG-27,47

Bolts
Par. UG-12
App. 2, Y

Alignment tolerance
longitudinal joint
Par. UW-33

Knuckle thickness
Toriconical head
Par. UG-32

Inspection opening
Par. UG-46

Skirt length
Par. UG-32,33
Fig. UW-13
Par. UHT-19

Fillet welds
Par. UW-18 and 36

Corrosion telltale
holes Par. UG-25,
UCL-25, ULW-76

Conical head
Par. UG-32
App. 1-5,L1,L2

Plus head
pressure on concave side
Par. UG-32
App. 1

Minus head
pressure on
convex side
Par. UG-33
App. L

Backing strip
Par. UW-2,12,
16,35

Threaded
connections
Par.UG-43,
Table UG-43

Staybolts Par. UW-19
Par. UG-14, 47-50

Ellipsoidal head = 1/4 inside dia.
Par. UG-32d, UCS-79
App. L

Skirt length
Par. UG-32
UW-13,
UHT-19

Fig. 2.1 Reference chart for *ASME Pressure Vessel Code*, Section VIII, Division 1.

MATERIAL CLASSIFICATION AND IDENTIFICATION

TABLE 2.1   Classes of materials

Material	Covering Code part	Applicable Code stress value tables	Remarks
Carbon and low-alloy steels	UCS	UCS-23 and UCS-27	Basis for establishing stress values — Code Appendix P Low-temperature operation requires use of notch-tough materials — Code Par. UCS-65, UCS-66, UCS-67, and UG-84 Corrosion allowance — Code Par. UCS-25 In high-temperature operation, creep strength is essential Design temperature — Code Par. UG-20 Design pressure — Code Par. UG-21, Footnote 8 Temperature above 800°F may cause carbide phase of carbon steel to convert to graphite
Nonferrous metals	UNF	UNF-23	Basis for establishing stress values — Code Appendix P Metal characteristics — Code Appendix NF Low-temperature operation — Code Par. UNF-65 Nonferrous castings — Code Par. UNF-8
High-alloy steels	UHA	UHA-23	Selection and treatment of austenitic chromium-nickel steels — Appendix HA

Material	Covering Code part	Applicable Code stress value tables	Remarks
High-alloy steels (cont.)			Inspection and tests — Code Par. UHA-34, UHA-50, UHA-51, and UHA-52 Fluid penetrant oil inspection required if shell thickness exceeds ¾ in — Code Par. UHA-34 Low-temperature operation Code Par. UHA-51 and UG-84 High-alloy castings — Code Par. UHA-8
Castings			Code Par. UG-11, UG-24, UCS-8 — Code Appendix 7
Cast iron	UCI	UCI-23	Vessels not permitted to contain lethal or flammable substances Inspection and tests — Code Par. UCI-90, UCI-99, and UCI-101
Dual cast iron	UCI		Code Par. UC-1, UCI-23, and UCI-29
Integrally clad plate or applied linings	UCL	(See Code Par. UCL-11 and UCL-23.)	Suggest careful study of entire UCL section Qualification of welding procedure — Code Par. UCL-40 to -46 Postweld heat treatment — Code Par. UCL-34 (including cautionary footnote) Inspection and test — Code Par. UCL-50, UCL-51, and UCL-52 Spot radiography required if cladding is included in computing required

TABLE 2.1   Classes of materials (Cont.)

Material	Covering Code part	Applicable Code stress value tables	Remarks
Integrally clad plate or applied linings (cont.)			thickness — Code Par. UCL-23(c) Use of linings — Code Par. UG-26 and Code Appendix F
Welded and seamless pipe (carbon and low-alloy steels)		UCS-23 and UCS-27	Thickness under internal pressure — Code Par. UG-27 Thickness under external pressure — Code Par. UG-28 Provide additional thickness when tubes are threaded and when corrosion, erosion, or wear caused by cleaning is expected — Code Par. UG-31 For calculating thickness required, minimum pipe wall thickness is 87.5 percent of nominal wall thickness. 30-in maximum on welded pipe made by open-hearth, basic oxygen, or electric-furnace process — Code Par. USC-27
Welded and seamless pipe (high-alloy steels)		UHA-23	
Forgings	UF		Materials — Code Par. UG-7, UG-11, and UF-6 Welding — Code Par. UF-32 (See also Section IX Code Par. QW-250 and Variables, Code QW-404.12, QW-406.3, QW-407.2, and QW-409.1 when welding forgings.)

TABLE 2.1 Classes of materials (Cont.)

Material	Covering Code part	Applicable Code stress value tables	Remarks
Low-temperature materials	ULT		Operation at very low temperatures requires use of notch-touch materials
Layered construction	ULW		Vessels having a shell and/or heads made up of two or more separate layers — Code Par. ULW-2
Ferritic steels with tensile properties enhanced by heat treatment	UHT	UHT-23	Scope — Code Par. UHT-1 Marking on plate or stamping, use "low-stress" stamps — Code Par. UHT-86

TABLE 2.2   Material and plate identification and inspection

Remarks	Code reference
General recommendations	Par. UG-4
Material must conform to a material specification given in Code Section II	Par. UG-6 Par. UG-10 Par. UG-11
One set of original identification marks should be visible, name of manufacturer, heat and slab numbers, quality and minimum of the range of tensile strength. If unavoidably cut out, one set must be transferred by vessel manufacturer. Inspector need not witness stamping but must check accuracy of transfer	Par. UG-77(a)
To guard against cracks in steel plate less than ¼ in and nonferrous plate less than ½ in, any method of transfer acceptable to inspector to identify material may be used.	Par. UG-77(b)
Plates given full heat treatment by the producing mill shall be stamped with the letters MT	Par. UG-85 Par. SA-20 Section II
When required heat treatments of material are not performed by the mill, they must be made under control of the fabricator who shall place the letter T following the letter G in the mill plate marking. To show that the heat treatments have been completed, the fabricator must show a supplement to the Mill Test Report	Par. UG-85 Par. UCS-85 Par. UHT-5 Par. UHT-81
Vessel manufacturer shall obtain Certified Test Report or Certificate of Compliance for plate material.	Par. UG-93
Inspector shall examine Certified Test Report	Par. UG-93
When stamping ferritic steels enhanced by heat treatment, use low-stress stamps	Par. UHT-86
Other markings in lieu of stamping	Par. UG-77, -93
For some product forms, marking of container or bundle is required	Par. ULT-86

TABLE 2.3   Nonidentified material

Remarks	Code reference
Material must conform to a material specification given in Code Section II except as otherwise permitted.	Par. UG-10 Par. UG-11

CYLINDRICAL AND SPHERICAL SHELLS

TABLE 2.4 Calculations for cylindrical and spherical shells under internal pressure

Equations	Remarks	Code reference
Cylindrical shells under internal pressure		
$$t = \frac{PR}{SE - 0.6P}$$ or $$P = \frac{SEt}{R + 0.6t}$$ in which: t = minimum thickness, in P = allowable pressure, psi S = allowable stress, psi E = joint efficiency, percent R = inside radius, in	Equations used when t is less than ½ R or P is less than 0.385 SE	Par. UG-27
	Circumferential joint formula in Code Par. UG-27 applies only if the joint efficiency is less than one-half the longitudinal joint efficiency or if the loadings described in Code Par UG-22 cause bending or tension (see examples in Code Appendix (L-2)	Par. UG-27 Par. UG-22 Appendix L-2 Appendix D Appendix G
	Thick cylindrical shells	Par. UA-2
	Formulas in terms of the outside radius may be used in place of those in Par. UG-27	Appendix 1
	Welded joint efficiencies	Par. UW-12 Table UW-12
	Stress values of materials (use appropriate table)	Table UCS-23 Table UCI-23 Table UHA-23 Table UNF-23 Table UCD-23 Table UHT-23 Table ULT-23
	Corrosion allowance for carbon-steel vessels	Par. UCS-25
Spherical shells under internal pressure		
$$t = \frac{PR}{2SE - 0.2P}$$ or $$P = \frac{2SEt}{R + 0.2t}$$	Equations used when t is less than 0.356 R or P is less than 0.665 SE (unknowns same as above)	Par. UG-27
	Thick spherical shells Stress values of materials	Appendix 1 Table UCS-23 Table UCI-23 Table UHA-23 Table UNF-23 Table UCD-23 Table UHT-23 Table ULT-23

USE OF INTERNAL-PRESSURE CYLINDRICAL SHELL THICKNESS TABLES (TABLES 2.4*a*–2.4*f*)

These tables may be applied to seamless shells (joint efficiency = 100 percent = 1.0). These tables may also be applied to shells with double-welded butt longitudinal joints that have been fully radiographed (joint efficiency = 1.0), spot-radiographed (joint efficiency = 0.85), or not radiographed (joint efficiency = 0.70). They may also be applied to single-welded butt joints with a backing strip that are fully radiographed (joint efficiency = 0.90), spot radiographed (joint efficiency = 0.80), or not radiographed (joint efficiency = 0.65). The tables may also be used when stress reduction factors of 85 percent (0.85) or 80 percent (0.80) are applied.

The required shell thickness may be determined by the following steps:

1. Determine the appropriate table corresponding to the desired joint efficiency. Note that the value of joint efficiency for each table is printed in the upper left-hand corner.

2. Locate the desired diameter. The diameter values are printed across the top of the page for each table.

3. Locate the desired pressure. The pressure values are located in the left column.

4. Read the thickness corresponding to the diameter and pressure.

Each table is based on an allowable stress value of 13,800 psi. For other stress values, multiply the thickness read from the table by either the following constant factors or factors read from Fig. 2.2.

Stress	11,300	12,500	13,800	15,000	16,300	17,500	18,800
Constant	1.224	1.105	1.000	0.919	0.846	0.787	0.733

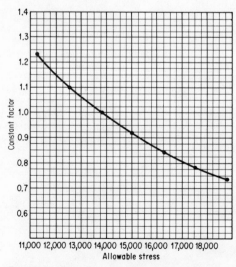

Fig. 2.2   Constant factor versus stress.

(Text continues on p. 35.)

TABLE 2.4a Shell thickness tables in inches: joint efficiency = 1.0

SHELL THICKNESS TABLES IN INCHES

JOINT EFF.=1.00   ALLOWABLE STRESS VALUE= 13800. PSI

INSIDE DIAMETER, IN.

PSI	12.	18.	24.	30.	36.	42.	48.	54.	60.	66.	72.	78.	84.	90.	96.	102.	108.	114.	120.	126.	132.
15.	0.007	0.010	0.013	0.016	0.020	0.023	0.026	0.029	0.033	0.036	0.039	0.042	0.046	0.049	0.052	0.055	0.059	0.062	0.065	0.069	0.072
20.	0.009	0.013	0.017	0.021	0.026	0.030	0.035	0.039	0.044	0.048	0.052	0.057	0.061	0.065	0.070	0.074	0.078	0.083	0.087	0.091	0.096
25.	0.011	0.016	0.022	0.027	0.033	0.038	0.044	0.049	0.054	0.060	0.065	0.071	0.076	0.082	0.087	0.092	0.098	0.103	0.109	0.114	0.120
30.	0.013	0.020	0.026	0.033	0.039	0.046	0.052	0.059	0.065	0.072	0.078	0.085	0.091	0.098	0.104	0.111	0.118	0.124	0.131	0.137	0.144
35.	0.015	0.023	0.030	0.038	0.046	0.053	0.061	0.069	0.076	0.084	0.091	0.099	0.107	0.114	0.122	0.130	0.137	0.145	0.152	0.160	0.168
40.	0.017	0.026	0.035	0.044	0.052	0.061	0.070	0.078	0.087	0.096	0.105	0.113	0.122	0.131	0.139	0.148	0.157	0.166	0.174	0.183	0.192
45.	0.020	0.029	0.039	0.049	0.059	0.069	0.078	0.088	0.098	0.108	0.118	0.127	0.137	0.147	0.157	0.167	0.176	0.186	0.196	0.206	0.216
50.	0.022	0.033	0.044	0.054	0.065	0.076	0.087	0.098	0.109	0.120	0.131	0.142	0.153	0.163	0.174	0.185	0.196	0.207	0.218	0.229	0.240
55.	0.024	0.036	0.048	0.060	0.072	0.084	0.096	0.108	0.120	0.132	0.144	0.156	0.168	0.180	0.192	0.204	0.216	0.228	0.240	0.252	0.264
60.	0.026	0.039	0.052	0.065	0.078	0.092	0.105	0.118	0.131	0.144	0.157	0.170	0.183	0.196	0.209	0.222	0.235	0.248	0.262	0.275	0.288
65.	0.028	0.043	0.057	0.071	0.085	0.099	0.113	0.128	0.142	0.156	0.170	0.184	0.198	0.213	0.227	0.241	0.255	0.269	0.283	0.298	0.312
70.	0.031	0.046	0.061	0.076	0.092	0.107	0.122	0.137	0.153	0.168	0.183	0.198	0.214	0.229	0.244	0.259	0.275	0.290	0.305	0.321	0.336
75.	0.033	0.049	0.065	0.082	0.098	0.115	0.131	0.147	0.164	0.180	0.196	0.213	0.229	0.245	0.262	0.278	0.294	0.311	0.327	0.344	0.360
80.	0.035	0.052	0.070	0.087	0.105	0.122	0.140	0.157	0.175	0.192	0.209	0.227	0.244	0.262	0.279	0.297	0.314	0.332	0.349	0.366	0.384
85.	0.037	0.056	0.074	0.093	0.111	0.130	0.148	0.167	0.185	0.204	0.223	0.241	0.260	0.278	0.297	0.315	0.334	0.352	0.371	0.389	0.408
90.	0.039	0.059	0.079	0.098	0.118	0.137	0.157	0.177	0.196	0.216	0.236	0.255	0.275	0.295	0.314	0.334	0.354	0.373	0.393	0.412	0.432
95.	0.041	0.062	0.083	0.104	0.124	0.145	0.166	0.187	0.207	0.228	0.249	0.270	0.290	0.311	0.332	0.353	0.373	0.394	0.415	0.435	0.456
100.	0.044	0.066	0.087	0.109	0.131	0.153	0.175	0.197	0.218	0.240	0.262	0.284	0.306	0.328	0.349	0.371	0.393	0.415	0.437	0.459	0.480
105.	0.046	0.069	0.092	0.115	0.138	0.161	0.183	0.206	0.229	0.252	0.275	0.298	0.321	0.344	0.367	0.390	0.413	0.436	0.459	0.482	0.504
110.	0.048	0.072	0.096	0.120	0.144	0.168	0.192	0.216	0.240	0.264	0.288	0.312	0.336	0.360	0.384	0.408	0.433	0.457	0.481	0.505	0.529
115.	0.050	0.075	0.101	0.126	0.151	0.176	0.201	0.226	0.251	0.276	0.302	0.327	0.352	0.377	0.402	0.427	0.452	0.477	0.503	0.528	0.553
120.	0.052	0.079	0.105	0.131	0.157	0.184	0.210	0.236	0.262	0.288	0.315	0.341	0.367	0.393	0.420	0.446	0.472	0.498	0.524	0.551	0.577
125.	0.055	0.082	0.109	0.137	0.164	0.191	0.219	0.246	0.273	0.301	0.328	0.355	0.383	0.410	0.437	0.464	0.492	0.519	0.546	0.574	0.601
130.	0.057	0.085	0.114	0.142	0.171	0.199	0.227	0.256	0.284	0.313	0.341	0.369	0.398	0.426	0.455	0.483	0.512	0.540	0.568	0.597	0.625
135.	0.059	0.089	0.118	0.148	0.177	0.207	0.236	0.266	0.295	0.325	0.354	0.384	0.413	0.443	0.472	0.502	0.531	0.561	0.590	0.620	0.649
140.	0.061	0.092	0.122	0.153	0.184	0.214	0.245	0.276	0.306	0.337	0.367	0.398	0.429	0.459	0.490	0.521	0.551	0.582	0.612	0.643	0.674
145.	0.063	0.095	0.127	0.159	0.190	0.222	0.254	0.285	0.317	0.349	0.381	0.412	0.444	0.476	0.508	0.539	0.571	0.603	0.634	0.666	0.698
150.	0.066	0.098	0.131	0.164	0.197	0.230	0.263	0.295	0.328	0.361	0.394	0.427	0.460	0.492	0.525	0.558	0.591	0.624	0.656	0.689	0.722
160.	0.070	0.105	0.140	0.175	0.210	0.245	0.280	0.315	0.350	0.385	0.420	0.455	0.490	0.525	0.560	0.595	0.630	0.665	0.701	0.736	0.771
170.	0.074	0.112	0.149	0.186	0.223	0.261	0.298	0.335	0.372	0.410	0.447	0.484	0.521	0.558	0.596	0.633	0.670	0.707	0.745	0.782	0.819
180.	0.079	0.118	0.158	0.197	0.237	0.276	0.316	0.355	0.394	0.434	0.473	0.513	0.552	0.592	0.631	0.670	0.710	0.749	0.789	0.828	0.868
190.	0.083	0.125	0.167	0.208	0.250	0.292	0.333	0.375	0.416	0.458	0.500	0.541	0.583	0.625	0.666	0.708	0.750	0.791	0.833	0.875	0.916
200.	0.088	0.132	0.175	0.219	0.263	0.307	0.351	0.395	0.439	0.482	0.526	0.570	0.614	0.658	0.702	0.746	0.789	0.833	0.877	0.921	0.965
210.	0.092	0.138	0.184	0.230	0.276	0.323	0.369	0.415	0.461	0.507	0.553	0.599	0.645	0.691	0.737	0.783	0.829	0.875	0.921	0.968	1.014
220.	0.097	0.145	0.193	0.241	0.290	0.338	0.386	0.435	0.483	0.531	0.579	0.628	0.676	0.724	0.773	0.821	0.869	0.917	0.966	1.014	1.062
230.	0.101	0.152	0.202	0.253	0.303	0.354	0.404	0.455	0.505	0.556	0.606	0.657	0.707	0.758	0.808	0.859	0.909	0.960	1.010	1.061	1.111
240.	0.105	0.158	0.211	0.264	0.316	0.369	0.422	0.475	0.527	0.580	0.633	0.685	0.738	0.791	0.844	0.896	0.949	1.002	1.054	1.107	1.160
250.	0.110	0.165	0.220	0.275	0.330	0.385	0.440	0.495	0.549	0.604	0.659	0.714	0.769	0.824	0.879	0.934	0.989	1.044	1.099	1.154	1.209
260.	0.114	0.172	0.229	0.286	0.343	0.400	0.457	0.515	0.572	0.629	0.686	0.743	0.800	0.858	0.915	0.972	1.029	1.086	1.143	1.201	1.258
270.	0.119	0.178	0.238	0.297	0.356	0.416	0.475	0.535	0.594	0.653	0.713	0.772	0.832	0.891	0.950	1.010	1.069	1.128	1.188	1.247	1.307
280.	0.123	0.185	0.246	0.308	0.370	0.431	0.493	0.555	0.616	0.678	0.739	0.801	0.863	0.924	0.986	1.048	1.109	1.171	1.232	1.294	1.356
290.	0.128	0.192	0.255	0.319	0.383	0.447	0.511	0.575	0.638	0.702	0.766	0.830	0.894	0.958	1.022	1.085	1.149	1.213	1.277	1.341	1.405
300.	0.132	0.198	0.264	0.330	0.396	0.463	0.529	0.595	0.661	0.727	0.793	0.859	0.925	0.991	1.057	1.123	1.189	1.256	1.322	1.388	1.454
310.	0.137	0.205	0.273	0.342	0.410	0.478	0.546	0.615	0.683	0.751	0.820	0.888	0.956	1.025	1.093	1.161	1.230	1.298	1.366	1.435	1.503
320.	0.141	0.212	0.282	0.353	0.423	0.494	0.564	0.635	0.705	0.776	0.847	0.917	0.988	1.058	1.129	1.199	1.270	1.340	1.411	1.481	1.552
330.	0.146	0.218	0.291	0.364	0.437	0.509	0.582	0.655	0.728	0.801	0.873	0.946	1.019	1.092	1.165	1.237	1.310	1.383	1.456	1.528	1.601
340.	0.150	0.225	0.300	0.375	0.450	0.525	0.600	0.675	0.750	0.825	0.900	0.975	1.050	1.125	1.200	1.275	1.350	1.425	1.500	1.575	1.650
350.	0.155	0.232	0.309	0.386	0.464	0.541	0.618	0.695	0.773	0.850	0.927	1.004	1.082	1.159	1.236	1.313	1.391	1.468	1.545	1.623	1.700
360.	0.159	0.239	0.318	0.398	0.477	0.557	0.636	0.716	0.795	0.875	0.954	1.034	1.113	1.193	1.272	1.352	1.431	1.511	1.590	1.670	1.749
370.	0.163	0.245	0.327	0.409	0.490	0.572	0.654	0.736	0.817	0.899	0.981	1.063	1.144	1.226	1.308	1.390	1.471	1.553	1.635	1.717	1.798

TABLE 2.4*b* Shell thickness tables in inches: joint efficiency = 0.90

SHELL THICKNESS TABLES IN INCHES

JOINT EFF.=0.90   ALLOWABLE STRESS VALUE= 13800. PSI

INSIDE DIAMETER,IN.

PSI	12.	18.	24.	30.	36.	42.	48.	54.	60.	66.	72.	78.	84.	90.	96.	102.	108.	114.	120.	126.	132.
15.	0.007	0.011	0.015	0.018	0.022	0.025	0.029	0.033	0.036	0.040	0.044	0.047	0.051	0.054	0.058	0.062	0.065	0.069	0.073	0.076	0.080
20.	0.010	0.015	0.019	0.024	0.029	0.034	0.039	0.044	0.048	0.053	0.058	0.063	0.068	0.073	0.077	0.082	0.087	0.092	0.097	0.102	0.106
25.	0.012	0.018	0.024	0.030	0.036	0.042	0.048	0.054	0.060	0.067	0.073	0.079	0.085	0.091	0.097	0.103	0.109	0.115	0.121	0.127	0.133
30.	0.015	0.022	0.029	0.036	0.044	0.051	0.058	0.065	0.073	0.080	0.087	0.094	0.102	0.109	0.116	0.123	0.131	0.138	0.145	0.152	0.160
35.	0.017	0.025	0.034	0.042	0.051	0.059	0.068	0.076	0.085	0.093	0.102	0.110	0.119	0.127	0.135	0.144	0.152	0.161	0.169	0.178	0.186
40.	0.019	0.029	0.039	0.048	0.058	0.068	0.077	0.087	0.097	0.106	0.116	0.126	0.136	0.145	0.155	0.165	0.174	0.184	0.194	0.203	0.213
45.	0.022	0.033	0.044	0.054	0.065	0.076	0.087	0.098	0.109	0.120	0.131	0.142	0.153	0.163	0.174	0.185	0.196	0.207	0.218	0.229	0.240
50.	0.024	0.036	0.048	0.061	0.073	0.085	0.097	0.109	0.121	0.133	0.145	0.157	0.169	0.182	0.194	0.206	0.218	0.230	0.242	0.254	0.266
55.	0.027	0.040	0.053	0.067	0.080	0.093	0.107	0.120	0.133	0.147	0.160	0.173	0.186	0.200	0.213	0.226	0.240	0.253	0.266	0.280	0.293
60.	0.029	0.044	0.058	0.073	0.087	0.102	0.116	0.131	0.145	0.160	0.174	0.189	0.203	0.218	0.233	0.247	0.262	0.276	0.291	0.305	0.320
65.	0.031	0.047	0.063	0.079	0.094	0.110	0.126	0.142	0.157	0.173	0.189	0.205	0.220	0.236	0.252	0.268	0.283	0.299	0.315	0.331	0.346
70.	0.034	0.051	0.068	0.085	0.102	0.119	0.136	0.153	0.170	0.187	0.204	0.221	0.238	0.254	0.271	0.288	0.305	0.322	0.339	0.356	0.373
75.	0.036	0.055	0.073	0.091	0.109	0.127	0.145	0.164	0.182	0.200	0.218	0.236	0.255	0.273	0.291	0.309	0.327	0.345	0.364	0.382	0.400
80.	0.039	0.058	0.078	0.097	0.116	0.136	0.155	0.175	0.194	0.213	0.233	0.252	0.272	0.291	0.310	0.330	0.349	0.369	0.388	0.407	0.427
85.	0.041	0.062	0.082	0.103	0.124	0.144	0.165	0.186	0.206	0.227	0.247	0.268	0.289	0.309	0.330	0.350	0.371	0.392	0.412	0.433	0.454
90.	0.044	0.066	0.087	0.109	0.131	0.153	0.175	0.197	0.218	0.240	0.262	0.284	0.306	0.328	0.349	0.371	0.393	0.415	0.437	0.459	0.480
95.	0.046	0.069	0.092	0.115	0.138	0.161	0.184	0.207	0.231	0.254	0.277	0.300	0.323	0.346	0.369	0.392	0.415	0.438	0.461	0.484	0.507
100.	0.049	0.073	0.097	0.121	0.146	0.170	0.194	0.218	0.243	0.267	0.291	0.316	0.340	0.364	0.388	0.413	0.437	0.461	0.485	0.510	0.534
105.	0.051	0.076	0.102	0.127	0.153	0.178	0.204	0.229	0.255	0.280	0.306	0.331	0.357	0.382	0.408	0.433	0.459	0.484	0.510	0.535	0.561
110.	0.053	0.080	0.107	0.134	0.160	0.187	0.214	0.240	0.267	0.294	0.321	0.347	0.374	0.401	0.427	0.454	0.481	0.508	0.534	0.561	0.588
115.	0.056	0.084	0.112	0.140	0.168	0.196	0.223	0.251	0.279	0.307	0.335	0.363	0.391	0.419	0.447	0.475	0.503	0.531	0.559	0.587	0.615
120.	0.058	0.087	0.117	0.146	0.175	0.204	0.233	0.262	0.292	0.321	0.350	0.379	0.408	0.437	0.466	0.496	0.525	0.554	0.583	0.612	0.641
125.	0.061	0.091	0.122	0.152	0.182	0.213	0.243	0.273	0.304	0.334	0.365	0.395	0.425	0.456	0.486	0.516	0.547	0.577	0.608	0.638	0.668
130.	0.063	0.095	0.126	0.158	0.190	0.221	0.253	0.284	0.316	0.348	0.379	0.411	0.442	0.474	0.506	0.537	0.569	0.600	0.632	0.664	0.695
135.	0.066	0.098	0.131	0.164	0.197	0.230	0.263	0.295	0.328	0.361	0.394	0.427	0.460	0.492	0.525	0.558	0.591	0.624	0.656	0.689	0.722
140.	0.068	0.102	0.136	0.170	0.204	0.238	0.272	0.306	0.340	0.375	0.409	0.443	0.477	0.511	0.545	0.579	0.613	0.647	0.681	0.715	0.749
145.	0.071	0.106	0.141	0.176	0.212	0.247	0.282	0.317	0.353	0.388	0.423	0.459	0.494	0.529	0.564	0.600	0.635	0.670	0.705	0.741	0.776
150.	0.073	0.109	0.146	0.182	0.219	0.255	0.292	0.328	0.365	0.401	0.438	0.474	0.511	0.547	0.584	0.620	0.657	0.693	0.730	0.766	0.803
155.	0.075	0.113	0.151	0.189	0.226	0.264	0.302	0.340	0.377	0.415	0.453	0.490	0.528	0.566	0.604	0.641	0.679	0.717	0.754	0.792	0.830
160.	0.078	0.117	0.156	0.195	0.234	0.273	0.312	0.351	0.389	0.428	0.467	0.506	0.545	0.584	0.623	0.662	0.701	0.740	0.779	0.818	0.857
165.	0.080	0.121	0.161	0.201	0.241	0.281	0.321	0.362	0.402	0.442	0.482	0.522	0.563	0.603	0.643	0.683	0.723	0.763	0.804	0.844	0.884
170.	0.083	0.124	0.166	0.207	0.248	0.290	0.331	0.373	0.414	0.455	0.497	0.538	0.580	0.621	0.662	0.704	0.745	0.787	0.828	0.869	0.911
175.	0.085	0.128	0.171	0.213	0.256	0.298	0.341	0.384	0.426	0.469	0.512	0.554	0.597	0.639	0.682	0.725	0.767	0.810	0.853	0.895	0.938
180.	0.088	0.132	0.175	0.219	0.263	0.307	0.351	0.395	0.439	0.482	0.526	0.570	0.614	0.658	0.702	0.746	0.789	0.833	0.877	0.921	0.965
185.	0.090	0.135	0.180	0.225	0.271	0.316	0.361	0.406	0.451	0.496	0.541	0.586	0.631	0.676	0.721	0.767	0.812	0.857	0.902	0.947	0.992
190.	0.093	0.139	0.185	0.232	0.278	0.324	0.371	0.417	0.463	0.510	0.556	0.602	0.648	0.695	0.741	0.787	0.834	0.880	0.926	0.973	1.019
195.	0.095	0.143	0.190	0.238	0.285	0.333	0.380	0.428	0.475	0.523	0.571	0.618	0.666	0.713	0.761	0.808	0.856	0.903	0.951	0.999	1.046
200.	0.098	0.146	0.195	0.244	0.293	0.341	0.390	0.439	0.488	0.537	0.585	0.634	0.683	0.732	0.780	0.829	0.878	0.927	0.976	1.024	1.073
210.	0.102	0.154	0.205	0.256	0.307	0.359	0.410	0.461	0.512	0.564	0.615	0.666	0.717	0.769	0.820	0.871	0.922	0.974	1.025	1.076	1.127
220.	0.107	0.161	0.215	0.269	0.322	0.376	0.430	0.483	0.537	0.591	0.645	0.698	0.752	0.806	0.859	0.913	0.967	1.021	1.074	1.128	1.182
230.	0.112	0.169	0.225	0.281	0.337	0.393	0.449	0.506	0.562	0.618	0.674	0.730	0.787	0.843	0.899	0.955	1.011	1.067	1.124	1.180	1.236
240.	0.117	0.176	0.235	0.293	0.352	0.411	0.469	0.528	0.587	0.645	0.704	0.762	0.821	0.880	0.938	0.997	1.056	1.114	1.173	1.232	1.290
250.	0.122	0.183	0.244	0.306	0.367	0.428	0.489	0.550	0.611	0.672	0.733	0.795	0.856	0.917	0.978	1.039	1.100	1.161	1.222	1.284	1.345
260.	0.127	0.191	0.254	0.318	0.382	0.445	0.509	0.572	0.636	0.700	0.763	0.827	0.890	0.954	1.018	1.081	1.145	1.208	1.272	1.336	1.399
270.	0.132	0.198	0.264	0.330	0.396	0.463	0.529	0.595	0.661	0.727	0.793	0.859	0.925	0.991	1.057	1.123	1.189	1.256	1.322	1.388	1.454
280.	0.137	0.206	0.274	0.343	0.411	0.480	0.548	0.617	0.686	0.754	0.823	0.891	0.960	1.028	1.097	1.166	1.234	1.303	1.371	1.440	1.508
290.	0.142	0.213	0.284	0.355	0.426	0.497	0.568	0.639	0.710	0.781	0.853	0.924	0.995	1.066	1.137	1.208	1.279	1.350	1.421	1.492	1.563
300.	0.147	0.221	0.294	0.368	0.441	0.515	0.588	0.662	0.735	0.809	0.882	0.956	1.029	1.103	1.176	1.250	1.324	1.397	1.471	1.544	1.618
310.	0.152	0.228	0.304	0.380	0.456	0.532	0.608	0.684	0.760	0.836	0.912	0.988	1.064	1.140	1.216	1.292	1.368	1.444	1.520	1.596	1.672
320.	0.157	0.236	0.314	0.393	0.471	0.550	0.628	0.707	0.785	0.864	0.942	1.021	1.099	1.178	1.256	1.335	1.413	1.492	1.570	1.649	1.727
330.	0.162	0.243	0.324	0.405	0.486	0.567	0.648	0.729	0.810	0.891	0.972	1.053	1.134	1.215	1.296	1.377	1.458	1.539	1.620	1.701	1.782
340.	0.167	0.250	0.334	0.417	0.501	0.584	0.668	0.751	0.835	0.918	1.002	1.085	1.169	1.252	1.336	1.419	1.503	1.586	1.670	1.753	1.837
350.	0.172	0.258	0.344	0.430	0.516	0.602	0.688	0.774	0.860	0.946	1.032	1.118	1.204	1.290	1.376	1.462	1.548	1.634	1.720	1.806	1.892
360.	0.177	0.265	0.354	0.442	0.531	0.619	0.708	0.796	0.885	0.973	1.062	1.150	1.239	1.327	1.416	1.504	1.593	1.681	1.770	1.858	1.947
370.	0.182	0.273	0.364	0.455	0.546	0.637	0.728	0.819	0.910	1.001	1.092	1.183	1.274	1.365	1.456	1.547	1.638	1.729	1.820	1.911	2.002

TABLE 2.4c Shell thickness tables in inches: joint efficiency = 0.85

SHELL THICKNESS TABLES IN INCHES

JOINT EFF.=0.85 ALLOWABLE STRESS VALUE= 13800. PSI

INSIDE DIAMETER, IN.

PSI	12.	18.	24.	30.	36.	42.	48.	54.	60.	66.	72.	78.	84.	90.	96.	102.	108.	114.	120.	126.	132.
15.	0.008	0.012	0.015	0.019	0.023	0.027	0.031	0.035	0.038	0.042	0.046	0.050	0.054	0.058	0.061	0.065	0.069	0.073	0.077	0.081	0.084
20.	0.010	0.015	0.020	0.026	0.031	0.036	0.041	0.046	0.051	0.056	0.061	0.067	0.072	0.077	0.082	0.087	0.092	0.097	0.102	0.108	0.113
25.	0.013	0.019	0.026	0.032	0.038	0.045	0.051	0.058	0.064	0.070	0.077	0.083	0.090	0.096	0.102	0.109	0.115	0.122	0.128	0.134	0.141
30.	0.015	0.023	0.031	0.038	0.046	0.054	0.061	0.069	0.077	0.085	0.092	0.100	0.108	0.115	0.123	0.131	0.138	0.146	0.154	0.161	0.169
35.	0.018	0.027	0.036	0.045	0.054	0.063	0.072	0.081	0.090	0.099	0.108	0.117	0.126	0.135	0.143	0.152	0.161	0.170	0.179	0.188	0.197
40.	0.021	0.031	0.041	0.051	0.062	0.072	0.082	0.092	0.103	0.113	0.123	0.133	0.144	0.154	0.164	0.174	0.185	0.195	0.205	0.215	0.226
45.	0.023	0.035	0.046	0.058	0.069	0.081	0.092	0.104	0.115	0.127	0.138	0.150	0.161	0.173	0.185	0.196	0.208	0.219	0.231	0.242	0.254
50.	0.026	0.038	0.051	0.064	0.077	0.090	0.103	0.115	0.128	0.141	0.154	0.167	0.179	0.192	0.205	0.218	0.231	0.244	0.256	0.269	0.282
55.	0.028	0.042	0.056	0.071	0.085	0.099	0.113	0.127	0.141	0.155	0.169	0.183	0.197	0.212	0.226	0.240	0.254	0.268	0.282	0.296	0.310
60.	0.031	0.046	0.062	0.077	0.092	0.108	0.123	0.139	0.154	0.169	0.185	0.200	0.215	0.231	0.246	0.262	0.277	0.292	0.308	0.323	0.339
65.	0.033	0.050	0.067	0.083	0.100	0.117	0.133	0.150	0.167	0.183	0.200	0.217	0.234	0.250	0.267	0.284	0.300	0.317	0.334	0.350	0.367
70.	0.036	0.054	0.072	0.090	0.108	0.126	0.144	0.162	0.180	0.198	0.216	0.234	0.252	0.270	0.287	0.305	0.323	0.341	0.359	0.377	0.395
75.	0.039	0.058	0.077	0.096	0.116	0.135	0.154	0.173	0.193	0.212	0.231	0.250	0.270	0.289	0.308	0.327	0.347	0.366	0.385	0.404	0.424
80.	0.041	0.062	0.082	0.103	0.123	0.144	0.164	0.185	0.205	0.226	0.247	0.267	0.288	0.308	0.329	0.349	0.370	0.390	0.411	0.431	0.452
85.	0.044	0.066	0.087	0.109	0.131	0.153	0.175	0.197	0.218	0.240	0.262	0.284	0.306	0.328	0.349	0.371	0.393	0.415	0.437	0.459	0.480
90.	0.046	0.069	0.092	0.116	0.139	0.162	0.185	0.208	0.231	0.254	0.277	0.301	0.324	0.347	0.370	0.393	0.416	0.439	0.462	0.486	0.509
95.	0.049	0.073	0.098	0.122	0.146	0.171	0.195	0.220	0.244	0.269	0.293	0.317	0.342	0.366	0.391	0.415	0.439	0.464	0.488	0.513	0.537
100.	0.051	0.077	0.103	0.129	0.154	0.180	0.206	0.231	0.257	0.283	0.308	0.334	0.360	0.386	0.411	0.437	0.463	0.488	0.514	0.540	0.566
105.	0.054	0.081	0.108	0.135	0.162	0.189	0.216	0.243	0.270	0.297	0.324	0.351	0.378	0.405	0.432	0.459	0.486	0.513	0.540	0.567	0.594
110.	0.057	0.085	0.113	0.141	0.170	0.198	0.226	0.255	0.283	0.311	0.340	0.368	0.396	0.424	0.453	0.481	0.509	0.538	0.566	0.594	0.622
115.	0.059	0.089	0.118	0.148	0.178	0.207	0.237	0.266	0.296	0.325	0.355	0.385	0.414	0.444	0.473	0.503	0.533	0.562	0.592	0.621	0.651
120.	0.062	0.093	0.124	0.154	0.185	0.216	0.247	0.278	0.309	0.340	0.371	0.402	0.432	0.463	0.494	0.525	0.556	0.587	0.618	0.648	0.679
125.	0.064	0.097	0.129	0.161	0.193	0.225	0.257	0.290	0.322	0.354	0.386	0.418	0.450	0.483	0.515	0.547	0.579	0.611	0.644	0.676	0.708
130.	0.067	0.100	0.134	0.167	0.201	0.234	0.268	0.301	0.335	0.368	0.402	0.435	0.469	0.502	0.536	0.569	0.602	0.636	0.669	0.703	0.736
135.	0.070	0.104	0.139	0.174	0.209	0.243	0.278	0.313	0.348	0.382	0.417	0.452	0.487	0.522	0.556	0.591	0.626	0.661	0.695	0.730	0.765
140.	0.072	0.108	0.144	0.180	0.216	0.252	0.289	0.325	0.361	0.397	0.433	0.469	0.505	0.541	0.577	0.613	0.649	0.685	0.721	0.757	0.793
145.	0.075	0.112	0.149	0.187	0.224	0.262	0.299	0.336	0.374	0.411	0.448	0.486	0.523	0.560	0.598	0.635	0.673	0.710	0.747	0.785	0.822
150.	0.077	0.116	0.155	0.193	0.232	0.271	0.309	0.348	0.387	0.425	0.464	0.503	0.541	0.580	0.619	0.657	0.696	0.735	0.773	0.812	0.851
160.	0.083	0.124	0.165	0.206	0.248	0.289	0.330	0.371	0.413	0.454	0.495	0.536	0.578	0.619	0.660	0.701	0.743	0.784	0.825	0.866	0.908
170.	0.088	0.132	0.175	0.219	0.263	0.307	0.351	0.395	0.439	0.482	0.526	0.570	0.614	0.658	0.702	0.746	0.789	0.833	0.877	0.921	0.965
180.	0.093	0.139	0.186	0.232	0.279	0.325	0.372	0.418	0.465	0.511	0.558	0.604	0.650	0.697	0.743	0.790	0.836	0.883	0.929	0.976	1.022
190.	0.098	0.147	0.196	0.245	0.294	0.343	0.393	0.442	0.491	0.540	0.589	0.638	0.687	0.736	0.785	0.834	0.883	0.932	0.981	1.030	1.080
200.	0.103	0.155	0.207	0.258	0.310	0.362	0.413	0.465	0.517	0.568	0.620	0.672	0.724	0.775	0.827	0.879	0.930	0.982	1.034	1.085	1.137
210.	0.108	0.163	0.217	0.271	0.326	0.380	0.434	0.489	0.543	0.597	0.651	0.706	0.760	0.814	0.869	0.923	0.977	1.032	1.086	1.140	1.194
220.	0.114	0.171	0.228	0.285	0.341	0.398	0.455	0.512	0.569	0.626	0.683	0.740	0.797	0.854	0.911	0.967	1.024	1.081	1.138	1.195	1.252
230.	0.119	0.179	0.238	0.298	0.357	0.417	0.476	0.536	0.595	0.655	0.714	0.774	0.833	0.893	0.952	1.012	1.071	1.131	1.190	1.250	1.310
240.	0.124	0.186	0.249	0.311	0.373	0.435	0.497	0.559	0.621	0.684	0.746	0.808	0.870	0.932	0.994	1.056	1.119	1.181	1.243	1.305	1.367
250.	0.130	0.194	0.259	0.324	0.389	0.453	0.518	0.583	0.648	0.712	0.777	0.842	0.907	0.972	1.036	1.101	1.166	1.231	1.295	1.360	1.425
260.	0.135	0.202	0.270	0.337	0.404	0.472	0.539	0.607	0.674	0.741	0.809	0.876	0.943	1.011	1.078	1.146	1.213	1.280	1.348	1.415	1.483
270.	0.140	0.210	0.280	0.350	0.420	0.490	0.560	0.630	0.700	0.770	0.840	0.910	0.980	1.050	1.120	1.190	1.260	1.330	1.400	1.470	1.540
280.	0.145	0.218	0.291	0.363	0.436	0.509	0.581	0.654	0.727	0.799	0.872	0.944	1.017	1.090	1.162	1.235	1.308	1.380	1.453	1.526	1.598
290.	0.151	0.226	0.301	0.376	0.452	0.527	0.602	0.678	0.753	0.828	0.903	0.979	1.054	1.129	1.205	1.280	1.355	1.430	1.506	1.581	1.656
300.	0.156	0.234	0.312	0.390	0.468	0.545	0.623	0.701	0.779	0.857	0.935	1.013	1.091	1.169	1.247	1.325	1.403	1.481	1.558	1.636	1.714
310.	0.161	0.242	0.322	0.403	0.483	0.564	0.644	0.725	0.806	0.886	0.967	1.047	1.128	1.208	1.289	1.370	1.450	1.531	1.611	1.692	1.772
320.	0.166	0.250	0.333	0.416	0.499	0.582	0.666	0.749	0.832	0.915	0.998	1.082	1.165	1.248	1.331	1.414	1.498	1.581	1.664	1.747	1.830
330.	0.172	0.258	0.343	0.429	0.515	0.601	0.687	0.773	0.858	0.944	1.030	1.116	1.202	1.288	1.374	1.459	1.545	1.631	1.717	1.803	1.889
340.	0.177	0.265	0.354	0.442	0.531	0.619	0.708	0.796	0.885	0.973	1.062	1.150	1.239	1.327	1.416	1.504	1.593	1.681	1.770	1.858	1.947
350.	0.182	0.273	0.365	0.456	0.547	0.638	0.729	0.820	0.911	1.003	1.094	1.185	1.276	1.367	1.458	1.549	1.641	1.732	1.823	1.914	2.005
360.	0.188	0.281	0.375	0.469	0.563	0.657	0.750	0.844	0.938	1.032	1.126	1.219	1.313	1.407	1.501	1.595	1.688	1.782	1.876	1.970	2.064
370.	0.193	0.289	0.386	0.482	0.579	0.675	0.772	0.868	0.965	1.061	1.157	1.254	1.350	1.447	1.543	1.640	1.736	1.833	1.929	2.026	2.122

TABLE 2.4d Shell thickness tables in inches: joint efficiency = 0.80

SHELL THICKNESS TABLES IN INCHES

JOINT EFF.=0.80 ALLOWABLE STRESS VALUE= 13800. PSI

INSIDE DIAMETER, IN.

PSI	12.	18.	24.	30.	36.	42.	48.	54.	60.	66.	72.	78.	84.	90.	96.	102.	108.	114.	120.	126.	132.
15.	0.008	0.012	0.016	0.020	0.024	0.029	0.033	0.037	0.041	0.045	0.049	0.053	0.057	0.061	0.065	0.069	0.073	0.078	0.082	0.086	0.090
20.	0.011	0.016	0.022	0.027	0.033	0.038	0.044	0.049	0.054	0.060	0.065	0.071	0.076	0.082	0.087	0.092	0.098	0.103	0.109	0.114	0.120
25.	0.014	0.020	0.027	0.034	0.041	0.048	0.054	0.061	0.068	0.075	0.082	0.088	0.095	0.102	0.109	0.116	0.122	0.129	0.136	0.143	0.150
30.	0.016	0.024	0.033	0.041	0.049	0.057	0.065	0.073	0.082	0.090	0.098	0.106	0.114	0.122	0.131	0.139	0.147	0.155	0.163	0.171	0.180
35.	0.019	0.029	0.038	0.048	0.057	0.067	0.076	0.086	0.095	0.105	0.114	0.124	0.133	0.143	0.152	0.162	0.172	0.181	0.191	0.200	0.210
40.	0.022	0.033	0.044	0.054	0.065	0.076	0.087	0.098	0.109	0.120	0.131	0.142	0.153	0.163	0.174	0.185	0.196	0.207	0.218	0.229	0.240
45.	0.025	0.037	0.049	0.061	0.074	0.086	0.098	0.110	0.123	0.135	0.147	0.159	0.172	0.184	0.196	0.208	0.221	0.233	0.245	0.257	0.270
50.	0.027	0.041	0.054	0.068	0.082	0.095	0.109	0.123	0.136	0.150	0.163	0.177	0.191	0.204	0.218	0.232	0.245	0.259	0.272	0.286	0.300
55.	0.030	0.045	0.060	0.075	0.090	0.105	0.120	0.135	0.150	0.165	0.180	0.195	0.210	0.225	0.240	0.255	0.270	0.285	0.300	0.315	0.330
60.	0.033	0.049	0.065	0.082	0.098	0.115	0.131	0.147	0.164	0.180	0.196	0.213	0.229	0.245	0.262	0.278	0.294	0.311	0.327	0.344	0.360
65.	0.035	0.053	0.071	0.089	0.106	0.124	0.142	0.160	0.177	0.195	0.213	0.230	0.248	0.266	0.284	0.301	0.319	0.337	0.355	0.372	0.390
70.	0.038	0.057	0.076	0.095	0.115	0.134	0.153	0.172	0.191	0.210	0.229	0.248	0.267	0.286	0.306	0.325	0.344	0.363	0.382	0.401	0.420
75.	0.041	0.061	0.082	0.102	0.123	0.143	0.164	0.184	0.205	0.225	0.246	0.266	0.286	0.307	0.327	0.348	0.368	0.389	0.409	0.430	0.450
80.	0.044	0.066	0.087	0.109	0.131	0.153	0.175	0.197	0.218	0.240	0.262	0.284	0.306	0.328	0.349	0.371	0.393	0.415	0.437	0.459	0.480
85.	0.046	0.070	0.093	0.116	0.139	0.162	0.186	0.209	0.232	0.255	0.278	0.302	0.325	0.348	0.371	0.394	0.418	0.441	0.464	0.487	0.511
90.	0.049	0.074	0.098	0.123	0.147	0.172	0.197	0.221	0.246	0.270	0.295	0.319	0.344	0.369	0.393	0.418	0.442	0.467	0.492	0.516	0.541
95.	0.052	0.078	0.104	0.130	0.156	0.182	0.208	0.234	0.259	0.285	0.311	0.337	0.363	0.389	0.415	0.441	0.467	0.493	0.519	0.545	0.571
100.	0.055	0.082	0.109	0.137	0.164	0.191	0.219	0.246	0.273	0.301	0.328	0.355	0.383	0.410	0.437	0.464	0.492	0.519	0.546	0.574	0.601
105.	0.057	0.086	0.115	0.143	0.172	0.201	0.230	0.258	0.287	0.316	0.344	0.373	0.402	0.430	0.459	0.488	0.517	0.545	0.574	0.603	0.631
110.	0.060	0.090	0.120	0.150	0.180	0.210	0.241	0.271	0.301	0.331	0.361	0.391	0.421	0.451	0.481	0.511	0.541	0.571	0.601	0.631	0.662
115.	0.063	0.094	0.126	0.157	0.189	0.220	0.252	0.283	0.314	0.346	0.377	0.409	0.440	0.472	0.503	0.535	0.566	0.597	0.629	0.660	0.692
120.	0.066	0.098	0.131	0.164	0.197	0.230	0.263	0.295	0.328	0.361	0.394	0.427	0.460	0.492	0.525	0.558	0.591	0.624	0.656	0.689	0.722
125.	0.068	0.103	0.137	0.171	0.205	0.239	0.274	0.308	0.342	0.376	0.410	0.445	0.479	0.513	0.547	0.581	0.616	0.650	0.684	0.718	0.752
130.	0.071	0.107	0.142	0.178	0.213	0.249	0.285	0.320	0.356	0.391	0.427	0.463	0.498	0.534	0.569	0.605	0.640	0.676	0.712	0.747	0.783
135.	0.074	0.111	0.148	0.185	0.222	0.259	0.296	0.333	0.370	0.407	0.443	0.480	0.517	0.554	0.591	0.628	0.665	0.702	0.739	0.776	0.813
140.	0.077	0.115	0.153	0.192	0.230	0.268	0.307	0.345	0.383	0.422	0.460	0.498	0.537	0.575	0.613	0.652	0.690	0.728	0.767	0.805	0.843
145.	0.079	0.119	0.159	0.199	0.238	0.278	0.318	0.357	0.397	0.437	0.476	0.516	0.556	0.596	0.635	0.675	0.715	0.755	0.794	0.834	0.874
150.	0.082	0.123	0.164	0.205	0.247	0.288	0.329	0.370	0.411	0.452	0.493	0.534	0.575	0.616	0.658	0.699	0.740	0.781	0.822	0.863	0.904
160.	0.088	0.132	0.175	0.219	0.263	0.307	0.351	0.395	0.439	0.482	0.526	0.570	0.614	0.658	0.702	0.746	0.789	0.833	0.877	0.921	0.965
170.	0.093	0.140	0.187	0.233	0.280	0.326	0.373	0.420	0.466	0.513	0.560	0.606	0.653	0.699	0.746	0.793	0.839	0.886	0.933	0.979	1.026
180.	0.099	0.148	0.198	0.247	0.296	0.346	0.395	0.445	0.494	0.543	0.593	0.642	0.692	0.741	0.790	0.840	0.889	0.939	0.988	1.037	1.087
190.	0.104	0.157	0.209	0.261	0.313	0.365	0.417	0.470	0.522	0.574	0.626	0.678	0.730	0.783	0.835	0.887	0.939	0.991	1.043	1.096	1.148
200.	0.110	0.165	0.220	0.275	0.330	0.385	0.440	0.495	0.549	0.604	0.659	0.714	0.769	0.824	0.879	0.934	0.989	1.044	1.099	1.154	1.209
210.	0.115	0.173	0.231	0.289	0.346	0.404	0.462	0.520	0.577	0.635	0.693	0.750	0.808	0.866	0.924	0.981	1.039	1.097	1.154	1.212	1.270
220.	0.121	0.182	0.242	0.303	0.363	0.424	0.484	0.545	0.605	0.666	0.726	0.787	0.847	0.908	0.968	1.029	1.089	1.150	1.210	1.271	1.331
230.	0.127	0.190	0.253	0.316	0.380	0.443	0.506	0.570	0.633	0.696	0.759	0.823	0.886	0.949	1.013	1.076	1.139	1.203	1.266	1.329	1.392
240.	0.132	0.198	0.264	0.330	0.396	0.463	0.529	0.595	0.661	0.727	0.793	0.859	0.925	0.991	1.057	1.123	1.189	1.256	1.322	1.388	1.454
250.	0.138	0.207	0.275	0.344	0.413	0.482	0.551	0.620	0.689	0.758	0.826	0.895	0.964	1.033	1.102	1.171	1.240	1.309	1.377	1.446	1.515
260.	0.143	0.215	0.287	0.358	0.430	0.502	0.573	0.645	0.717	0.788	0.860	0.932	1.003	1.075	1.147	1.218	1.290	1.362	1.433	1.505	1.577
270.	0.149	0.223	0.298	0.372	0.447	0.521	0.596	0.670	0.745	0.819	0.894	0.968	1.042	1.117	1.191	1.266	1.340	1.415	1.489	1.564	1.638
280.	0.155	0.232	0.309	0.386	0.464	0.541	0.618	0.695	0.773	0.850	0.927	1.004	1.082	1.159	1.236	1.313	1.391	1.468	1.545	1.623	1.700
290.	0.160	0.240	0.320	0.400	0.480	0.560	0.641	0.721	0.801	0.881	0.961	1.042	1.117	1.201	1.281	1.361	1.441	1.521	1.601	1.681	1.761
300.	0.166	0.249	0.331	0.414	0.497	0.580	0.663	0.746	0.829	0.912	0.994	1.077	1.160	1.243	1.326	1.409	1.492	1.575	1.657	1.740	1.823
310.	0.171	0.257	0.343	0.428	0.514	0.600	0.685	0.771	0.857	0.943	1.028	1.114	1.199	1.285	1.371	1.457	1.542	1.628	1.714	1.799	1.885
320.	0.177	0.265	0.354	0.442	0.531	0.619	0.708	0.796	0.885	0.973	1.062	1.150	1.239	1.327	1.416	1.504	1.593	1.681	1.770	1.858	1.947
330.	0.185	0.274	0.365	0.457	0.548	0.639	0.730	0.822	0.913	1.004	1.096	1.187	1.278	1.370	1.461	1.552	1.644	1.735	1.826	1.918	2.009
340.	0.188	0.282	0.377	0.471	0.565	0.659	0.753	0.847	0.941	1.035	1.130	1.224	1.318	1.412	1.506	1.600	1.694	1.789	1.883	1.977	2.071
350.	0.194	0.291	0.388	0.485	0.582	0.679	0.776	0.873	0.970	1.066	1.163	1.260	1.357	1.454	1.551	1.648	1.745	1.842	1.939	2.036	2.133
360.	0.200	0.299	0.399	0.499	0.599	0.698	0.798	0.898	0.998	1.098	1.197	1.297	1.397	1.497	1.596	1.696	1.796	1.896	1.996	2.095	2.195
370.	0.205	0.308	0.410	0.513	0.616	0.718	0.821	0.923	1.026	1.129	1.231	1.334	1.436	1.539	1.642	1.744	1.847	1.950	2.052	2.155	2.257

TABLE 2.4*e* Shell thickness tables in inches: joint efficiency = 0.70

SHELL THICKNESS TABLES IN INCHES

JOINT EFF.=0.70   ALLOWABLE STRESS VALUE= 13800. PSI

INSIDE DIAMETER,IN.

PSI	12.	18.	24.	30.	36.	42.	48.	54.	60.	66.	72.	78.	84.	90.	96.	102.	108.	114.	120.	126.	132.
15.	0.009	0.014	0.019	0.023	0.028	0.033	0.037	0.042	0.047	0.051	0.056	0.061	0.065	0.070	0.075	0.079	0.084	0.089	0.093	0.098	0.103
20.	0.012	0.019	0.025	0.031	0.037	0.044	0.050	0.056	0.062	0.068	0.075	0.081	0.087	0.093	0.100	0.106	0.112	0.118	0.124	0.131	0.137
25.	0.016	0.023	0.031	0.039	0.047	0.054	0.062	0.070	0.078	0.086	0.093	0.101	0.109	0.117	0.124	0.132	0.140	0.148	0.156	0.163	0.171
30.	0.019	0.028	0.037	0.047	0.056	0.065	0.075	0.084	0.093	0.103	0.112	0.121	0.131	0.140	0.149	0.159	0.168	0.177	0.187	0.196	0.205
35.	0.022	0.033	0.044	0.054	0.065	0.076	0.087	0.098	0.109	0.120	0.131	0.142	0.153	0.163	0.174	0.185	0.196	0.207	0.218	0.229	0.240
40.	0.025	0.037	0.050	0.062	0.075	0.087	0.100	0.112	0.125	0.137	0.149	0.162	0.174	0.187	0.199	0.212	0.224	0.237	0.249	0.262	0.274
45.	0.028	0.042	0.056	0.070	0.084	0.098	0.112	0.126	0.140	0.154	0.168	0.182	0.196	0.210	0.224	0.238	0.252	0.266	0.280	0.294	0.308
50.	0.031	0.047	0.062	0.078	0.093	0.109	0.125	0.140	0.156	0.171	0.187	0.202	0.218	0.234	0.249	0.265	0.280	0.296	0.312	0.327	0.343
55.	0.034	0.051	0.069	0.086	0.103	0.120	0.137	0.154	0.171	0.189	0.206	0.223	0.240	0.257	0.274	0.291	0.309	0.326	0.343	0.360	0.377
60.	0.037	0.056	0.075	0.094	0.112	0.131	0.150	0.168	0.187	0.206	0.224	0.243	0.262	0.281	0.299	0.318	0.337	0.355	0.374	0.393	0.411
65.	0.041	0.061	0.081	0.101	0.122	0.142	0.162	0.182	0.203	0.223	0.243	0.263	0.284	0.304	0.324	0.345	0.365	0.385	0.405	0.426	0.446
70.	0.044	0.066	0.087	0.109	0.131	0.153	0.175	0.197	0.218	0.240	0.262	0.284	0.306	0.328	0.349	0.371	0.393	0.415	0.437	0.459	0.480
75.	0.047	0.070	0.094	0.117	0.140	0.164	0.187	0.211	0.234	0.257	0.281	0.304	0.328	0.351	0.374	0.398	0.421	0.445	0.468	0.491	0.515
80.	0.050	0.075	0.100	0.125	0.150	0.175	0.200	0.225	0.250	0.275	0.300	0.325	0.350	0.375	0.400	0.424	0.449	0.474	0.499	0.524	0.549
85.	0.053	0.080	0.106	0.133	0.159	0.186	0.212	0.239	0.265	0.292	0.318	0.345	0.372	0.398	0.425	0.451	0.478	0.504	0.531	0.557	0.584
90.	0.056	0.084	0.112	0.141	0.169	0.197	0.225	0.253	0.281	0.309	0.337	0.365	0.394	0.422	0.450	0.478	0.506	0.534	0.562	0.590	0.618
95.	0.059	0.089	0.119	0.148	0.178	0.208	0.237	0.267	0.297	0.326	0.356	0.386	0.415	0.445	0.475	0.505	0.534	0.564	0.594	0.623	0.653
100.	0.063	0.094	0.125	0.156	0.188	0.219	0.250	0.281	0.313	0.344	0.375	0.406	0.438	0.469	0.500	0.531	0.563	0.594	0.625	0.656	0.688
105.	0.066	0.098	0.131	0.164	0.197	0.230	0.263	0.295	0.328	0.361	0.394	0.427	0.460	0.492	0.525	0.558	0.591	0.624	0.656	0.689	0.722
110.	0.069	0.103	0.138	0.172	0.206	0.241	0.275	0.310	0.344	0.378	0.413	0.447	0.482	0.516	0.550	0.585	0.619	0.654	0.688	0.722	0.757
115.	0.072	0.108	0.144	0.180	0.216	0.252	0.288	0.324	0.360	0.396	0.432	0.468	0.504	0.540	0.576	0.612	0.647	0.683	0.719	0.755	0.791
120.	0.075	0.113	0.150	0.188	0.225	0.263	0.300	0.338	0.375	0.413	0.451	0.488	0.526	0.563	0.601	0.638	0.676	0.713	0.751	0.788	0.826
125.	0.078	0.117	0.156	0.196	0.235	0.274	0.313	0.352	0.391	0.430	0.469	0.509	0.548	0.587	0.626	0.665	0.704	0.743	0.782	0.822	0.861
130.	0.081	0.122	0.163	0.204	0.245	0.285	0.326	0.366	0.407	0.448	0.488	0.529	0.570	0.611	0.651	0.692	0.733	0.773	0.814	0.855	0.895
135.	0.085	0.127	0.169	0.211	0.254	0.296	0.338	0.381	0.423	0.465	0.507	0.550	0.592	0.634	0.676	0.719	0.761	0.803	0.846	0.888	0.930
140.	0.088	0.132	0.175	0.219	0.263	0.307	0.351	0.395	0.439	0.482	0.526	0.570	0.614	0.658	0.702	0.746	0.789	0.833	0.877	0.921	0.965
145.	0.091	0.136	0.182	0.227	0.273	0.318	0.364	0.409	0.454	0.500	0.545	0.591	0.636	0.682	0.727	0.772	0.818	0.863	0.909	0.954	1.000
150.	0.094	0.141	0.188	0.235	0.282	0.329	0.376	0.423	0.470	0.517	0.564	0.611	0.658	0.705	0.752	0.799	0.846	0.893	0.940	0.987	1.034
160.	0.100	0.151	0.201	0.251	0.301	0.351	0.402	0.452	0.502	0.552	0.602	0.652	0.703	0.753	0.803	0.853	0.903	0.954	1.004	1.054	1.104
170.	0.107	0.160	0.213	0.267	0.320	0.374	0.427	0.480	0.534	0.587	0.640	0.694	0.747	0.800	0.854	0.907	0.960	1.014	1.067	1.121	1.174
180.	0.113	0.170	0.226	0.283	0.339	0.396	0.452	0.509	0.565	0.622	0.678	0.735	0.791	0.848	0.905	0.961	1.018	1.074	1.131	1.187	1.244
190.	0.119	0.179	0.239	0.299	0.358	0.418	0.478	0.537	0.597	0.657	0.717	0.776	0.836	0.896	0.955	1.015	1.075	1.135	1.194	1.254	1.314
200.	0.126	0.189	0.252	0.314	0.377	0.440	0.503	0.566	0.629	0.692	0.755	0.818	0.881	0.943	1.006	1.069	1.132	1.195	1.258	1.321	1.384
210.	0.132	0.198	0.264	0.330	0.396	0.463	0.529	0.595	0.661	0.727	0.793	0.859	0.925	0.991	1.057	1.123	1.189	1.256	1.322	1.388	1.454
220.	0.139	0.208	0.277	0.346	0.416	0.485	0.554	0.623	0.693	0.762	0.831	0.901	0.970	1.039	1.108	1.178	1.247	1.316	1.385	1.455	1.524
230.	0.145	0.217	0.290	0.362	0.435	0.507	0.580	0.652	0.725	0.797	0.870	0.942	1.014	1.087	1.159	1.232	1.304	1.376	1.449	1.521	1.594
240.	0.151	0.227	0.303	0.378	0.454	0.530	0.605	0.681	0.757	0.832	0.908	0.984	1.059	1.135	1.211	1.286	1.362	1.438	1.513	1.589	1.665
250.	0.158	0.237	0.315	0.394	0.473	0.552	0.631	0.710	0.789	0.868	0.946	1.025	1.104	1.183	1.262	1.341	1.420	1.498	1.577	1.656	1.735
260.	0.164	0.246	0.328	0.410	0.492	0.574	0.657	0.739	0.821	0.903	0.985	1.067	1.149	1.231	1.313	1.395	1.477	1.559	1.641	1.723	1.806
270.	0.171	0.256	0.341	0.426	0.512	0.597	0.682	0.768	0.853	0.938	1.023	1.109	1.194	1.279	1.364	1.450	1.535	1.620	1.706	1.791	1.876
280.	0.177	0.265	0.354	0.442	0.531	0.619	0.708	0.796	0.885	0.973	1.062	1.150	1.239	1.327	1.416	1.504	1.593	1.681	1.770	1.858	1.947
290.	0.183	0.275	0.367	0.459	0.550	0.642	0.734	0.825	0.917	1.009	1.101	1.192	1.284	1.376	1.467	1.559	1.651	1.743	1.834	1.926	2.018
300.	0.190	0.285	0.380	0.475	0.570	0.665	0.759	0.854	0.949	1.044	1.139	1.234	1.329	1.424	1.519	1.614	1.709	1.804	1.899	1.994	2.089
310.	0.196	0.294	0.393	0.491	0.589	0.687	0.785	0.883	0.982	1.080	1.178	1.276	1.374	1.472	1.571	1.669	1.767	1.865	1.963	2.061	2.160
320.	0.203	0.304	0.406	0.507	0.608	0.710	0.811	0.913	1.014	1.115	1.217	1.318	1.420	1.521	1.622	1.724	1.825	1.926	2.028	2.129	2.231
330.	0.209	0.314	0.419	0.523	0.628	0.732	0.837	0.942	1.046	1.151	1.256	1.360	1.465	1.569	1.674	1.779	1.883	1.988	2.093	2.197	2.302
340.	0.216	0.324	0.431	0.539	0.647	0.755	0.863	0.971	1.079	1.187	1.294	1.402	1.510	1.618	1.726	1.834	1.942	2.049	2.157	2.265	2.373
350.	0.222	0.333	0.444	0.556	0.667	0.778	0.889	1.000	1.111	1.222	1.333	1.444	1.556	1.667	1.778	1.889	2.000	2.111	2.222	2.333	2.444
360.	0.229	0.343	0.457	0.572	0.686	0.801	0.915	1.029	1.144	1.258	1.372	1.487	1.601	1.715	1.830	1.944	2.058	2.173	2.287	2.402	2.516
370.	0.235	0.353	0.470	0.588	0.706	0.823	0.941	1.058	1.176	1.294	1.411	1.529	1.647	1.764	1.882	1.999	2.117	2.235	2.352	2.470	2.587

TABLE 2.4f Shell thickness tables in inches; joint efficiency = 0.65

SHELL THICKNESS TABLES IN INCHES

JOINT EFF.=0.65   ALLOWABLE STRESS VALUE= 13800. PSI

INSIDE DIAMETER, IN.

SHELL THICKNESS TABLES IN INCHES

PSI	12.	18.	24.	30.	36.	42.	48.	54.	60.	66.	72.	78.	84.	90.	96.	102.	108.	114.	120.	126.	132.
15.	0.010	0.015	0.020	0.025	0.030	0.035	0.040	0.045	0.050	0.055	0.060	0.065	0.070	0.075	0.080	0.085	0.090	0.095	0.100	0.105	0.110
20.	0.013	0.020	0.027	0.033	0.040	0.047	0.054	0.060	0.067	0.074	0.080	0.087	0.094	0.100	0.107	0.114	0.121	0.127	0.134	0.141	0.147
25.	0.017	0.025	0.034	0.042	0.050	0.059	0.067	0.075	0.084	0.092	0.101	0.109	0.117	0.126	0.134	0.142	0.151	0.159	0.168	0.176	0.184
30.	0.020	0.030	0.040	0.050	0.060	0.070	0.080	0.090	0.101	0.111	0.121	0.131	0.141	0.151	0.161	0.171	0.181	0.191	0.201	0.211	0.221
35.	0.023	0.035	0.047	0.059	0.070	0.082	0.094	0.106	0.117	0.129	0.141	0.153	0.164	0.176	0.188	0.199	0.211	0.223	0.235	0.246	0.258
40.	0.027	0.040	0.054	0.067	0.080	0.094	0.107	0.121	0.134	0.148	0.161	0.174	0.188	0.201	0.215	0.228	0.241	0.255	0.268	0.282	0.295
45.	0.030	0.045	0.060	0.075	0.091	0.106	0.121	0.136	0.151	0.166	0.181	0.196	0.211	0.226	0.242	0.257	0.272	0.287	0.302	0.317	0.332
50.	0.034	0.050	0.067	0.084	0.101	0.117	0.134	0.151	0.168	0.185	0.201	0.218	0.235	0.252	0.268	0.285	0.302	0.319	0.336	0.352	0.369
55.	0.037	0.055	0.074	0.092	0.111	0.129	0.148	0.166	0.185	0.203	0.222	0.240	0.258	0.277	0.295	0.314	0.332	0.351	0.369	0.388	0.406
60.	0.040	0.060	0.081	0.101	0.121	0.141	0.161	0.181	0.201	0.222	0.242	0.262	0.282	0.302	0.322	0.343	0.363	0.383	0.403	0.423	0.443
65.	0.044	0.066	0.087	0.109	0.131	0.153	0.175	0.197	0.218	0.240	0.262	0.284	0.306	0.328	0.349	0.371	0.393	0.415	0.437	0.459	0.480
70.	0.047	0.071	0.094	0.118	0.141	0.165	0.188	0.212	0.235	0.259	0.282	0.306	0.329	0.353	0.376	0.400	0.423	0.447	0.470	0.494	0.517
75.	0.050	0.076	0.101	0.126	0.151	0.176	0.202	0.227	0.252	0.277	0.303	0.328	0.353	0.378	0.403	0.429	0.454	0.479	0.504	0.529	0.555
80.	0.054	0.081	0.108	0.134	0.161	0.188	0.215	0.242	0.269	0.296	0.323	0.350	0.377	0.403	0.430	0.457	0.484	0.511	0.538	0.565	0.592
85.	0.057	0.086	0.114	0.143	0.172	0.200	0.229	0.257	0.286	0.314	0.343	0.372	0.400	0.429	0.457	0.486	0.515	0.543	0.572	0.600	0.629
90.	0.061	0.091	0.121	0.151	0.182	0.212	0.242	0.273	0.303	0.333	0.363	0.394	0.424	0.454	0.485	0.515	0.545	0.575	0.606	0.636	0.666
95.	0.064	0.096	0.128	0.160	0.192	0.224	0.256	0.288	0.320	0.352	0.384	0.416	0.448	0.480	0.512	0.544	0.576	0.608	0.640	0.671	0.703
100.	0.067	0.101	0.135	0.168	0.202	0.236	0.269	0.303	0.337	0.370	0.404	0.438	0.471	0.505	0.539	0.572	0.606	0.640	0.673	0.707	0.741
105.	0.071	0.106	0.141	0.177	0.212	0.248	0.283	0.318	0.354	0.389	0.424	0.460	0.495	0.530	0.566	0.601	0.637	0.672	0.707	0.743	0.778
110.	0.074	0.111	0.148	0.185	0.222	0.259	0.296	0.334	0.371	0.408	0.445	0.482	0.519	0.556	0.593	0.630	0.667	0.704	0.741	0.778	0.815
115.	0.078	0.116	0.155	0.194	0.233	0.271	0.310	0.349	0.388	0.426	0.465	0.504	0.543	0.581	0.620	0.659	0.698	0.736	0.775	0.814	0.853
120.	0.081	0.121	0.162	0.202	0.243	0.283	0.324	0.364	0.405	0.445	0.486	0.526	0.566	0.607	0.647	0.688	0.728	0.769	0.809	0.850	0.890
125.	0.084	0.126	0.169	0.211	0.253	0.295	0.337	0.379	0.422	0.464	0.506	0.548	0.590	0.632	0.675	0.717	0.759	0.801	0.843	0.885	0.927
130.	0.088	0.132	0.175	0.219	0.263	0.307	0.351	0.395	0.439	0.482	0.526	0.570	0.614	0.658	0.702	0.746	0.789	0.833	0.877	0.921	0.965
135.	0.091	0.137	0.182	0.228	0.273	0.319	0.364	0.410	0.456	0.501	0.547	0.592	0.638	0.683	0.729	0.775	0.820	0.866	0.911	0.957	1.002
140.	0.095	0.142	0.189	0.236	0.284	0.331	0.378	0.425	0.473	0.520	0.567	0.614	0.662	0.709	0.756	0.804	0.851	0.898	0.945	0.993	1.040
145.	0.098	0.147	0.196	0.245	0.294	0.343	0.392	0.441	0.490	0.539	0.588	0.637	0.686	0.735	0.784	0.832	0.881	0.930	0.979	1.028	1.077
150.	0.101	0.152	0.203	0.253	0.304	0.355	0.405	0.456	0.507	0.557	0.608	0.659	0.709	0.760	0.811	0.861	0.912	0.963	1.014	1.064	1.115
160.	0.108	0.162	0.216	0.270	0.325	0.379	0.433	0.487	0.541	0.595	0.649	0.703	0.757	0.811	0.865	0.920	0.974	1.028	1.082	1.136	1.190
170.	0.115	0.173	0.230	0.288	0.345	0.403	0.460	0.518	0.575	0.633	0.690	0.748	0.805	0.863	0.920	0.978	1.035	1.093	1.150	1.208	1.265
180.	0.122	0.183	0.244	0.305	0.366	0.427	0.487	0.548	0.609	0.670	0.731	0.792	0.853	0.914	0.975	1.036	1.097	1.158	1.219	1.280	1.341
190.	0.129	0.193	0.257	0.322	0.386	0.451	0.515	0.579	0.644	0.708	0.772	0.837	0.901	0.965	1.030	1.094	1.159	1.223	1.287	1.352	1.416
200.	0.136	0.203	0.271	0.339	0.407	0.475	0.542	0.610	0.678	0.746	0.814	0.881	0.949	1.017	1.085	1.153	1.220	1.288	1.356	1.424	1.492
210.	0.142	0.214	0.285	0.356	0.427	0.499	0.570	0.641	0.712	0.784	0.855	0.926	0.997	1.069	1.140	1.211	1.282	1.353	1.425	1.496	1.567
220.	0.149	0.224	0.299	0.373	0.448	0.523	0.597	0.672	0.747	0.821	0.896	0.971	1.045	1.120	1.195	1.270	1.344	1.419	1.494	1.568	1.643
230.	0.156	0.234	0.313	0.391	0.469	0.547	0.625	0.703	0.781	0.859	0.938	1.016	1.094	1.172	1.250	1.328	1.406	1.484	1.563	1.641	1.719
240.	0.163	0.245	0.326	0.408	0.489	0.571	0.653	0.734	0.816	0.897	0.979	1.061	1.142	1.224	1.305	1.387	1.468	1.550	1.632	1.713	1.795
250.	0.170	0.255	0.340	0.425	0.510	0.595	0.680	0.765	0.850	0.935	1.020	1.105	1.190	1.276	1.361	1.446	1.531	1.616	1.701	1.786	1.871
260.	0.177	0.265	0.354	0.442	0.531	0.619	0.708	0.796	0.885	0.973	1.062	1.150	1.239	1.327	1.416	1.504	1.593	1.681	1.770	1.858	1.947
270.	0.184	0.276	0.368	0.460	0.552	0.644	0.736	0.828	0.920	1.012	1.104	1.196	1.287	1.379	1.471	1.563	1.655	1.747	1.839	1.931	2.023
280.	0.191	0.286	0.382	0.477	0.573	0.668	0.763	0.859	0.954	1.050	1.145	1.241	1.336	1.431	1.527	1.622	1.718	1.813	1.909	2.004	2.100
290.	0.198	0.297	0.396	0.495	0.593	0.692	0.791	0.890	0.989	1.088	1.187	1.286	1.385	1.484	1.583	1.681	1.780	1.879	1.978	2.077	2.176
300.	0.205	0.307	0.410	0.512	0.614	0.717	0.819	0.922	1.024	1.126	1.229	1.331	1.433	1.536	1.638	1.741	1.843	1.945	2.048	2.150	2.253
310.	0.212	0.318	0.423	0.529	0.635	0.741	0.847	0.953	1.059	1.165	1.270	1.376	1.482	1.588	1.694	1.800	1.906	2.012	2.117	2.223	2.329
320.	0.219	0.328	0.437	0.547	0.656	0.766	0.875	0.984	1.094	1.203	1.312	1.422	1.531	1.640	1.750	1.859	1.969	2.078	2.187	2.297	2.406
330.	0.226	0.339	0.451	0.564	0.677	0.790	0.903	1.016	1.129	1.241	1.354	1.467	1.580	1.693	1.806	1.919	2.031	2.144	2.257	2.370	2.483
340.	0.233	0.349	0.465	0.582	0.698	0.815	0.931	1.047	1.164	1.280	1.396	1.513	1.629	1.745	1.862	1.978	2.094	2.211	2.327	2.444	2.560
350.	0.240	0.360	0.479	0.599	0.719	0.839	0.959	1.079	1.199	1.318	1.438	1.558	1.678	1.798	1.918	2.038	2.158	2.277	2.397	2.517	2.637
360.	0.247	0.370	0.493	0.617	0.740	0.864	0.987	1.110	1.234	1.357	1.480	1.604	1.727	1.851	1.974	2.097	2.221	2.344	2.467	2.591	2.714
370.	0.254	0.381	0.508	0.634	0.761	0.888	1.015	1.142	1.269	1.396	1.523	1.650	1.776	1.903	2.030	2.157	2.284	2.411	2.538	2.665	2.791

Thickness equations for cylindrical shells may be found in Table 2.4. For corrosion allowance in spot-radiographed or fully radiograhed carbon- and low-alloy steel vessels containing air, steam, or water, less than ¼-in thick, add ⅙ of the calculated plate thickness (see Code Par. UCS-25). No corrosion allowance is necessary for vessels that have not been radiograhed (see Code Table UW-12, column C).

USE OF SIMPLIFIED EXTERNAL-PRESSURE CYLINDRICAL SHELL THICKNESS CHARTS (FIGS. 2.3 AND 2.4)

These charts give at a glance the shell thickness required for carbon steel and type 304 stainless steel vessels under external pressure, thereby eliminating the seven to fourteen steps required by Code Par. UG-28, Code Fig. 5, UGO-28.0, and Table 5 — UCS-28.2: Code Fig. 5 — UCS-28.2 for carbon steel with a minimum yield strength of 30,000 psi and type 405 and type 410 stainless steels; and for stainless steel type 304 also using Code Fig. 5 — UHA-28.1.

The required thickness can be obtained by (1) placing a rule horizontally at the intersection of the applicable L/D_o and pressure lines on the left side of the chart and (2) reading the thickness where the rule crosses the vertical D_o line on the right side of the chart.

Other common problems may also be easily solved. The allowable external pressure of an existing (or proposed) vessel may be determined by (1) entering the chart on the right side at the proper thickness and diameter and (2) moving horizontally to the left and reading the pressure at the vertical L/D_o line.

A similar procedure may be used to find the maximum allowable effective length L for spacing of stiffening rings at a specified pressure. (See Figs. 2.3 and 2.4.)

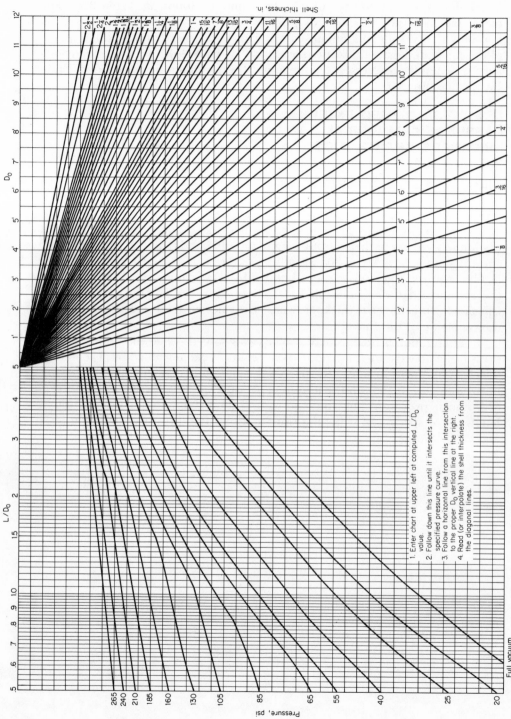

Fig. 2.3   Simplified external-pressure shell thickness chart: *Carbon-steel yield strength is 30,000 to 38,000 psi; temperature, 300°F.*

36

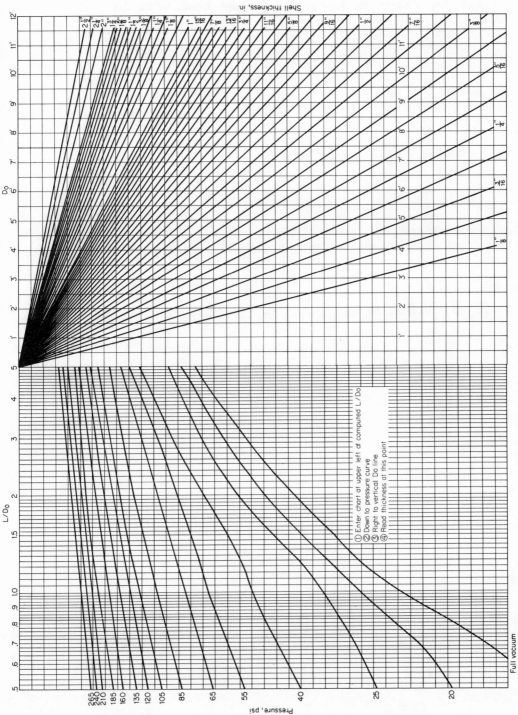

Fig. 2.4 Simplified External-Pressure Shell Thickness Chart: *Stainless steel (18cr-8Ni) Type 304 temperature, 400° F.*

Shell thickness, in.

Pressure, psi

Full vacuum

① Enter chart at upper left at computed L/Do
② Down to pressure curve
③ Right to vertical Do line
④ Read thickness at this point

37

TABLE 2.5   Calculations for cylindrical and spherical shells under external pressure

Equations	Remarks	Code reference
Cylindrical and spherical shells under external pressure		
Vessels intended for service under external working pressure of 15 psi or less		Par. UG-28(f)

Use equations in Code Par. UG-28(c) and (d) and thickness charts for the applicable material (figure numbers of charts listed under Code references)	Carbon-steel shells	Par. UG-28
		Par. UCS-28
		Fig. UG-28
		Fig. UG-29.1 and UG-29.2
		Appendix 5
		Fig. UGO-28.0,
		   Fig. UCS-28.1,
		   UCS-28.2, and
		   UCS-28.4

For stiffening rings, use equations in Code Par. UG-29 and UG-30	Nonferrous metal shells	Par. UG-28 Par. UNF-28 Fig. UGO-28.0 Fig. UNF-28.1 to UNF-28.32
	High-alloy steel shells	Par. UG-28 Par. UHA-28 Fig. UGO-28.0 Fig. UHA-28.1 to UHA-28.4 Fig. UCS-28.2
	Cast-iron shells	Par. UG-28 Par. UCI-28 Fig. UCI-28
	Clad-steel shells	Par. UG-28 Par. UCL-26
	Corrugated shells	Par. UCS-28(c)*
	See Code Appendix L for explanation of thickness charts for various materials (introductory note) and for applications of Code formulas and rules	Appendix L
	Basis for establishing external pressure charts	Appendix Q

* Rules in Par. PFT-19 of Power Boiler Code may also be used.

Equations	Remarks	Code reference
Stiffening rings for cylindrical shells under external pressure		
	Avoid stress concentrations	Par. UG-28 and UG-29 Fig. UG-29.1 and UG-29.2 Appendix G
	Stiffening rings may be placed outside or inside a vessel	Par. UG-30 Fig. UG-29.1 and UG-30 Fig. UG-29.2
	External material charts and tabular values	Appendix 5
	Stiffening rings may be attached to shell by continuous or intermittent welding	Par UG-30
	Total length of intermittent welding on each side of stiffening ring shall be: (a) Rings on outside Not less than one-half outside circumference of the vessel (b) Rings on inside Not less than one-third of the circumference of the vessel Maximum spacing of intermittent weld $8 \times$ thickness of shell for external rings $12 \times$ thickness of shell for internal rings	Fig. UG-30
	Drainage — Opening at bottom for drainage	Fig. UG-29.1
	Vent — Opening at top for vent	Fig. UG-29.1
	Maximum arc of shell left unsupported	Fig. UG-29.1 and UG-29.2

ELLIPSOIDAL AND TORISPHERICAL HEADS

TABLE 2.6 Calculations for ellipsoidal and torispherical flanged and dished heads under internal pressure

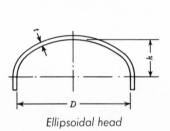

Ellipsoidal head

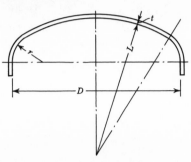

Torispherical head

Equations	Remarks	Code reference
	Ellipsoidal heads under internal pressure	
$t = \dfrac{PD}{2SE - 0.2P}$ or $P = \dfrac{2\,SEt}{D + 0.2t}$	Equations to be used when ratio of the major axis, D, to the minor axis, h, is $2:1$ (see illustration above). For other ratios, see Appendix 1	Par. UG-32(d) Appendix 1 (also footnotes)
in which: t = minimum thickness, in P = internal pressure, psi S = allowable stress, psi E = minimum joint efficiency, percent D = inside diameter of head skirt, in	Minimum thickness Allowable percentages of stress values and joint efficiencies for material depend upon type of radiography — full, partial, spot, or no radiography	Par. UG-32(b), Par. UW-11 and UW-12
	Required skirt length	Par. UG-32(l) and (m)
	Attachment of heads to shells	Fig. UW-13
	Cold-formed heads of P-1, Group 1 and 2 material may require heat treatment	Par. UCS-79

Equations	Remarks	Code reference
Torispherical heads under internal pressure		
$t = \dfrac{0.885\,PL}{SE - 0.1P}$ or $P = \dfrac{SEt}{0.885\,L + 0.1t}$	ASME head is one in which the knuckle radius, r, is 6 percent of the inside crown radius, L	Par. UG-32(e) Appendix 1 (also footnotes)
	Minimum thickness	Par. UG-32(a) and (b)
in which L equals the inside crown radius (in inches) and other unknowns are same as above	Allowable percentages of stress values and joint efficiencies for material depend upon type of radiography — full, partial, spot, or no radiography	Par. UW-11 Par. UW-12
	Required skirt length	Par. UG-32(l) and (m)
	Attachment of heads to shells	Fig. UW-13
	Cold-formed heads of P-1, Group 1 and 2 material may require heat treatment	Par. UCS-79

USE OF INTERNAL-PRESSURE THICKNESS TABLES FOR ELLIPSOIDAL AND TORISPHERICAL HEADS (TABLES 2.6a–2.6l)

The head thickness tables may be applied to ellipsoidal 2 : 1 heads or torispherical heads. They may be applied to seamless heads (joint efficiency = 1.0), double-welded butt heads that have been fully radiographed (joint efficiency = 1.0), spot-radiographed (joint efficiency = 0.85), or not radiographed (joint efficiency = 0.70). They may also be applied to single-welded butt joints with a backing strip that are fully radiographed (joint efficiency = 0.90), spot radiographed (joint efficiency = 0.80), or not radiographed (joint efficiency = 0.65). The tables may also be used when stress reduction factors of 85 percent (0.85) or 80 percent (0.80) are applied.

The required head thickness may be determined by the following steps.

1. Determine the appropriate table corresponding to the desired joint efficiency. Note that the value of joint efficiency for each table is printed in the upper left-hand corner.

(Text continues on p. 54.)

TABLE 2.6a Torispherical head thickness tables in inches: joint efficiency = 1.0

TORISPHERICAL HEAD THICKNESS TABLES IN INCHES

JOINT EFF.=1.00 ALLOWABLE STRESS VALUE =13800. PSI

INSIDE CROWN RADIUS, INCHES

PSI	12.	18.	24.	30.	36.	42.	48.	54.	60.	66.	72.	78.	84.	90.	96.	102.	108.	114.	120.	126.	132.
15.	0.012	0.017	0.023	0.029	0.035	0.040	0.046	0.052	0.058	0.063	0.069	0.075	0.081	0.087	0.092	0.098	0.104	0.110	0.115	0.121	0.127
20.	0.015	0.023	0.031	0.038	0.046	0.054	0.062	0.069	0.077	0.085	0.092	0.100	0.108	0.115	0.123	0.131	0.139	0.146	0.154	0.162	0.169
25.	0.019	0.029	0.038	0.048	0.058	0.067	0.077	0.087	0.096	0.106	0.115	0.125	0.135	0.144	0.154	0.164	0.173	0.183	0.192	0.202	0.212
30.	0.023	0.035	0.046	0.058	0.069	0.081	0.092	0.104	0.115	0.127	0.139	0.150	0.162	0.173	0.185	0.196	0.208	0.219	0.231	0.242	0.254
35.	0.027	0.040	0.054	0.067	0.081	0.094	0.108	0.121	0.135	0.148	0.162	0.175	0.189	0.202	0.216	0.229	0.242	0.256	0.269	0.283	0.296
40.	0.031	0.046	0.062	0.077	0.092	0.108	0.123	0.139	0.154	0.169	0.185	0.200	0.216	0.231	0.246	0.262	0.277	0.293	0.308	0.323	0.339
45.	0.035	0.052	0.069	0.087	0.104	0.121	0.139	0.156	0.173	0.191	0.208	0.225	0.242	0.260	0.277	0.294	0.312	0.329	0.346	0.364	0.381
50.	0.038	0.058	0.077	0.096	0.115	0.135	0.154	0.173	0.192	0.212	0.231	0.250	0.269	0.289	0.308	0.327	0.346	0.366	0.385	0.404	0.423
55.	0.042	0.064	0.085	0.106	0.127	0.148	0.169	0.191	0.212	0.233	0.254	0.275	0.296	0.318	0.339	0.360	0.381	0.402	0.423	0.445	0.466
60.	0.046	0.069	0.092	0.115	0.139	0.162	0.185	0.208	0.231	0.254	0.277	0.300	0.323	0.346	0.370	0.393	0.416	0.439	0.462	0.485	0.508
65.	0.050	0.075	0.100	0.125	0.150	0.175	0.200	0.225	0.250	0.275	0.300	0.325	0.350	0.375	0.400	0.425	0.450	0.475	0.500	0.525	0.550
70.	0.054	0.081	0.108	0.135	0.162	0.189	0.216	0.243	0.269	0.296	0.323	0.350	0.377	0.404	0.431	0.458	0.485	0.512	0.539	0.566	0.593
75.	0.058	0.087	0.115	0.144	0.173	0.202	0.231	0.260	0.289	0.318	0.346	0.375	0.404	0.433	0.462	0.491	0.520	0.549	0.577	0.606	0.635
80.	0.062	0.092	0.123	0.154	0.185	0.216	0.246	0.277	0.308	0.339	0.370	0.400	0.431	0.462	0.493	0.524	0.554	0.585	0.616	0.647	0.678
85.	0.065	0.098	0.131	0.164	0.196	0.229	0.262	0.295	0.327	0.360	0.393	0.425	0.458	0.491	0.524	0.556	0.589	0.622	0.655	0.687	0.720
90.	0.069	0.104	0.139	0.173	0.208	0.243	0.277	0.312	0.347	0.381	0.416	0.450	0.485	0.520	0.554	0.589	0.624	0.658	0.693	0.728	0.762
95.	0.073	0.110	0.146	0.183	0.219	0.256	0.293	0.329	0.366	0.402	0.439	0.476	0.512	0.549	0.585	0.622	0.658	0.695	0.732	0.768	0.805
100.	0.077	0.116	0.154	0.193	0.231	0.270	0.308	0.347	0.385	0.424	0.462	0.501	0.539	0.578	0.616	0.655	0.693	0.732	0.770	0.809	0.847
105.	0.081	0.121	0.162	0.202	0.243	0.283	0.323	0.364	0.404	0.445	0.485	0.526	0.566	0.606	0.647	0.687	0.728	0.768	0.809	0.849	0.890
110.	0.085	0.127	0.169	0.212	0.254	0.297	0.339	0.381	0.424	0.466	0.508	0.551	0.593	0.635	0.678	0.720	0.762	0.805	0.847	0.890	0.932
115.	0.089	0.133	0.177	0.221	0.266	0.310	0.354	0.399	0.443	0.487	0.531	0.576	0.620	0.664	0.709	0.753	0.797	0.841	0.886	0.930	0.974
120.	0.092	0.139	0.185	0.231	0.277	0.323	0.370	0.416	0.462	0.508	0.555	0.601	0.647	0.693	0.739	0.786	0.832	0.878	0.924	0.970	1.017
125.	0.096	0.144	0.193	0.241	0.289	0.337	0.385	0.433	0.481	0.530	0.578	0.626	0.674	0.722	0.770	0.818	0.867	0.915	0.963	1.011	1.059
130.	0.100	0.150	0.200	0.250	0.300	0.350	0.401	0.451	0.501	0.551	0.601	0.651	0.701	0.751	0.801	0.851	0.901	0.951	1.001	1.051	1.102
135.	0.104	0.156	0.208	0.260	0.312	0.364	0.416	0.468	0.520	0.572	0.624	0.676	0.728	0.780	0.832	0.884	0.936	0.988	1.040	1.092	1.144
140.	0.108	0.162	0.216	0.270	0.324	0.377	0.431	0.485	0.539	0.593	0.647	0.701	0.755	0.809	0.863	0.917	0.971	1.025	1.078	1.132	1.186
145.	0.112	0.168	0.223	0.279	0.335	0.391	0.447	0.503	0.559	0.614	0.670	0.726	0.782	0.838	0.894	0.949	1.005	1.061	1.117	1.173	1.229
150.	0.116	0.173	0.231	0.289	0.347	0.404	0.462	0.520	0.578	0.636	0.693	0.751	0.809	0.867	0.924	0.982	1.040	1.098	1.156	1.214	1.271
160.	0.123	0.185	0.247	0.308	0.370	0.431	0.493	0.555	0.616	0.678	0.740	0.801	0.863	0.925	0.986	1.048	1.109	1.171	1.233	1.294	1.356
170.	0.131	0.196	0.262	0.327	0.393	0.458	0.524	0.589	0.655	0.720	0.786	0.851	0.917	0.982	1.048	1.113	1.179	1.244	1.310	1.375	1.441
180.	0.139	0.208	0.277	0.347	0.416	0.485	0.555	0.624	0.694	0.763	0.832	0.902	0.971	1.040	1.110	1.179	1.248	1.318	1.387	1.456	1.526
190.	0.146	0.220	0.293	0.366	0.439	0.512	0.586	0.659	0.732	0.805	0.879	0.952	1.025	1.098	1.171	1.245	1.318	1.391	1.464	1.537	1.611
200.	0.154	0.231	0.308	0.385	0.462	0.539	0.617	0.694	0.771	0.848	0.925	1.002	1.079	1.156	1.233	1.310	1.387	1.464	1.541	1.618	1.696
210.	0.162	0.243	0.324	0.405	0.486	0.566	0.647	0.728	0.809	0.890	0.971	1.052	1.133	1.214	1.295	1.376	1.457	1.538	1.619	1.699	1.780
220.	0.170	0.254	0.339	0.424	0.509	0.594	0.678	0.763	0.848	0.933	1.018	1.102	1.187	1.272	1.357	1.441	1.526	1.611	1.696	1.781	1.865
230.	0.177	0.266	0.355	0.443	0.532	0.621	0.709	0.798	0.886	0.975	1.064	1.152	1.241	1.330	1.418	1.507	1.596	1.684	1.773	1.862	1.950
240.	0.185	0.277	0.370	0.462	0.555	0.648	0.740	0.833	0.925	1.018	1.110	1.203	1.295	1.388	1.480	1.573	1.665	1.758	1.850	1.943	2.035
250.	0.193	0.289	0.385	0.482	0.578	0.675	0.771	0.867	0.964	1.060	1.156	1.253	1.349	1.446	1.542	1.638	1.735	1.831	1.927	2.024	2.120
260.	0.200	0.301	0.401	0.501	0.601	0.702	0.802	0.902	1.002	1.103	1.203	1.303	1.403	1.503	1.604	1.704	1.804	1.904	2.005	2.105	2.205
270.	0.208	0.312	0.416	0.520	0.625	0.729	0.833	0.937	1.041	1.145	1.249	1.353	1.457	1.561	1.666	1.770	1.874	1.978	2.082	2.186	2.290
280.	0.216	0.324	0.432	0.540	0.648	0.756	0.864	0.972	1.080	1.188	1.295	1.403	1.511	1.619	1.727	1.835	1.943	2.051	2.159	2.267	2.375
290.	0.224	0.335	0.447	0.559	0.671	0.783	0.895	1.006	1.118	1.230	1.342	1.454	1.566	1.677	1.789	1.901	2.013	2.125	2.236	2.348	2.460
300.	0.231	0.347	0.463	0.578	0.694	0.810	0.925	1.041	1.157	1.273	1.388	1.504	1.620	1.735	1.851	1.967	2.082	2.198	2.314	2.429	2.545
310.	0.239	0.359	0.478	0.598	0.717	0.837	0.956	1.076	1.196	1.315	1.435	1.554	1.674	1.793	1.913	2.032	2.152	2.271	2.391	2.511	2.630
320.	0.247	0.370	0.494	0.617	0.740	0.864	0.987	1.111	1.234	1.358	1.481	1.604	1.728	1.851	1.975	2.098	2.221	2.345	2.468	2.592	2.715
330.	0.255	0.382	0.509	0.636	0.764	0.891	1.018	1.146	1.273	1.400	1.527	1.655	1.782	1.909	2.037	2.164	2.291	2.418	2.546	2.673	2.800
340.	0.262	0.393	0.525	0.656	0.787	0.918	1.049	1.180	1.311	1.443	1.574	1.705	1.836	1.967	2.098	2.230	2.361	2.492	2.623	2.754	2.885
350.	0.270	0.405	0.540	0.675	0.810	0.945	1.080	1.215	1.350	1.485	1.620	1.755	1.890	2.025	2.160	2.295	2.430	2.565	2.700	2.835	2.970
360.	0.278	0.417	0.556	0.694	0.833	0.972	1.111	1.250	1.389	1.528	1.667	1.805	1.944	2.083	2.222	2.361	2.500	2.639	2.778	2.917	3.055
370.	0.286	0.428	0.571	0.714	0.857	0.999	1.142	1.285	1.428	1.570	1.713	1.856	1.999	2.141	2.284	2.427	2.570	2.712	2.855	2.998	3.141

TABLE 2.6*b* Torispherical head thickness tables in inches: joint efficiency = 0.90

TORISPHERICAL HEAD THICKNESS TABLES IN INCHES

JOINT EFF.=0.90   ALLOWABLE STRESS VALUE =13800. PSI

INSIDE CROWN RADIUS,INCHES

PSI	12.	18.	24.	30.	36.	42.	48.	54.	60.	66.	72.	78.	84.	90.	96.	102.	108.	114.	120.	126.	132.
15.	0.013	0.019	0.026	0.032	0.038	0.045	0.051	0.058	0.064	0.071	0.077	0.083	0.090	0.096	0.103	0.109	0.115	0.122	0.128	0.135	0.141
20.	0.017	0.026	0.034	0.043	0.051	0.060	0.068	0.077	0.086	0.094	0.103	0.111	0.120	0.128	0.137	0.145	0.154	0.162	0.171	0.180	0.188
25.	0.021	0.032	0.043	0.053	0.064	0.075	0.086	0.096	0.107	0.118	0.128	0.139	0.150	0.160	0.171	0.182	0.192	0.203	0.214	0.225	0.235
30.	0.026	0.038	0.051	0.064	0.077	0.090	0.103	0.115	0.128	0.141	0.154	0.167	0.180	0.192	0.205	0.218	0.231	0.244	0.257	0.269	0.282
35.	0.030	0.045	0.060	0.075	0.090	0.105	0.120	0.135	0.150	0.165	0.180	0.195	0.210	0.225	0.239	0.254	0.269	0.284	0.299	0.314	0.329
40.	0.034	0.051	0.068	0.086	0.103	0.120	0.137	0.154	0.171	0.188	0.205	0.222	0.239	0.257	0.274	0.291	0.308	0.325	0.342	0.359	0.376
45.	0.038	0.058	0.077	0.096	0.115	0.135	0.154	0.173	0.192	0.212	0.231	0.250	0.269	0.289	0.308	0.327	0.346	0.366	0.385	0.404	0.423
50.	0.043	0.064	0.086	0.107	0.128	0.150	0.171	0.192	0.214	0.235	0.257	0.278	0.299	0.321	0.342	0.364	0.385	0.406	0.428	0.449	0.470
55.	0.047	0.071	0.094	0.118	0.141	0.165	0.188	0.212	0.235	0.259	0.282	0.306	0.329	0.353	0.376	0.400	0.423	0.447	0.470	0.494	0.518
60.	0.051	0.077	0.103	0.128	0.154	0.180	0.205	0.231	0.257	0.282	0.308	0.334	0.359	0.385	0.411	0.436	0.462	0.488	0.513	0.539	0.565
65.	0.056	0.083	0.111	0.139	0.167	0.195	0.222	0.250	0.278	0.306	0.334	0.361	0.389	0.417	0.445	0.473	0.500	0.528	0.556	0.584	0.612
70.	0.060	0.090	0.120	0.150	0.180	0.210	0.240	0.269	0.299	0.329	0.359	0.389	0.419	0.449	0.479	0.509	0.539	0.569	0.599	0.629	0.659
75.	0.064	0.096	0.128	0.160	0.193	0.225	0.257	0.289	0.321	0.353	0.385	0.417	0.449	0.481	0.513	0.545	0.578	0.610	0.642	0.674	0.706
80.	0.068	0.103	0.137	0.171	0.205	0.240	0.274	0.308	0.342	0.376	0.411	0.445	0.479	0.513	0.548	0.582	0.616	0.650	0.684	0.719	0.753
85.	0.073	0.109	0.145	0.182	0.218	0.255	0.291	0.327	0.364	0.400	0.436	0.473	0.509	0.545	0.582	0.618	0.655	0.691	0.727	0.764	0.800
90.	0.077	0.116	0.154	0.193	0.231	0.270	0.308	0.347	0.385	0.424	0.462	0.501	0.539	0.578	0.616	0.655	0.693	0.732	0.770	0.800	0.847
95.	0.081	0.122	0.163	0.203	0.244	0.285	0.325	0.366	0.406	0.447	0.488	0.528	0.569	0.610	0.650	0.691	0.732	0.772	0.813	0.854	0.894
100.	0.086	0.128	0.171	0.214	0.257	0.300	0.342	0.385	0.428	0.471	0.513	0.556	0.599	0.642	0.685	0.727	0.770	0.813	0.856	0.899	0.941
105.	0.090	0.135	0.180	0.225	0.270	0.315	0.359	0.404	0.449	0.494	0.539	0.584	0.629	0.674	0.719	0.764	0.809	0.854	0.899	0.944	0.988
110.	0.094	0.141	0.188	0.235	0.282	0.329	0.377	0.424	0.471	0.518	0.565	0.612	0.659	0.706	0.753	0.800	0.847	0.894	0.941	0.988	1.036
115.	0.098	0.148	0.197	0.246	0.295	0.344	0.394	0.443	0.492	0.541	0.591	0.640	0.689	0.738	0.787	0.837	0.886	0.935	0.984	1.033	1.083
120.	0.103	0.154	0.205	0.257	0.308	0.359	0.411	0.462	0.514	0.565	0.616	0.668	0.719	0.770	0.822	0.873	0.924	0.976	1.027	1.078	1.130
125.	0.107	0.160	0.214	0.267	0.321	0.374	0.428	0.481	0.535	0.588	0.642	0.695	0.749	0.802	0.856	0.909	0.963	1.016	1.070	1.123	1.177
130.	0.111	0.167	0.223	0.278	0.334	0.389	0.445	0.501	0.556	0.612	0.668	0.723	0.779	0.835	0.899	0.946	1.001	1.057	1.113	1.168	1.224
135.	0.116	0.173	0.231	0.289	0.347	0.404	0.462	0.520	0.578	0.636	0.693	0.751	0.809	0.867	0.924	0.982	1.040	1.098	1.156	1.213	1.271
140.	0.120	0.180	0.240	0.300	0.360	0.419	0.479	0.539	0.599	0.659	0.719	0.779	0.839	0.899	0.959	1.019	1.079	1.139	1.198	1.258	1.318
145.	0.124	0.186	0.248	0.310	0.372	0.434	0.497	0.559	0.621	0.683	0.745	0.807	0.869	0.931	0.993	1.055	1.117	1.179	1.241	1.303	1.365
150.	0.128	0.193	0.257	0.321	0.385	0.449	0.514	0.578	0.642	0.706	0.770	0.835	0.899	0.963	1.027	1.092	1.156	1.220	1.284	1.348	1.413
160.	0.137	0.205	0.274	0.342	0.411	0.479	0.547	0.616	0.685	0.753	0.822	0.890	0.959	1.027	1.096	1.164	1.233	1.301	1.370	1.438	1.507
170.	0.146	0.218	0.291	0.364	0.437	0.509	0.582	0.655	0.728	0.801	0.873	0.946	1.019	1.092	1.164	1.237	1.310	1.383	1.456	1.528	1.601
180.	0.154	0.231	0.308	0.385	0.462	0.539	0.617	0.694	0.771	0.848	0.925	1.002	1.079	1.156	1.233	1.310	1.387	1.464	1.541	1.618	1.696
190.	0.163	0.244	0.325	0.407	0.488	0.569	0.651	0.732	0.814	0.895	0.976	1.058	1.139	1.220	1.302	1.383	1.464	1.546	1.627	1.708	1.790
200.	0.171	0.257	0.343	0.428	0.514	0.600	0.685	0.771	0.856	0.942	1.028	1.113	1.199	1.285	1.370	1.456	1.542	1.627	1.713	1.799	1.884
210.	0.180	0.270	0.360	0.450	0.540	0.630	0.719	0.809	0.899	0.989	1.079	1.169	1.259	1.349	1.439	1.529	1.619	1.709	1.799	1.889	1.979
220.	0.188	0.283	0.377	0.471	0.565	0.660	0.754	0.848	0.942	1.036	1.131	1.225	1.319	1.413	1.508	1.602	1.696	1.790	1.884	1.979	2.073
230.	0.197	0.296	0.394	0.493	0.591	0.690	0.788	0.887	0.985	1.084	1.182	1.281	1.379	1.478	1.576	1.675	1.773	1.872	1.970	2.069	2.167
240.	0.206	0.308	0.411	0.514	0.617	0.720	0.823	0.925	1.028	1.131	1.234	1.336	1.439	1.542	1.645	1.748	1.851	1.953	2.056	2.159	2.262
250.	0.214	0.321	0.428	0.535	0.643	0.750	0.857	0.964	1.071	1.178	1.285	1.392	1.499	1.606	1.714	1.821	1.928	2.035	2.142	2.249	2.356
260.	0.223	0.334	0.446	0.557	0.668	0.780	0.891	1.003	1.114	1.225	1.337	1.448	1.560	1.671	1.782	1.894	2.005	2.116	2.228	2.339	2.451
270.	0.231	0.347	0.463	0.578	0.694	0.810	0.925	1.041	1.157	1.273	1.388	1.504	1.620	1.735	1.851	1.967	2.082	2.198	2.314	2.429	2.545
280.	0.240	0.360	0.480	0.600	0.720	0.840	0.960	1.080	1.200	1.320	1.440	1.560	1.680	1.800	1.920	2.040	2.160	2.280	2.400	2.520	2.640
290.	0.249	0.373	0.497	0.621	0.746	0.870	0.994	1.118	1.243	1.367	1.491	1.616	1.740	1.864	1.988	2.113	2.237	2.361	2.486	2.610	2.734
300.	0.257	0.386	0.514	0.643	0.771	0.900	1.029	1.157	1.286	1.414	1.543	1.671	1.800	1.929	2.057	2.186	2.314	2.443	2.571	2.700	2.829
310.	0.266	0.399	0.531	0.664	0.797	0.930	1.063	1.196	1.329	1.462	1.594	1.727	1.860	1.993	2.126	2.259	2.392	2.524	2.657	2.790	2.923
320.	0.274	0.411	0.549	0.686	0.823	0.960	1.097	1.234	1.372	1.509	1.646	1.783	1.920	2.057	2.195	2.332	2.469	2.606	2.743	2.880	3.018
330.	0.283	0.424	0.566	0.707	0.849	0.990	1.132	1.273	1.415	1.556	1.698	1.839	1.980	2.122	2.263	2.405	2.546	2.688	2.829	2.971	3.112
340.	0.292	0.437	0.583	0.729	0.875	1.020	1.166	1.312	1.458	1.603	1.749	1.895	2.041	2.186	2.332	2.478	2.624	2.769	2.915	3.061	3.207
350.	0.300	0.450	0.600	0.750	0.900	1.050	1.200	1.351	1.501	1.651	1.801	1.951	2.101	2.251	2.401	2.551	2.701	2.851	3.001	3.151	3.301
360.	0.309	0.463	0.617	0.772	0.926	1.081	1.235	1.389	1.544	1.698	1.852	2.007	2.161	2.315	2.470	2.624	2.778	2.933	3.087	3.242	3.396
370.	0.317	0.476	0.635	0.793	0.952	1.111	1.269	1.428	1.587	1.745	1.904	2.063	2.221	2.380	2.539	2.697	2.856	3.015	3.173	3.332	3.491

TABLE 2.6c Torispherical head thickness tables in inches; joint efficiency = 0.85

TORISPHERICAL HEAD THICKNESS TABLES IN INCHES

JOINT EFF.=0.85   ALLOWABLE STRESS VALUE =13800. PSI

INSIDE CROWN RADIUS,INCHES

PSI	12.	18.	24.	30.	36.	42.	48.	54.	60.	66.	72.	78.	84.	90.	96.	102.	108.	114.	120.	126.	132.
15.	0.014	0.020	0.027	0.034	0.041	0.048	0.054	0.061	0.068	0.075	0.081	0.088	0.095	0.102	0.109	0.115	0.122	0.129	0.136	0.143	0.149
20.	0.018	0.027	0.036	0.045	0.054	0.063	0.072	0.081	0.091	0.100	0.109	0.118	0.127	0.136	0.145	0.154	0.163	0.172	0.181	0.190	0.199
25.	0.023	0.034	0.045	0.057	0.068	0.079	0.091	0.102	0.113	0.125	0.136	0.147	0.158	0.170	0.181	0.192	0.204	0.215	0.226	0.238	0.249
30.	0.027	0.041	0.054	0.068	0.082	0.095	0.109	0.122	0.136	0.149	0.163	0.177	0.190	0.204	0.217	0.231	0.245	0.258	0.272	0.285	0.299
35.	0.032	0.048	0.063	0.079	0.095	0.111	0.127	0.143	0.158	0.174	0.190	0.206	0.222	0.238	0.254	0.269	0.285	0.301	0.317	0.333	0.349
40.	0.036	0.054	0.072	0.091	0.109	0.127	0.145	0.163	0.181	0.199	0.217	0.235	0.254	0.272	0.290	0.308	0.326	0.344	0.362	0.380	0.398
45.	0.041	0.061	0.081	0.102	0.122	0.143	0.163	0.183	0.204	0.224	0.245	0.265	0.285	0.306	0.326	0.346	0.367	0.387	0.408	0.428	0.448
50.	0.045	0.068	0.091	0.113	0.136	0.159	0.181	0.204	0.226	0.249	0.272	0.294	0.317	0.340	0.362	0.385	0.408	0.430	0.453	0.476	0.498
55.	0.050	0.075	0.100	0.125	0.149	0.174	0.199	0.224	0.249	0.274	0.299	0.324	0.349	0.374	0.399	0.423	0.448	0.473	0.498	0.523	0.548
60.	0.054	0.082	0.109	0.136	0.163	0.190	0.217	0.245	0.272	0.299	0.326	0.353	0.380	0.408	0.435	0.462	0.489	0.516	0.544	0.571	0.598
65.	0.059	0.088	0.118	0.147	0.177	0.206	0.236	0.265	0.294	0.324	0.353	0.383	0.412	0.442	0.471	0.500	0.530	0.559	0.589	0.618	0.648
70.	0.063	0.095	0.127	0.159	0.190	0.222	0.254	0.285	0.317	0.349	0.380	0.412	0.444	0.476	0.507	0.539	0.571	0.602	0.634	0.666	0.698
75.	0.068	0.102	0.136	0.170	0.204	0.238	0.272	0.306	0.340	0.374	0.408	0.442	0.476	0.510	0.544	0.578	0.612	0.645	0.679	0.713	0.747
80.	0.072	0.109	0.145	0.181	0.217	0.254	0.290	0.326	0.362	0.399	0.435	0.471	0.507	0.544	0.580	0.616	0.652	0.689	0.725	0.761	0.797
85.	0.077	0.116	0.154	0.193	0.231	0.270	0.308	0.347	0.385	0.424	0.462	0.501	0.539	0.578	0.616	0.655	0.693	0.732	0.770	0.809	0.847
90.	0.082	0.122	0.163	0.204	0.245	0.285	0.326	0.367	0.408	0.449	0.489	0.530	0.571	0.612	0.652	0.693	0.734	0.775	0.815	0.856	0.897
95.	0.086	0.129	0.172	0.215	0.258	0.301	0.344	0.387	0.430	0.473	0.516	0.560	0.603	0.646	0.689	0.732	0.775	0.818	0.861	0.904	0.947
100.	0.091	0.136	0.181	0.227	0.272	0.317	0.362	0.408	0.453	0.498	0.544	0.589	0.634	0.680	0.725	0.770	0.816	0.861	0.906	0.951	0.997
105.	0.095	0.143	0.190	0.238	0.285	0.333	0.381	0.428	0.476	0.523	0.571	0.618	0.666	0.714	0.761	0.809	0.856	0.904	0.951	0.999	1.047
110.	0.100	0.150	0.199	0.249	0.299	0.349	0.399	0.449	0.498	0.548	0.598	0.648	0.698	0.748	0.797	0.847	0.897	0.947	0.997	1.047	1.097
115.	0.104	0.156	0.208	0.261	0.313	0.365	0.417	0.469	0.521	0.573	0.625	0.677	0.730	0.782	0.834	0.886	0.938	0.990	1.042	1.094	1.146
120.	0.109	0.163	0.218	0.272	0.326	0.381	0.435	0.489	0.544	0.598	0.653	0.707	0.761	0.816	0.870	0.924	0.979	1.033	1.088	1.142	1.196
125.	0.113	0.170	0.227	0.283	0.340	0.397	0.453	0.510	0.566	0.623	0.680	0.736	0.793	0.850	0.906	0.963	1.020	1.076	1.133	1.190	1.246
130.	0.118	0.177	0.236	0.295	0.353	0.412	0.471	0.530	0.589	0.648	0.707	0.766	0.825	0.884	0.943	1.002	1.060	1.119	1.178	1.237	1.296
135.	0.122	0.184	0.245	0.306	0.367	0.428	0.489	0.551	0.612	0.673	0.734	0.795	0.857	0.918	0.979	1.040	1.101	1.162	1.224	1.285	1.346
140.	0.127	0.190	0.254	0.317	0.381	0.444	0.508	0.571	0.635	0.698	0.761	0.825	0.888	0.952	1.015	1.079	1.142	1.206	1.269	1.332	1.396
145.	0.131	0.197	0.263	0.329	0.394	0.460	0.526	0.591	0.657	0.723	0.789	0.854	0.920	0.986	1.052	1.117	1.183	1.249	1.314	1.380	1.446
150.	0.136	0.204	0.272	0.340	0.408	0.476	0.544	0.612	0.680	0.748	0.816	0.884	0.952	1.020	1.088	1.156	1.224	1.292	1.360	1.428	1.496
160.	0.145	0.218	0.290	0.363	0.435	0.508	0.580	0.653	0.725	0.798	0.870	0.943	1.015	1.088	1.160	1.233	1.306	1.378	1.451	1.523	1.596
170.	0.154	0.231	0.308	0.385	0.462	0.539	0.617	0.694	0.771	0.848	0.925	1.002	1.079	1.156	1.233	1.310	1.387	1.464	1.541	1.618	1.696
180.	0.163	0.245	0.326	0.408	0.490	0.571	0.653	0.734	0.816	0.898	0.979	1.061	1.143	1.224	1.306	1.387	1.469	1.551	1.632	1.714	1.795
190.	0.172	0.258	0.345	0.431	0.517	0.603	0.689	0.775	0.861	0.948	1.034	1.120	1.206	1.292	1.378	1.465	1.551	1.637	1.723	1.809	1.895
200.	0.181	0.272	0.363	0.453	0.544	0.635	0.726	0.816	0.907	0.998	1.088	1.179	1.270	1.360	1.451	1.542	1.632	1.723	1.814	1.905	1.995
210.	0.190	0.286	0.381	0.476	0.571	0.667	0.762	0.857	0.952	1.048	1.143	1.238	1.333	1.429	1.524	1.619	1.714	1.809	1.905	2.000	2.095
220.	0.200	0.299	0.399	0.499	0.598	0.698	0.798	0.898	0.998	1.097	1.197	1.297	1.397	1.497	1.596	1.696	1.796	1.896	1.996	2.095	2.195
230.	0.209	0.313	0.417	0.522	0.626	0.730	0.835	0.939	1.043	1.148	1.252	1.356	1.461	1.565	1.669	1.773	1.878	1.982	2.086	2.191	2.295
240.	0.218	0.327	0.435	0.544	0.653	0.762	0.871	0.980	1.089	1.198	1.306	1.415	1.524	1.633	1.742	1.851	1.960	2.068	2.177	2.286	2.395
250.	0.227	0.340	0.453	0.566	0.680	0.794	0.907	1.021	1.134	1.248	1.361	1.474	1.588	1.701	1.815	1.928	2.041	2.155	2.268	2.382	2.495
260.	0.236	0.354	0.472	0.590	0.708	0.826	0.944	1.062	1.180	1.298	1.416	1.533	1.651	1.769	1.887	2.005	2.123	2.241	2.359	2.477	2.595
270.	0.245	0.368	0.490	0.613	0.735	0.858	0.980	1.103	1.225	1.348	1.470	1.593	1.715	1.838	1.960	2.083	2.205	2.328	2.450	2.573	2.695
280.	0.254	0.381	0.508	0.635	0.762	0.889	1.016	1.143	1.271	1.398	1.525	1.652	1.779	1.906	2.033	2.160	2.287	2.414	2.541	2.668	2.795
290.	0.263	0.395	0.526	0.658	0.790	0.921	1.053	1.184	1.316	1.448	1.579	1.711	1.842	1.974	2.106	2.237	2.369	2.500	2.632	2.764	2.895
300.	0.272	0.408	0.545	0.681	0.817	0.953	1.089	1.225	1.362	1.498	1.634	1.770	1.906	2.042	2.178	2.315	2.451	2.587	2.723	2.859	2.995
310.	0.281	0.422	0.563	0.704	0.844	0.985	1.126	1.266	1.407	1.548	1.688	1.829	1.970	2.111	2.251	2.392	2.533	2.673	2.814	2.955	3.095
320.	0.291	0.436	0.581	0.726	0.872	1.017	1.162	1.307	1.453	1.598	1.743	1.888	2.034	2.179	2.324	2.469	2.615	2.760	2.905	3.050	3.196
330.	0.300	0.449	0.599	0.749	0.899	1.049	1.198	1.348	1.498	1.648	1.798	1.947	2.097	2.247	2.397	2.547	2.697	2.846	2.996	3.146	3.296
340.	0.309	0.463	0.617	0.772	0.926	1.081	1.235	1.389	1.544	1.698	1.852	2.007	2.161	2.315	2.470	2.624	2.778	2.933	3.087	3.242	3.396
350.	0.318	0.477	0.636	0.795	0.953	1.112	1.271	1.430	1.589	1.748	1.907	2.065	2.225	2.384	2.543	2.702	2.860	3.019	3.178	3.337	3.496
360.	0.327	0.490	0.654	0.817	0.981	1.144	1.308	1.471	1.635	1.798	1.962	2.125	2.289	2.452	2.615	2.779	2.942	3.106	3.269	3.433	3.596
370.	0.336	0.504	0.672	0.840	1.008	1.176	1.344	1.512	1.680	1.848	2.016	2.184	2.352	2.520	2.688	2.856	3.024	3.192	3.360	3.528	3.697

TABLE 2.6d Torispherical head thickness tables in inches: joint efficiency = 0.80

TORISPHERICAL HEAD THICKNESS TABLES IN INCHES

JOINT EFF.=0.80 ALLOWABLE STRESS VALUE =13800. PSI

INSIDE CROWN RADIUS, INCHES

PSI	12.	18.	24.	30.	36.	42.	48.	54.	60.	66.	72.	78.	84.	90.	96.	102.	108.	114.	120.	126.	132.
15.	0.014	0.022	0.029	0.036	0.043	0.051	0.058	0.065	0.072	0.079	0.087	0.094	0.101	0.108	0.115	0.123	0.130	0.137	0.144	0.152	0.159
20.	0.019	0.029	0.038	0.048	0.058	0.067	0.077	0.087	0.096	0.106	0.115	0.125	0.135	0.144	0.154	0.164	0.173	0.183	0.192	0.202	0.212
25.	0.024	0.036	0.048	0.060	0.072	0.084	0.096	0.108	0.120	0.132	0.144	0.156	0.168	0.180	0.192	0.204	0.216	0.229	0.241	0.253	0.265
30.	0.029	0.043	0.058	0.072	0.087	0.101	0.115	0.130	0.144	0.159	0.173	0.188	0.202	0.216	0.231	0.245	0.260	0.274	0.289	0.303	0.318
35.	0.034	0.051	0.067	0.084	0.101	0.118	0.135	0.152	0.168	0.185	0.202	0.219	0.236	0.253	0.269	0.286	0.303	0.320	0.337	0.354	0.370
40.	0.038	0.058	0.077	0.096	0.115	0.135	0.154	0.173	0.192	0.212	0.231	0.250	0.269	0.289	0.308	0.327	0.346	0.366	0.385	0.404	0.423
45.	0.043	0.065	0.087	0.108	0.130	0.152	0.173	0.195	0.217	0.238	0.260	0.281	0.303	0.325	0.346	0.368	0.390	0.411	0.433	0.455	0.476
50.	0.048	0.072	0.096	0.120	0.144	0.168	0.192	0.217	0.241	0.265	0.289	0.313	0.337	0.361	0.385	0.409	0.433	0.457	0.481	0.505	0.529
55.	0.053	0.079	0.106	0.132	0.159	0.185	0.212	0.238	0.265	0.291	0.318	0.344	0.371	0.397	0.423	0.450	0.476	0.503	0.529	0.556	0.582
60.	0.058	0.087	0.115	0.144	0.173	0.202	0.231	0.260	0.289	0.318	0.346	0.375	0.404	0.433	0.462	0.491	0.520	0.549	0.577	0.606	0.635
65.	0.063	0.094	0.125	0.156	0.188	0.219	0.250	0.282	0.313	0.344	0.375	0.407	0.438	0.469	0.501	0.532	0.563	0.594	0.626	0.657	0.688
70.	0.067	0.101	0.135	0.168	0.202	0.236	0.270	0.303	0.337	0.371	0.404	0.438	0.472	0.505	0.539	0.573	0.606	0.640	0.674	0.707	0.741
75.	0.072	0.108	0.144	0.180	0.217	0.253	0.289	0.325	0.361	0.397	0.433	0.469	0.505	0.541	0.578	0.614	0.650	0.686	0.722	0.758	0.794
80.	0.077	0.116	0.154	0.193	0.231	0.270	0.308	0.347	0.385	0.424	0.462	0.501	0.539	0.578	0.616	0.655	0.693	0.732	0.770	0.809	0.847
85.	0.082	0.123	0.164	0.205	0.245	0.286	0.327	0.368	0.409	0.449	0.491	0.532	0.573	0.614	0.655	0.696	0.736	0.777	0.818	0.859	0.900
90.	0.087	0.130	0.173	0.217	0.260	0.303	0.347	0.390	0.433	0.477	0.520	0.563	0.607	0.650	0.693	0.736	0.780	0.823	0.866	0.910	0.953
95.	0.091	0.137	0.183	0.229	0.274	0.320	0.366	0.412	0.457	0.503	0.549	0.595	0.640	0.686	0.732	0.777	0.823	0.869	0.915	0.960	1.006
100.	0.096	0.144	0.193	0.241	0.289	0.337	0.385	0.433	0.481	0.530	0.578	0.626	0.674	0.722	0.770	0.818	0.867	0.915	0.963	1.011	1.059
105.	0.101	0.152	0.202	0.253	0.303	0.354	0.404	0.455	0.506	0.556	0.607	0.657	0.708	0.758	0.809	0.859	0.910	0.960	1.011	1.062	1.112
110.	0.106	0.159	0.212	0.265	0.318	0.371	0.424	0.477	0.530	0.583	0.636	0.688	0.741	0.794	0.847	0.900	0.953	1.006	1.059	1.112	1.165
115.	0.111	0.166	0.221	0.277	0.332	0.388	0.443	0.498	0.554	0.609	0.664	0.720	0.775	0.831	0.886	0.941	0.997	1.052	1.107	1.163	1.218
120.	0.116	0.173	0.231	0.289	0.347	0.404	0.462	0.520	0.578	0.636	0.693	0.751	0.809	0.867	0.924	0.982	1.040	1.098	1.156	1.213	1.271
125.	0.120	0.181	0.241	0.301	0.361	0.421	0.482	0.542	0.602	0.662	0.722	0.782	0.843	0.903	0.963	1.023	1.083	1.144	1.204	1.264	1.324
130.	0.125	0.188	0.250	0.313	0.376	0.438	0.501	0.563	0.626	0.689	0.751	0.814	0.876	0.939	1.002	1.064	1.127	1.189	1.252	1.315	1.377
135.	0.130	0.195	0.260	0.325	0.390	0.455	0.520	0.585	0.650	0.715	0.780	0.845	0.910	0.975	1.040	1.105	1.170	1.235	1.300	1.365	1.430
140.	0.135	0.202	0.270	0.337	0.405	0.472	0.539	0.607	0.674	0.742	0.809	0.876	0.944	1.011	1.079	1.146	1.214	1.281	1.348	1.416	1.483
145.	0.140	0.210	0.279	0.349	0.419	0.489	0.559	0.629	0.698	0.768	0.838	0.908	0.978	1.048	1.117	1.187	1.257	1.327	1.397	1.467	1.536
150.	0.144	0.217	0.289	0.361	0.433	0.506	0.578	0.650	0.722	0.795	0.867	0.939	1.011	1.084	1.156	1.228	1.300	1.373	1.445	1.517	1.589
160.	0.154	0.231	0.308	0.385	0.462	0.539	0.617	0.694	0.771	0.848	0.925	1.002	1.079	1.156	1.233	1.310	1.387	1.464	1.541	1.618	1.696
170.	0.164	0.246	0.328	0.409	0.491	0.573	0.655	0.737	0.819	0.901	0.983	1.065	1.146	1.228	1.310	1.392	1.474	1.556	1.638	1.720	1.802
180.	0.173	0.260	0.347	0.434	0.520	0.607	0.694	0.780	0.867	0.954	1.041	1.127	1.214	1.301	1.387	1.474	1.561	1.648	1.734	1.821	1.908
190.	0.183	0.275	0.366	0.458	0.549	0.641	0.732	0.824	0.915	1.007	1.098	1.190	1.282	1.373	1.465	1.556	1.648	1.739	1.831	1.922	2.014
200.	0.193	0.289	0.385	0.482	0.578	0.675	0.771	0.867	0.964	1.060	1.156	1.253	1.349	1.446	1.542	1.638	1.735	1.831	1.927	2.024	2.120
210.	0.202	0.304	0.405	0.506	0.607	0.708	0.810	0.911	1.012	1.113	1.214	1.316	1.417	1.518	1.619	1.720	1.822	1.923	2.024	2.125	2.226
220.	0.212	0.318	0.424	0.530	0.636	0.742	0.848	0.954	1.060	1.166	1.272	1.378	1.484	1.590	1.696	1.802	1.908	2.015	2.121	2.227	2.333
230.	0.221	0.335	0.443	0.554	0.665	0.776	0.887	0.998	1.109	1.219	1.330	1.441	1.552	1.663	1.774	1.885	1.995	2.106	2.217	2.328	2.439
240.	0.231	0.347	0.463	0.578	0.694	0.810	0.925	1.041	1.157	1.273	1.388	1.504	1.620	1.735	1.851	1.967	2.082	2.198	2.314	2.429	2.545
250.	0.241	0.362	0.482	0.603	0.723	0.844	0.964	1.085	1.205	1.326	1.446	1.567	1.687	1.808	1.928	2.049	2.169	2.290	2.410	2.531	2.651
260.	0.251	0.376	0.501	0.627	0.752	0.877	1.003	1.128	1.253	1.379	1.504	1.630	1.755	1.880	2.006	2.131	2.256	2.382	2.507	2.632	2.758
270.	0.260	0.391	0.521	0.651	0.781	0.911	1.041	1.172	1.302	1.432	1.562	1.692	1.823	1.953	2.083	2.213	2.343	2.473	2.604	2.734	2.864
280.	0.270	0.405	0.540	0.675	0.810	0.945	1.080	1.215	1.350	1.485	1.620	1.755	1.890	2.025	2.160	2.295	2.430	2.565	2.700	2.835	2.970
290.	0.280	0.420	0.559	0.699	0.839	0.979	1.119	1.259	1.399	1.538	1.678	1.818	1.958	2.098	2.238	2.377	2.517	2.657	2.797	2.937	3.077
300.	0.289	0.434	0.579	0.723	0.868	1.013	1.157	1.302	1.447	1.592	1.736	1.881	2.026	2.170	2.315	2.460	2.604	2.749	2.894	3.038	3.183
310.	0.299	0.449	0.598	0.748	0.897	1.047	1.196	1.346	1.495	1.645	1.794	1.944	2.093	2.243	2.392	2.542	2.691	2.841	2.990	3.140	3.290
320.	0.309	0.463	0.617	0.772	0.926	1.081	1.235	1.389	1.544	1.698	1.852	2.007	2.161	2.315	2.470	2.624	2.778	2.933	3.087	3.242	3.396
330.	0.318	0.478	0.637	0.796	0.955	1.114	1.274	1.433	1.592	1.751	1.910	2.070	2.229	2.388	2.547	2.706	2.866	3.025	3.184	3.343	3.502
340.	0.328	0.492	0.656	0.820	0.984	1.148	1.312	1.476	1.640	1.804	1.968	2.132	2.297	2.461	2.625	2.789	2.953	3.117	3.281	3.445	3.609
350.	0.338	0.507	0.676	0.844	1.013	1.182	1.351	1.520	1.689	1.858	2.027	2.195	2.364	2.533	2.702	2.871	3.040	3.209	3.378	3.546	3.715
360.	0.347	0.521	0.695	0.869	1.042	1.216	1.390	1.563	1.737	1.911	2.085	2.258	2.432	2.606	2.779	2.953	3.127	3.301	3.474	3.648	3.822
370.	0.357	0.536	0.714	0.893	1.071	1.250	1.428	1.607	1.786	1.964	2.143	2.321	2.500	2.678	2.857	3.036	3.214	3.393	3.571	3.750	3.928

TABLE 2.6e Torispherical head thickness tables in inches: joint efficiency = 0.70

TORISPHERICAL HEAD THICKNESS TABLES IN INCHES

JOINT EFF.=0.70 ALLOWABLE STRESS VALUE =13800. PSI

INSIDE CROWN RADIUS, INCHES

PSI	12.	18.	24.	30.	36.	42.	48.	54.	60.	66.	72.	78.	84.	90.	96.	102.	108.	114.	120.	126.	132.
15.	0.016	0.025	0.033	0.041	0.049	0.058	0.066	0.074	0.082	0.091	0.099	0.107	0.115	0.124	0.132	0.140	0.148	0.157	0.165	0.173	0.181
20.	0.022	0.033	0.044	0.055	0.066	0.077	0.088	0.099	0.110	0.121	0.132	0.143	0.154	0.165	0.176	0.187	0.198	0.209	0.220	0.231	0.242
25.	0.027	0.041	0.055	0.069	0.082	0.096	0.110	0.124	0.137	0.151	0.165	0.179	0.192	0.206	0.220	0.234	0.247	0.261	0.275	0.289	0.302
30.	0.033	0.049	0.066	0.082	0.099	0.115	0.132	0.148	0.165	0.181	0.198	0.214	0.231	0.247	0.264	0.280	0.297	0.313	0.330	0.346	0.363
35.	0.038	0.058	0.077	0.096	0.115	0.135	0.154	0.173	0.192	0.212	0.231	0.250	0.269	0.289	0.308	0.327	0.346	0.366	0.385	0.404	0.423
40.	0.044	0.066	0.088	0.110	0.132	0.154	0.176	0.198	0.220	0.242	0.264	0.286	0.308	0.330	0.352	0.374	0.396	0.418	0.440	0.462	0.484
45.	0.049	0.074	0.099	0.124	0.148	0.173	0.198	0.223	0.247	0.272	0.297	0.322	0.346	0.371	0.396	0.421	0.445	0.470	0.495	0.520	0.544
50.	0.055	0.082	0.110	0.137	0.165	0.192	0.220	0.247	0.275	0.302	0.330	0.357	0.385	0.412	0.440	0.467	0.495	0.522	0.550	0.577	0.605
55.	0.060	0.091	0.121	0.151	0.182	0.212	0.242	0.272	0.303	0.333	0.363	0.393	0.424	0.454	0.484	0.514	0.545	0.575	0.605	0.635	0.666
60.	0.066	0.099	0.132	0.165	0.198	0.231	0.264	0.297	0.330	0.363	0.396	0.429	0.462	0.495	0.528	0.561	0.594	0.627	0.660	0.693	0.726
65.	0.072	0.107	0.143	0.179	0.215	0.250	0.286	0.322	0.358	0.393	0.429	0.465	0.501	0.536	0.572	0.608	0.644	0.679	0.715	0.751	0.787
70.	0.077	0.116	0.154	0.193	0.231	0.270	0.308	0.347	0.385	0.424	0.462	0.501	0.539	0.578	0.616	0.655	0.693	0.732	0.770	0.809	0.847
75.	0.083	0.124	0.165	0.206	0.248	0.289	0.330	0.371	0.413	0.454	0.495	0.536	0.578	0.619	0.660	0.701	0.743	0.784	0.825	0.866	0.908
80.	0.088	0.132	0.176	0.220	0.264	0.308	0.352	0.396	0.440	0.484	0.528	0.572	0.616	0.660	0.704	0.748	0.792	0.836	0.880	0.924	0.968
85.	0.094	0.140	0.187	0.234	0.281	0.327	0.374	0.421	0.468	0.514	0.561	0.608	0.655	0.701	0.748	0.795	0.842	0.889	0.935	0.982	1.029
90.	0.099	0.148	0.198	0.248	0.297	0.347	0.396	0.446	0.495	0.545	0.594	0.644	0.693	0.743	0.792	0.842	0.891	0.941	0.990	1.040	1.089
95.	0.105	0.157	0.209	0.261	0.314	0.366	0.418	0.470	0.523	0.575	0.627	0.680	0.732	0.784	0.836	0.889	0.941	0.993	1.045	1.098	1.150
100.	0.110	0.165	0.220	0.275	0.330	0.385	0.440	0.495	0.550	0.605	0.660	0.715	0.770	0.826	0.880	0.935	0.990	1.045	1.101	1.156	1.211
105.	0.116	0.173	0.231	0.289	0.347	0.404	0.462	0.520	0.578	0.636	0.693	0.751	0.809	0.867	0.924	0.982	1.040	1.098	1.156	1.213	1.271
110.	0.121	0.182	0.242	0.303	0.363	0.424	0.484	0.545	0.605	0.666	0.726	0.787	0.847	0.908	0.969	1.029	1.090	1.150	1.211	1.271	1.332
115.	0.127	0.190	0.253	0.316	0.380	0.443	0.506	0.570	0.633	0.696	0.759	0.823	0.886	0.949	1.013	1.076	1.139	1.203	1.266	1.329	1.392
120.	0.132	0.198	0.264	0.330	0.396	0.462	0.528	0.594	0.660	0.726	0.793	0.859	0.925	0.991	1.057	1.123	1.189	1.255	1.321	1.387	1.453
125.	0.138	0.206	0.275	0.344	0.413	0.482	0.550	0.619	0.688	0.757	0.826	0.894	0.963	1.032	1.101	1.170	1.238	1.307	1.376	1.445	1.514
130.	0.143	0.215	0.286	0.358	0.429	0.501	0.572	0.644	0.716	0.787	0.859	0.930	1.002	1.073	1.145	1.216	1.288	1.360	1.431	1.503	1.574
135.	0.149	0.223	0.297	0.372	0.446	0.520	0.594	0.669	0.743	0.817	0.892	0.966	1.040	1.115	1.189	1.263	1.338	1.412	1.486	1.561	1.635
140.	0.154	0.231	0.308	0.385	0.462	0.539	0.617	0.694	0.771	0.848	0.925	1.002	1.079	1.156	1.233	1.310	1.387	1.464	1.541	1.618	1.696
145.	0.160	0.239	0.319	0.399	0.479	0.559	0.639	0.718	0.798	0.878	0.958	1.038	1.118	1.197	1.277	1.357	1.437	1.517	1.596	1.676	1.756
150.	0.165	0.248	0.330	0.413	0.495	0.578	0.661	0.743	0.826	0.908	0.991	1.074	1.156	1.239	1.321	1.404	1.486	1.569	1.652	1.734	1.817
160.	0.176	0.264	0.352	0.440	0.529	0.617	0.705	0.793	0.881	0.969	1.057	1.145	1.233	1.321	1.410	1.498	1.586	1.674	1.762	1.850	1.938
170.	0.187	0.281	0.374	0.468	0.562	0.655	0.749	0.843	0.936	1.030	1.123	1.217	1.311	1.404	1.498	1.591	1.685	1.779	1.872	1.966	2.059
180.	0.198	0.297	0.396	0.496	0.595	0.694	0.793	0.892	0.991	1.090	1.190	1.289	1.388	1.487	1.586	1.685	1.784	1.883	1.983	2.082	2.181
190.	0.209	0.314	0.419	0.523	0.628	0.733	0.837	0.942	1.046	1.151	1.256	1.360	1.465	1.570	1.674	1.779	1.884	1.988	2.093	2.198	2.302
200.	0.220	0.330	0.441	0.551	0.661	0.771	0.881	0.991	1.102	1.212	1.322	1.432	1.542	1.652	1.763	1.873	1.983	2.093	2.203	2.314	2.424
210.	0.231	0.347	0.463	0.578	0.694	0.810	0.925	1.041	1.157	1.273	1.388	1.504	1.620	1.735	1.851	1.967	2.082	2.198	2.314	2.429	2.545
220.	0.242	0.364	0.485	0.606	0.727	0.848	0.970	1.091	1.212	1.333	1.455	1.576	1.697	1.818	1.939	2.061	2.182	2.303	2.424	2.545	2.667
230.	0.253	0.380	0.507	0.634	0.760	0.887	1.014	1.141	1.267	1.394	1.521	1.647	1.774	1.901	2.028	2.154	2.281	2.408	2.535	2.661	2.788
240.	0.265	0.397	0.529	0.661	0.794	0.926	1.058	1.190	1.323	1.455	1.587	1.719	1.852	1.984	2.116	2.248	2.381	2.513	2.645	2.777	2.910
250.	0.276	0.413	0.551	0.688	0.827	0.964	1.102	1.240	1.378	1.516	1.655	1.791	1.929	2.067	2.204	2.342	2.480	2.618	2.756	2.893	3.031
260.	0.287	0.430	0.573	0.717	0.860	1.003	1.146	1.290	1.433	1.576	1.720	1.863	2.006	2.150	2.293	2.436	2.579	2.723	2.866	3.009	3.153
270.	0.298	0.446	0.595	0.744	0.893	1.042	1.191	1.339	1.488	1.637	1.786	1.935	2.084	2.232	2.381	2.530	2.679	2.828	2.977	3.125	3.274
280.	0.309	0.463	0.617	0.772	0.926	1.081	1.235	1.389	1.544	1.698	1.852	2.007	2.161	2.315	2.470	2.624	2.778	2.933	3.087	3.242	3.396
290.	0.320	0.480	0.640	0.799	0.959	1.119	1.279	1.439	1.599	1.759	1.919	2.079	2.238	2.398	2.558	2.718	2.878	3.038	3.198	3.358	3.518
300.	0.331	0.496	0.662	0.827	0.993	1.158	1.323	1.489	1.654	1.820	1.985	2.150	2.316	2.481	2.647	2.812	2.978	3.143	3.308	3.474	3.639
310.	0.342	0.513	0.684	0.855	1.026	1.197	1.368	1.539	1.710	1.880	2.051	2.222	2.393	2.564	2.735	2.906	3.077	3.248	3.419	3.590	3.761
320.	0.353	0.529	0.706	0.882	1.059	1.235	1.412	1.588	1.765	1.941	2.118	2.294	2.471	2.647	2.824	3.000	3.177	3.353	3.530	3.706	3.883
330.	0.364	0.546	0.728	0.910	1.092	1.274	1.456	1.638	1.820	2.002	2.184	2.366	2.548	2.730	2.912	3.094	3.276	3.458	3.640	3.822	4.004
340.	0.375	0.563	0.750	0.938	1.125	1.313	1.500	1.688	1.876	2.063	2.251	2.438	2.626	2.813	3.001	3.188	3.376	3.564	3.751	3.939	4.126
350.	0.386	0.579	0.772	0.965	1.159	1.352	1.545	1.738	1.931	2.124	2.317	2.510	2.703	2.896	3.089	3.283	3.476	3.669	3.862	4.055	4.248
360.	0.397	0.596	0.795	0.993	1.192	1.390	1.589	1.788	1.986	2.185	2.384	2.582	2.781	2.979	3.178	3.377	3.575	3.774	3.973	4.171	4.370
370.	0.408	0.613	0.817	1.021	1.225	1.429	1.633	1.838	2.042	2.246	2.450	2.654	2.858	3.063	3.267	3.471	3.675	3.879	4.083	4.288	4.492

TABLE 2.6f Torispherical head thickness tables in inches: joint efficiency = 0.65

TORISPHERICAL HEAD THICKNESS TABLES IN INCHES

JOINT EFF.=0.65   ALLOWABLE STRESS VALUE =13800. PSI

INSIDE CROWN RADIUS, INCHES

PSI	12.	18.	24.	30.	36.	42.	48.	54.	60.	66.	72.	78.	84.	90.	96.	102.	108.	114.	120.	126.	132.
15.	0.018	0.027	0.036	0.044	0.053	0.062	0.071	0.080	0.089	0.098	0.107	0.115	0.124	0.133	0.142	0.151	0.160	0.169	0.178	0.187	0.195
20.	0.024	0.036	0.047	0.059	0.071	0.083	0.095	0.107	0.118	0.130	0.142	0.154	0.166	0.178	0.189	0.201	0.213	0.225	0.237	0.249	0.261
25.	0.030	0.044	0.059	0.074	0.089	0.104	0.118	0.133	0.148	0.163	0.178	0.192	0.207	0.222	0.237	0.252	0.266	0.281	0.296	0.311	0.326
30.	0.036	0.053	0.071	0.089	0.107	0.124	0.142	0.160	0.178	0.195	0.213	0.231	0.249	0.266	0.284	0.302	0.320	0.338	0.355	0.373	0.391
35.	0.041	0.062	0.083	0.104	0.124	0.145	0.166	0.187	0.207	0.228	0.249	0.269	0.290	0.311	0.332	0.352	0.373	0.394	0.415	0.435	0.456
40.	0.047	0.071	0.095	0.118	0.142	0.166	0.190	0.213	0.237	0.261	0.284	0.308	0.332	0.355	0.379	0.403	0.426	0.450	0.474	0.497	0.521
45.	0.053	0.080	0.107	0.133	0.160	0.187	0.213	0.240	0.267	0.293	0.320	0.346	0.373	0.400	0.426	0.453	0.480	0.506	0.533	0.560	0.586
50.	0.059	0.089	0.118	0.148	0.178	0.207	0.237	0.267	0.296	0.326	0.355	0.385	0.415	0.444	0.474	0.503	0.533	0.563	0.592	0.622	0.652
55.	0.065	0.098	0.130	0.163	0.195	0.228	0.261	0.293	0.326	0.358	0.391	0.424	0.456	0.489	0.521	0.554	0.586	0.619	0.652	0.684	0.717
60.	0.071	0.107	0.142	0.178	0.213	0.249	0.284	0.320	0.355	0.391	0.427	0.462	0.498	0.533	0.569	0.604	0.640	0.675	0.711	0.746	0.782
65.	0.077	0.116	0.154	0.193	0.231	0.270	0.308	0.347	0.385	0.424	0.462	0.501	0.539	0.578	0.616	0.655	0.693	0.732	0.770	0.809	0.847
70.	0.083	0.124	0.166	0.207	0.249	0.290	0.332	0.373	0.415	0.456	0.498	0.539	0.581	0.622	0.664	0.705	0.746	0.788	0.829	0.871	0.912
75.	0.089	0.133	0.178	0.222	0.267	0.311	0.355	0.400	0.444	0.489	0.533	0.578	0.622	0.667	0.711	0.755	0.800	0.844	0.889	0.933	0.978
80.	0.095	0.142	0.190	0.237	0.284	0.332	0.379	0.427	0.474	0.521	0.569	0.616	0.664	0.711	0.758	0.806	0.853	0.901	0.948	0.995	1.043
85.	0.101	0.151	0.201	0.252	0.302	0.352	0.403	0.453	0.504	0.554	0.604	0.655	0.705	0.755	0.806	0.856	0.907	0.957	1.007	1.058	1.108
90.	0.107	0.160	0.213	0.267	0.320	0.373	0.427	0.480	0.533	0.587	0.640	0.693	0.747	0.800	0.853	0.907	0.960	1.013	1.067	1.120	1.173
95.	0.113	0.169	0.225	0.281	0.338	0.394	0.450	0.507	0.563	0.619	0.676	0.732	0.788	0.844	0.901	0.957	1.013	1.070	1.126	1.182	1.239
100.	0.119	0.178	0.237	0.296	0.356	0.415	0.474	0.533	0.593	0.652	0.711	0.770	0.830	0.889	0.948	1.007	1.067	1.126	1.185	1.245	1.304
105.	0.124	0.187	0.249	0.311	0.373	0.436	0.498	0.560	0.622	0.685	0.747	0.809	0.871	0.933	0.996	1.058	1.120	1.182	1.245	1.307	1.369
110.	0.130	0.196	0.261	0.326	0.391	0.456	0.522	0.587	0.652	0.717	0.782	0.848	0.913	0.978	1.043	1.108	1.174	1.239	1.304	1.369	1.434
115.	0.136	0.204	0.273	0.341	0.409	0.477	0.545	0.613	0.682	0.750	0.818	0.886	0.954	1.022	1.091	1.159	1.227	1.295	1.363	1.431	1.500
120.	0.142	0.213	0.285	0.356	0.427	0.498	0.569	0.640	0.711	0.782	0.854	0.925	0.996	1.067	1.138	1.209	1.280	1.352	1.423	1.494	1.565
125.	0.148	0.222	0.296	0.370	0.445	0.519	0.593	0.667	0.741	0.815	0.889	0.963	1.037	1.111	1.186	1.260	1.334	1.408	1.482	1.556	1.630
130.	0.154	0.231	0.308	0.385	0.462	0.539	0.617	0.694	0.771	0.848	0.925	1.002	1.079	1.156	1.233	1.310	1.387	1.464	1.541	1.618	1.696
135.	0.160	0.240	0.320	0.400	0.480	0.560	0.640	0.720	0.800	0.880	0.960	1.040	1.121	1.201	1.281	1.361	1.441	1.521	1.601	1.681	1.761
140.	0.166	0.249	0.332	0.415	0.498	0.581	0.664	0.747	0.830	0.913	0.996	1.079	1.162	1.245	1.328	1.411	1.494	1.577	1.660	1.743	1.826
145.	0.172	0.258	0.344	0.430	0.516	0.602	0.688	0.774	0.860	0.946	1.032	1.118	1.204	1.290	1.376	1.462	1.548	1.634	1.720	1.805	1.891
150.	0.178	0.267	0.356	0.445	0.534	0.623	0.712	0.801	0.889	0.978	1.067	1.156	1.245	1.334	1.423	1.512	1.601	1.690	1.779	1.868	1.957
160.	0.190	0.285	0.380	0.474	0.569	0.664	0.759	0.854	0.949	1.044	1.139	1.234	1.328	1.423	1.518	1.613	1.708	1.803	1.898	1.993	2.087
170.	0.202	0.302	0.403	0.504	0.605	0.706	0.807	0.907	1.008	1.109	1.210	1.311	1.412	1.512	1.613	1.714	1.815	1.916	2.017	2.117	2.218
180.	0.214	0.320	0.427	0.534	0.641	0.747	0.854	0.961	1.068	1.174	1.281	1.388	1.495	1.602	1.708	1.815	1.922	2.029	2.135	2.242	2.349
190.	0.225	0.338	0.451	0.564	0.676	0.789	0.902	1.014	1.127	1.240	1.353	1.465	1.578	1.691	1.803	1.916	2.029	2.142	2.254	2.367	2.480
200.	0.237	0.356	0.475	0.593	0.712	0.831	0.949	1.068	1.187	1.305	1.424	1.543	1.661	1.780	1.899	2.017	2.136	2.255	2.373	2.492	2.611
210.	0.249	0.374	0.498	0.623	0.748	0.872	0.997	1.121	1.246	1.371	1.495	1.620	1.744	1.869	1.994	2.118	2.243	2.368	2.492	2.617	2.741
220.	0.261	0.392	0.522	0.653	0.783	0.914	1.044	1.175	1.306	1.436	1.567	1.697	1.828	1.958	2.089	2.219	2.350	2.481	2.611	2.742	2.872
230.	0.273	0.410	0.546	0.683	0.819	0.956	1.092	1.229	1.365	1.502	1.638	1.775	1.911	2.048	2.184	2.321	2.457	2.594	2.730	2.867	3.003
240.	0.285	0.427	0.570	0.712	0.855	0.997	1.140	1.282	1.425	1.567	1.709	1.852	1.994	2.137	2.279	2.422	2.564	2.707	2.849	2.992	3.134
250.	0.297	0.445	0.594	0.742	0.890	1.039	1.187	1.336	1.484	1.632	1.781	1.929	2.078	2.226	2.375	2.523	2.671	2.820	2.968	3.117	3.265
260.	0.309	0.463	0.617	0.772	0.926	1.081	1.235	1.389	1.544	1.698	1.852	2.007	2.161	2.315	2.470	2.624	2.778	2.933	3.087	3.242	3.396
270.	0.321	0.481	0.641	0.802	0.962	1.122	1.283	1.443	1.603	1.763	1.924	2.084	2.244	2.405	2.565	2.725	2.886	3.046	3.206	3.367	3.527
280.	0.333	0.499	0.665	0.831	0.998	1.164	1.330	1.496	1.663	1.829	1.995	2.161	2.328	2.494	2.660	2.827	2.993	3.159	3.325	3.492	3.658
290.	0.344	0.517	0.689	0.861	1.033	1.206	1.378	1.550	1.722	1.895	2.067	2.239	2.411	2.583	2.756	2.928	3.100	3.272	3.445	3.617	3.789
300.	0.356	0.535	0.713	0.891	1.069	1.247	1.426	1.604	1.782	1.960	2.138	2.317	2.495	2.673	2.851	3.029	3.207	3.386	3.564	3.742	3.920
310.	0.368	0.552	0.737	0.921	1.105	1.289	1.473	1.657	1.841	2.026	2.210	2.394	2.578	2.762	2.946	3.131	3.315	3.499	3.683	3.867	4.051
320.	0.380	0.570	0.760	0.951	1.141	1.331	1.521	1.711	1.901	2.091	2.281	2.471	2.662	2.852	3.042	3.232	3.422	3.612	3.802	3.992	4.182
330.	0.392	0.588	0.784	0.980	1.176	1.373	1.569	1.765	1.961	2.157	2.353	2.549	2.745	2.941	3.137	3.333	3.529	3.725	3.921	4.118	4.314
340.	0.404	0.606	0.808	1.010	1.212	1.414	1.616	1.818	2.020	2.222	2.424	2.626	2.829	3.031	3.233	3.435	3.637	3.839	4.041	4.243	4.445
350.	0.416	0.624	0.832	1.040	1.248	1.456	1.664	1.872	2.080	2.288	2.496	2.704	2.912	3.120	3.328	3.536	3.744	3.952	4.160	4.368	4.576
360.	0.428	0.642	0.856	1.070	1.284	1.498	1.712	1.926	2.140	2.354	2.568	2.782	2.996	3.210	3.424	3.637	3.851	4.065	4.279	4.493	4.707
370.	0.440	0.660	0.880	1.100	1.320	1.540	1.759	1.979	2.199	2.419	2.639	2.859	3.079	3.299	3.519	3.739	3.959	4.179	4.399	4.619	4.839

TABLE 2.6g Ellipsoidal head thickness tables in inches: joint efficiency = 1.0

ELLIPSOIDAL HEAD THICKNESS TABLES IN INCHES   MAJOR TO MINOR AXIS RATIO=2:1

JOINT EFF.=1.00   ALLOWABLE STRESS VALUE =13800. PSI

INSIDE DIAMETER,IN.

PSI	12.	18.	24.	30.	36.	42.	48.	54.	60.	66.	72.	78.	84.	90.	96.	102.	108.	114.	120.	126.	132.
15.	0.007	0.010	0.013	0.016	0.020	0.023	0.026	0.029	0.033	0.036	0.039	0.042	0.046	0.049	0.052	0.055	0.059	0.062	0.065	0.068	0.072
20.	0.009	0.013	0.017	0.022	0.026	0.030	0.035	0.039	0.043	0.048	0.052	0.057	0.061	0.065	0.070	0.074	0.078	0.083	0.087	0.091	0.096
25.	0.011	0.016	0.022	0.027	0.033	0.038	0.043	0.049	0.054	0.060	0.065	0.071	0.076	0.082	0.087	0.092	0.098	0.103	0.109	0.114	0.120
30.	0.013	0.020	0.026	0.033	0.039	0.046	0.052	0.059	0.065	0.072	0.078	0.085	0.091	0.098	0.104	0.111	0.117	0.124	0.130	0.137	0.144
35.	0.015	0.023	0.030	0.038	0.046	0.053	0.061	0.068	0.076	0.084	0.091	0.099	0.107	0.114	0.122	0.129	0.137	0.145	0.152	0.160	0.167
40.	0.017	0.026	0.035	0.043	0.052	0.061	0.070	0.078	0.087	0.096	0.104	0.113	0.122	0.130	0.139	0.148	0.157	0.165	0.174	0.183	0.191
45.	0.020	0.029	0.039	0.049	0.059	0.069	0.078	0.088	0.098	0.108	0.117	0.127	0.137	0.147	0.157	0.166	0.176	0.186	0.196	0.206	0.215
50.	0.022	0.033	0.043	0.054	0.065	0.076	0.087	0.098	0.109	0.120	0.130	0.141	0.152	0.163	0.174	0.185	0.196	0.207	0.217	0.228	0.239
55.	0.024	0.036	0.048	0.060	0.072	0.084	0.096	0.108	0.120	0.132	0.144	0.155	0.167	0.179	0.191	0.203	0.215	0.227	0.239	0.251	0.263
60.	0.026	0.039	0.052	0.065	0.078	0.091	0.104	0.117	0.130	0.144	0.157	0.170	0.183	0.196	0.209	0.222	0.235	0.248	0.261	0.274	0.287
65.	0.028	0.042	0.057	0.071	0.085	0.099	0.113	0.127	0.141	0.156	0.170	0.184	0.198	0.212	0.226	0.240	0.254	0.269	0.283	0.297	0.311
70.	0.030	0.046	0.061	0.076	0.091	0.107	0.122	0.137	0.152	0.167	0.183	0.198	0.213	0.228	0.244	0.259	0.274	0.289	0.305	0.320	0.335
75.	0.033	0.049	0.065	0.082	0.098	0.114	0.131	0.147	0.163	0.179	0.196	0.212	0.228	0.245	0.261	0.277	0.294	0.310	0.326	0.343	0.359
80.	0.035	0.052	0.070	0.087	0.104	0.122	0.139	0.157	0.174	0.191	0.209	0.226	0.244	0.261	0.278	0.296	0.313	0.331	0.348	0.365	0.383
85.	0.037	0.055	0.074	0.092	0.111	0.129	0.148	0.166	0.185	0.203	0.222	0.240	0.259	0.277	0.296	0.314	0.333	0.351	0.370	0.388	0.407
90.	0.039	0.059	0.078	0.098	0.117	0.137	0.157	0.176	0.196	0.215	0.235	0.255	0.274	0.294	0.313	0.333	0.352	0.372	0.392	0.411	0.431
95.	0.041	0.062	0.083	0.103	0.124	0.145	0.165	0.186	0.207	0.227	0.248	0.269	0.289	0.310	0.331	0.351	0.372	0.393	0.413	0.434	0.455
100.	0.044	0.065	0.087	0.109	0.131	0.152	0.174	0.196	0.218	0.239	0.261	0.283	0.305	0.326	0.348	0.370	0.392	0.413	0.435	0.457	0.479
105.	0.046	0.069	0.091	0.114	0.137	0.160	0.183	0.206	0.228	0.251	0.274	0.297	0.320	0.343	0.365	0.388	0.411	0.434	0.457	0.480	0.503
110.	0.048	0.072	0.096	0.120	0.144	0.168	0.191	0.215	0.239	0.263	0.287	0.311	0.335	0.359	0.383	0.407	0.431	0.455	0.479	0.503	0.527
115.	0.050	0.075	0.100	0.125	0.150	0.175	0.200	0.225	0.250	0.275	0.300	0.325	0.350	0.375	0.400	0.425	0.450	0.475	0.500	0.525	0.550
120.	0.052	0.078	0.104	0.131	0.157	0.183	0.209	0.235	0.261	0.287	0.313	0.339	0.366	0.392	0.418	0.444	0.470	0.496	0.522	0.548	0.574
125.	0.054	0.082	0.109	0.136	0.163	0.190	0.218	0.245	0.272	0.299	0.326	0.354	0.381	0.408	0.435	0.462	0.490	0.517	0.544	0.571	0.598
130.	0.057	0.085	0.113	0.141	0.170	0.198	0.226	0.255	0.283	0.311	0.339	0.368	0.396	0.424	0.453	0.481	0.509	0.537	0.566	0.594	0.622
135.	0.059	0.088	0.118	0.147	0.176	0.206	0.235	0.264	0.294	0.323	0.353	0.382	0.411	0.441	0.470	0.499	0.529	0.558	0.588	0.617	0.646
140.	0.061	0.091	0.122	0.152	0.183	0.213	0.244	0.274	0.305	0.335	0.366	0.396	0.427	0.457	0.487	0.518	0.548	0.579	0.609	0.640	0.670
145.	0.063	0.095	0.126	0.158	0.189	0.221	0.252	0.284	0.316	0.347	0.379	0.410	0.442	0.473	0.505	0.536	0.568	0.600	0.631	0.663	0.694
150.	0.065	0.098	0.131	0.163	0.196	0.229	0.261	0.294	0.326	0.359	0.392	0.424	0.457	0.490	0.522	0.555	0.588	0.620	0.653	0.686	0.718
160.	0.070	0.104	0.139	0.174	0.209	0.244	0.279	0.313	0.348	0.383	0.418	0.453	0.488	0.522	0.557	0.592	0.627	0.662	0.696	0.731	0.766
170.	0.074	0.111	0.148	0.185	0.222	0.259	0.296	0.333	0.370	0.407	0.444	0.481	0.518	0.555	0.592	0.629	0.666	0.703	0.740	0.777	0.814
180.	0.078	0.118	0.157	0.196	0.235	0.274	0.313	0.353	0.392	0.431	0.470	0.509	0.549	0.588	0.627	0.666	0.705	0.744	0.784	0.823	0.862
190.	0.083	0.124	0.165	0.207	0.248	0.290	0.331	0.372	0.414	0.455	0.496	0.538	0.579	0.620	0.662	0.703	0.745	0.786	0.827	0.869	0.910
200.	0.087	0.131	0.174	0.218	0.261	0.305	0.348	0.392	0.435	0.479	0.522	0.566	0.610	0.653	0.697	0.740	0.784	0.827	0.871	0.914	0.958
210.	0.091	0.137	0.183	0.229	0.274	0.320	0.366	0.411	0.457	0.503	0.549	0.594	0.640	0.686	0.732	0.777	0.823	0.869	0.914	0.960	1.006
220.	0.096	0.144	0.192	0.240	0.287	0.335	0.383	0.431	0.479	0.527	0.575	0.623	0.671	0.719	0.766	0.814	0.862	0.910	0.958	1.006	1.054
230.	0.100	0.150	0.200	0.250	0.301	0.351	0.401	0.451	0.501	0.551	0.601	0.651	0.701	0.751	0.801	0.851	0.902	0.952	1.002	1.052	1.102
240.	0.105	0.157	0.209	0.261	0.314	0.366	0.418	0.470	0.523	0.575	0.627	0.679	0.732	0.784	0.836	0.889	0.941	0.993	1.045	1.098	1.150
250.	0.109	0.163	0.218	0.272	0.327	0.381	0.436	0.490	0.544	0.599	0.653	0.708	0.762	0.817	0.871	0.926	0.980	1.034	1.089	1.143	1.198
260.	0.113	0.170	0.227	0.283	0.340	0.396	0.453	0.510	0.566	0.623	0.680	0.736	0.793	0.849	0.906	0.963	1.019	1.076	1.133	1.189	1.246
270.	0.118	0.176	0.235	0.294	0.353	0.412	0.470	0.529	0.588	0.647	0.706	0.765	0.823	0.882	0.941	1.000	1.059	1.117	1.176	1.235	1.294
280.	0.122	0.183	0.244	0.305	0.366	0.427	0.488	0.549	0.610	0.671	0.732	0.793	0.854	0.915	0.976	1.037	1.098	1.159	1.220	1.281	1.342
290.	0.126	0.190	0.253	0.316	0.379	0.442	0.505	0.569	0.632	0.695	0.758	0.821	0.884	0.948	1.011	1.074	1.137	1.200	1.264	1.327	1.390
300.	0.131	0.196	0.261	0.327	0.392	0.458	0.523	0.588	0.654	0.719	0.784	0.850	0.915	0.980	1.046	1.111	1.176	1.242	1.307	1.373	1.438
310.	0.135	0.203	0.270	0.338	0.405	0.473	0.540	0.608	0.675	0.743	0.811	0.878	0.946	1.013	1.081	1.148	1.216	1.283	1.351	1.418	1.486
320.	0.139	0.209	0.279	0.349	0.418	0.488	0.558	0.628	0.697	0.767	0.837	0.906	0.976	1.046	1.116	1.185	1.255	1.325	1.395	1.464	1.534
330.	0.144	0.216	0.288	0.360	0.431	0.503	0.575	0.647	0.719	0.791	0.863	0.935	1.007	1.079	1.151	1.222	1.294	1.366	1.438	1.510	1.582
340.	0.148	0.222	0.296	0.370	0.445	0.519	0.593	0.667	0.741	0.815	0.889	0.963	1.037	1.111	1.186	1.260	1.334	1.408	1.482	1.556	1.630
350.	0.153	0.229	0.305	0.381	0.458	0.534	0.610	0.687	0.763	0.839	0.915	0.992	1.068	1.144	1.220	1.297	1.373	1.449	1.526	1.602	1.678
360.	0.157	0.235	0.314	0.392	0.471	0.549	0.628	0.706	0.785	0.863	0.942	1.020	1.099	1.177	1.255	1.334	1.412	1.491	1.569	1.648	1.726
370.	0.161	0.242	0.323	0.403	0.484	0.565	0.645	0.726	0.807	0.887	0.968	1.048	1.129	1.210	1.290	1.371	1.452	1.532	1.613	1.694	1.774

TABLE 2.6h Ellipsoidal head thickness tables in inches: joint efficiency = 0.90

ELLIPSOIDAL HEAD THICKNESS TABLES IN INCHES   MAJOR TO MINOR AXIS RATIO=2:1

JOINT EFF.=0.90   ALLOWABLE STRESS VALUE =13800. PSI

PSI	INSIDE DIAMETER,IN. 12.	18.	24.	30.	36.	42.	48.	54.	60.	66.	72.	78.	84.	90.	96.	102.	108.	114.	120.	126.	132.
15.	0.007	0.011	0.014	0.018	0.022	0.025	0.029	0.033	0.036	0.040	0.043	0.047	0.051	0.054	0.058	0.062	0.065	0.069	0.072	0.076	0.080
20.	0.010	0.014	0.019	0.024	0.029	0.034	0.039	0.043	0.048	0.053	0.058	0.063	0.068	0.072	0.077	0.082	0.087	0.092	0.097	0.101	0.106
25.	0.012	0.018	0.024	0.030	0.036	0.042	0.048	0.054	0.060	0.066	0.072	0.079	0.085	0.091	0.097	0.103	0.109	0.115	0.121	0.127	0.133
30.	0.014	0.022	0.029	0.036	0.043	0.051	0.058	0.065	0.072	0.080	0.087	0.094	0.101	0.109	0.116	0.123	0.130	0.138	0.145	0.152	0.159
35.	0.017	0.025	0.034	0.042	0.051	0.059	0.068	0.076	0.085	0.093	0.101	0.110	0.118	0.127	0.135	0.144	0.152	0.161	0.169	0.178	0.186
40.	0.019	0.029	0.039	0.048	0.058	0.068	0.077	0.087	0.097	0.106	0.116	0.126	0.135	0.145	0.155	0.164	0.174	0.184	0.193	0.203	0.213
45.	0.022	0.033	0.043	0.054	0.065	0.076	0.087	0.098	0.109	0.120	0.130	0.141	0.152	0.163	0.174	0.185	0.196	0.207	0.217	0.228	0.239
50.	0.024	0.036	0.048	0.060	0.072	0.085	0.097	0.109	0.121	0.133	0.145	0.157	0.169	0.181	0.193	0.205	0.217	0.230	0.242	0.254	0.266
55.	0.027	0.040	0.053	0.066	0.080	0.093	0.106	0.120	0.133	0.146	0.159	0.173	0.186	0.199	0.213	0.226	0.239	0.253	0.266	0.279	0.292
60.	0.029	0.043	0.058	0.072	0.087	0.101	0.116	0.130	0.145	0.159	0.174	0.188	0.203	0.217	0.232	0.246	0.261	0.275	0.290	0.304	0.319
65.	0.031	0.047	0.063	0.079	0.094	0.110	0.126	0.141	0.157	0.173	0.189	0.204	0.220	0.236	0.251	0.267	0.283	0.298	0.314	0.330	0.346
70.	0.034	0.051	0.068	0.085	0.102	0.118	0.135	0.152	0.169	0.186	0.203	0.220	0.237	0.254	0.271	0.288	0.305	0.321	0.338	0.355	0.372
75.	0.036	0.054	0.073	0.091	0.109	0.127	0.145	0.163	0.181	0.199	0.218	0.236	0.254	0.272	0.290	0.308	0.326	0.344	0.363	0.381	0.399
80.	0.039	0.058	0.077	0.097	0.116	0.135	0.155	0.174	0.193	0.213	0.232	0.251	0.271	0.290	0.309	0.329	0.348	0.367	0.387	0.406	0.425
85.	0.041	0.062	0.082	0.103	0.123	0.144	0.164	0.185	0.205	0.226	0.247	0.267	0.288	0.308	0.329	0.349	0.370	0.390	0.411	0.431	0.452
90.	0.044	0.065	0.087	0.109	0.131	0.152	0.174	0.196	0.218	0.239	0.261	0.283	0.305	0.326	0.348	0.370	0.392	0.413	0.435	0.457	0.479
95.	0.046	0.069	0.092	0.115	0.138	0.161	0.184	0.207	0.230	0.253	0.276	0.299	0.322	0.344	0.367	0.390	0.413	0.436	0.459	0.482	0.505
100.	0.048	0.073	0.097	0.121	0.145	0.169	0.193	0.218	0.242	0.266	0.290	0.314	0.338	0.363	0.387	0.411	0.435	0.459	0.483	0.508	0.532
105.	0.051	0.076	0.102	0.127	0.152	0.178	0.203	0.228	0.254	0.279	0.305	0.330	0.355	0.381	0.406	0.432	0.457	0.482	0.508	0.533	0.558
110.	0.053	0.080	0.106	0.133	0.160	0.186	0.213	0.239	0.266	0.293	0.319	0.346	0.372	0.399	0.425	0.452	0.479	0.505	0.532	0.558	0.585
115.	0.056	0.083	0.111	0.139	0.167	0.195	0.222	0.250	0.278	0.306	0.334	0.361	0.389	0.417	0.445	0.473	0.500	0.528	0.556	0.584	0.612
120.	0.058	0.087	0.116	0.145	0.174	0.203	0.232	0.261	0.290	0.319	0.348	0.377	0.406	0.435	0.464	0.493	0.522	0.551	0.580	0.609	0.638
125.	0.060	0.091	0.121	0.151	0.181	0.212	0.242	0.272	0.302	0.332	0.363	0.393	0.423	0.453	0.484	0.514	0.544	0.574	0.604	0.635	0.665
130.	0.063	0.094	0.126	0.157	0.189	0.220	0.251	0.283	0.314	0.346	0.377	0.409	0.440	0.472	0.503	0.534	0.566	0.597	0.629	0.660	0.692
135.	0.065	0.098	0.131	0.163	0.196	0.229	0.261	0.294	0.326	0.359	0.392	0.424	0.457	0.490	0.522	0.555	0.588	0.620	0.653	0.686	0.718
140.	0.068	0.102	0.135	0.169	0.203	0.237	0.271	0.305	0.339	0.372	0.406	0.440	0.474	0.508	0.542	0.576	0.609	0.643	0.677	0.711	0.745
145.	0.070	0.105	0.140	0.175	0.210	0.245	0.281	0.316	0.351	0.386	0.421	0.456	0.491	0.526	0.561	0.597	0.631	0.666	0.701	0.736	0.771
150.	0.073	0.109	0.145	0.181	0.218	0.254	0.290	0.326	0.363	0.399	0.435	0.472	0.508	0.544	0.580	0.617	0.653	0.689	0.726	0.762	0.798
160.	0.077	0.116	0.155	0.193	0.232	0.271	0.310	0.348	0.387	0.426	0.464	0.503	0.542	0.580	0.619	0.658	0.697	0.735	0.774	0.813	0.851
170.	0.082	0.123	0.164	0.206	0.247	0.288	0.329	0.370	0.411	0.452	0.493	0.535	0.576	0.617	0.658	0.699	0.740	0.781	0.822	0.864	0.905
180.	0.087	0.131	0.174	0.218	0.261	0.305	0.348	0.392	0.435	0.479	0.522	0.566	0.610	0.653	0.697	0.740	0.784	0.827	0.871	0.914	0.958
190.	0.092	0.138	0.184	0.230	0.276	0.322	0.368	0.414	0.460	0.506	0.552	0.598	0.643	0.689	0.735	0.781	0.827	0.873	0.919	0.965	1.011
200.	0.097	0.145	0.194	0.242	0.290	0.339	0.387	0.435	0.484	0.532	0.581	0.629	0.677	0.726	0.774	0.823	0.871	0.919	0.968	1.016	1.065
210.	0.102	0.152	0.203	0.254	0.305	0.356	0.406	0.457	0.508	0.559	0.610	0.661	0.711	0.762	0.813	0.864	0.915	0.965	1.016	1.067	1.118
220.	0.106	0.160	0.213	0.266	0.319	0.373	0.426	0.479	0.532	0.586	0.639	0.692	0.745	0.799	0.852	0.905	0.958	1.011	1.065	1.118	1.171
230.	0.111	0.167	0.223	0.278	0.334	0.390	0.445	0.501	0.557	0.612	0.668	0.724	0.779	0.835	0.891	0.946	1.002	1.058	1.113	1.169	1.224
240.	0.116	0.174	0.232	0.290	0.348	0.407	0.465	0.523	0.581	0.639	0.697	0.755	0.813	0.871	0.929	0.987	1.045	1.104	1.162	1.220	1.278
250.	0.121	0.182	0.242	0.303	0.363	0.424	0.484	0.545	0.605	0.666	0.726	0.787	0.847	0.908	0.968	1.029	1.089	1.150	1.210	1.271	1.331
260.	0.126	0.189	0.252	0.315	0.378	0.441	0.503	0.566	0.629	0.692	0.755	0.818	0.881	0.944	1.007	1.070	1.133	1.196	1.259	1.322	1.385
270.	0.131	0.196	0.261	0.327	0.392	0.457	0.523	0.588	0.654	0.719	0.784	0.850	0.915	0.980	1.046	1.111	1.176	1.242	1.307	1.373	1.438
280.	0.136	0.203	0.271	0.339	0.407	0.474	0.542	0.610	0.678	0.746	0.813	0.881	0.949	1.017	1.085	1.152	1.220	1.288	1.356	1.423	1.491
290.	0.140	0.211	0.281	0.351	0.421	0.491	0.562	0.632	0.702	0.772	0.843	0.913	0.983	1.053	1.123	1.194	1.264	1.334	1.404	1.474	1.545
300.	0.145	0.218	0.291	0.363	0.436	0.508	0.581	0.653	0.726	0.799	0.872	0.944	1.017	1.090	1.162	1.235	1.308	1.380	1.453	1.525	1.598
310.	0.150	0.225	0.300	0.375	0.450	0.525	0.601	0.676	0.751	0.826	0.901	0.976	1.051	1.126	1.201	1.276	1.351	1.426	1.501	1.576	1.651
320.	0.155	0.232	0.310	0.387	0.465	0.542	0.620	0.697	0.775	0.852	0.930	1.007	1.085	1.162	1.240	1.317	1.395	1.472	1.550	1.627	1.705
330.	0.160	0.240	0.320	0.400	0.480	0.559	0.639	0.719	0.799	0.879	0.959	1.039	1.119	1.199	1.279	1.359	1.439	1.519	1.598	1.678	1.758
340.	0.165	0.247	0.329	0.412	0.494	0.576	0.659	0.741	0.824	0.906	0.988	1.071	1.153	1.235	1.318	1.400	1.482	1.565	1.647	1.729	1.812
350.	0.170	0.254	0.339	0.424	0.509	0.593	0.678	0.763	0.848	0.933	1.017	1.102	1.187	1.272	1.356	1.441	1.526	1.611	1.696	1.780	1.865
360.	0.174	0.262	0.349	0.436	0.523	0.610	0.698	0.785	0.872	0.959	1.047	1.134	1.221	1.308	1.395	1.483	1.570	1.657	1.744	1.831	1.919
370.	0.179	0.269	0.359	0.448	0.538	0.627	0.717	0.807	0.896	0.986	1.076	1.165	1.255	1.345	1.434	1.524	1.614	1.703	1.793	1.882	1.972

TABLE 2.6i Ellipsoidal head thickness tables in inches: joint efficiency = 0.85

ELLIPSOIDAL HEAD THICKNESS TABLES IN INCHES   MAJOR TO MINOR AXIS RATIO=2:1

JOINT EFF.=0.85   ALLOWABLE STRESS VALUE =13800. PSI

PSI	INSIDE DIAMETER, IN.																				
	12.	18.	24.	30.	36.	42.	48.	54.	60.	66.	72.	78.	84.	90.	96.	102.	108.	114.	120.	126.	132.
15.	0.008	0.012	0.015	0.019	0.023	0.027	0.031	0.035	0.038	0.042	0.046	0.050	0.054	0.058	0.061	0.065	0.069	0.073	0.077	0.081	0.084
20.	0.010	0.015	0.020	0.026	0.031	0.036	0.041	0.046	0.051	0.056	0.061	0.067	0.072	0.077	0.082	0.087	0.092	0.097	0.102	0.107	0.113
25.	0.013	0.019	0.026	0.032	0.038	0.045	0.051	0.058	0.064	0.070	0.077	0.083	0.090	0.096	0.102	0.109	0.115	0.122	0.128	0.134	0.141
30.	0.015	0.023	0.031	0.038	0.046	0.054	0.061	0.069	0.077	0.084	0.092	0.100	0.107	0.115	0.123	0.130	0.138	0.146	0.153	0.161	0.169
35.	0.018	0.027	0.036	0.045	0.054	0.063	0.072	0.081	0.090	0.098	0.107	0.116	0.125	0.134	0.143	0.152	0.161	0.170	0.179	0.188	0.197
40.	0.020	0.031	0.041	0.051	0.061	0.072	0.082	0.092	0.102	0.113	0.123	0.133	0.143	0.154	0.164	0.174	0.184	0.194	0.205	0.215	0.225
45.	0.023	0.035	0.046	0.058	0.069	0.081	0.092	0.104	0.115	0.127	0.138	0.150	0.161	0.173	0.184	0.196	0.207	0.219	0.230	0.242	0.253
50.	0.026	0.038	0.051	0.064	0.077	0.090	0.102	0.115	0.128	0.141	0.154	0.166	0.179	0.192	0.205	0.217	0.230	0.243	0.256	0.269	0.281
55.	0.028	0.042	0.056	0.070	0.084	0.099	0.113	0.127	0.141	0.155	0.169	0.183	0.197	0.211	0.225	0.239	0.253	0.267	0.281	0.296	0.310
60.	0.031	0.046	0.061	0.077	0.092	0.107	0.123	0.138	0.154	0.169	0.184	0.200	0.215	0.230	0.246	0.261	0.276	0.292	0.307	0.322	0.338
65.	0.033	0.050	0.067	0.083	0.100	0.116	0.133	0.150	0.166	0.183	0.200	0.216	0.233	0.249	0.266	0.283	0.299	0.316	0.333	0.349	0.366
70.	0.036	0.054	0.072	0.090	0.107	0.125	0.143	0.161	0.179	0.197	0.215	0.233	0.251	0.269	0.287	0.305	0.322	0.340	0.358	0.376	0.394
75.	0.038	0.058	0.077	0.096	0.116	0.134	0.154	0.173	0.192	0.211	0.230	0.250	0.269	0.288	0.307	0.326	0.345	0.365	0.384	0.403	0.422
80.	0.041	0.061	0.082	0.102	0.123	0.143	0.164	0.184	0.205	0.225	0.246	0.266	0.287	0.307	0.328	0.348	0.369	0.389	0.409	0.430	0.450
85.	0.044	0.065	0.087	0.109	0.131	0.152	0.174	0.196	0.218	0.239	0.261	0.283	0.305	0.326	0.348	0.370	0.392	0.413	0.435	0.457	0.479
90.	0.046	0.069	0.092	0.115	0.138	0.161	0.184	0.207	0.230	0.253	0.276	0.299	0.322	0.346	0.369	0.392	0.415	0.438	0.461	0.484	0.507
95.	0.049	0.073	0.097	0.122	0.146	0.170	0.195	0.219	0.243	0.267	0.292	0.316	0.340	0.365	0.389	0.413	0.438	0.462	0.486	0.511	0.535
100.	0.051	0.077	0.102	0.128	0.154	0.179	0.205	0.230	0.256	0.282	0.307	0.333	0.358	0.384	0.410	0.435	0.461	0.486	0.512	0.538	0.563
105.	0.054	0.081	0.108	0.134	0.161	0.188	0.215	0.242	0.269	0.296	0.323	0.349	0.376	0.403	0.430	0.457	0.484	0.511	0.538	0.564	0.591
110.	0.056	0.084	0.113	0.141	0.169	0.197	0.225	0.253	0.282	0.310	0.338	0.366	0.394	0.422	0.451	0.479	0.507	0.535	0.563	0.591	0.620
115.	0.059	0.088	0.118	0.147	0.176	0.206	0.236	0.265	0.294	0.324	0.353	0.383	0.412	0.442	0.471	0.500	0.530	0.559	0.589	0.618	0.648
120.	0.061	0.092	0.123	0.154	0.184	0.215	0.246	0.276	0.307	0.338	0.369	0.399	0.430	0.461	0.492	0.522	0.553	0.584	0.614	0.645	0.676
125.	0.064	0.096	0.128	0.160	0.192	0.224	0.256	0.288	0.320	0.352	0.384	0.416	0.448	0.480	0.512	0.544	0.576	0.608	0.640	0.672	0.704
130.	0.067	0.100	0.133	0.166	0.200	0.233	0.266	0.300	0.333	0.366	0.399	0.433	0.466	0.499	0.533	0.566	0.599	0.632	0.666	0.699	0.732
135.	0.069	0.104	0.138	0.173	0.207	0.242	0.277	0.311	0.346	0.380	0.415	0.449	0.484	0.518	0.553	0.588	0.622	0.657	0.691	0.726	0.760
140.	0.072	0.108	0.143	0.179	0.215	0.251	0.287	0.323	0.358	0.394	0.430	0.466	0.502	0.538	0.574	0.609	0.645	0.681	0.717	0.753	0.789
145.	0.074	0.111	0.149	0.186	0.223	0.260	0.297	0.334	0.371	0.408	0.446	0.483	0.520	0.557	0.594	0.631	0.668	0.705	0.743	0.780	0.817
150.	0.077	0.115	0.154	0.192	0.230	0.269	0.307	0.346	0.384	0.423	0.461	0.499	0.538	0.576	0.615	0.653	0.691	0.730	0.768	0.807	0.845
160.	0.082	0.123	0.164	0.205	0.246	0.287	0.328	0.369	0.410	0.451	0.492	0.533	0.574	0.615	0.656	0.697	0.738	0.779	0.820	0.861	0.901
170.	0.087	0.131	0.174	0.218	0.261	0.305	0.348	0.392	0.435	0.479	0.522	0.566	0.610	0.653	0.697	0.740	0.784	0.827	0.871	0.914	0.958
180.	0.092	0.138	0.184	0.231	0.277	0.323	0.369	0.415	0.461	0.507	0.553	0.599	0.645	0.692	0.738	0.784	0.830	0.876	0.922	0.968	1.014
190.	0.097	0.146	0.195	0.243	0.292	0.341	0.389	0.438	0.487	0.535	0.584	0.633	0.681	0.730	0.779	0.827	0.876	0.925	0.973	1.022	1.071
200.	0.102	0.154	0.205	0.256	0.307	0.359	0.410	0.461	0.512	0.564	0.615	0.666	0.717	0.769	0.820	0.871	0.922	0.974	1.025	1.076	1.127
210.	0.108	0.161	0.215	0.269	0.323	0.377	0.430	0.484	0.538	0.592	0.646	0.699	0.753	0.807	0.861	0.915	0.968	1.022	1.076	1.130	1.184
220.	0.113	0.169	0.225	0.282	0.338	0.395	0.451	0.507	0.564	0.620	0.676	0.733	0.789	0.846	0.902	0.958	1.015	1.071	1.127	1.184	1.240
230.	0.118	0.177	0.236	0.295	0.354	0.413	0.472	0.530	0.589	0.648	0.707	0.766	0.825	0.884	0.943	1.002	1.061	1.120	1.179	1.238	1.297
240.	0.123	0.185	0.246	0.308	0.369	0.431	0.492	0.554	0.615	0.677	0.738	0.800	0.861	0.923	0.984	1.046	1.107	1.169	1.230	1.292	1.353
250.	0.128	0.192	0.256	0.320	0.384	0.449	0.513	0.577	0.641	0.705	0.769	0.833	0.897	0.961	1.025	1.089	1.153	1.217	1.282	1.346	1.410
260.	0.133	0.200	0.267	0.333	0.400	0.467	0.533	0.600	0.666	0.733	0.800	0.866	0.933	1.000	1.066	1.133	1.200	1.266	1.333	1.400	1.466
270.	0.138	0.208	0.277	0.346	0.415	0.484	0.554	0.623	0.692	0.761	0.831	0.900	0.969	1.038	1.107	1.177	1.246	1.315	1.384	1.453	1.523
280.	0.144	0.215	0.287	0.359	0.431	0.502	0.574	0.646	0.718	0.790	0.861	0.933	1.005	1.077	1.149	1.220	1.292	1.364	1.436	1.507	1.579
290.	0.149	0.223	0.297	0.372	0.446	0.520	0.595	0.669	0.744	0.818	0.892	0.967	1.041	1.115	1.190	1.264	1.338	1.413	1.487	1.561	1.636
300.	0.154	0.231	0.308	0.385	0.462	0.539	0.615	0.692	0.769	0.846	0.923	1.000	1.077	1.154	1.231	1.308	1.385	1.462	1.538	1.615	1.692
310.	0.159	0.238	0.318	0.397	0.477	0.556	0.636	0.715	0.795	0.874	0.954	1.033	1.113	1.192	1.272	1.351	1.431	1.510	1.590	1.669	1.749
320.	0.164	0.246	0.328	0.410	0.492	0.574	0.657	0.739	0.821	0.903	0.985	1.067	1.149	1.231	1.313	1.395	1.477	1.559	1.641	1.723	1.805
330.	0.169	0.254	0.339	0.423	0.508	0.592	0.677	0.762	0.846	0.931	1.016	1.100	1.185	1.270	1.354	1.439	1.523	1.608	1.693	1.777	1.862
340.	0.174	0.262	0.349	0.436	0.523	0.610	0.698	0.785	0.872	0.959	1.047	1.134	1.221	1.308	1.395	1.483	1.570	1.657	1.744	1.831	1.919
350.	0.180	0.269	0.359	0.449	0.539	0.628	0.718	0.808	0.898	0.988	1.077	1.167	1.257	1.347	1.437	1.526	1.616	1.706	1.796	1.885	1.975
360.	0.185	0.277	0.369	0.462	0.554	0.646	0.739	0.831	0.924	1.016	1.108	1.201	1.293	1.385	1.478	1.570	1.662	1.755	1.847	1.939	2.032
370.	0.190	0.285	0.380	0.475	0.570	0.665	0.759	0.854	0.949	1.044	1.139	1.234	1.329	1.424	1.519	1.614	1.709	1.804	1.899	1.994	2.088

TABLE 2.6j Ellipsoidal head thickness tables in inches: joint efficiency = 0.80

ELLIPSOIDAL HEAD THICKNESS TABLES IN INCHES   MAJOR TO MINOR AXIS RATIO=2:1

JOINT EFF.=0.80   ALLOWABLE STRESS VALUE =13800. PSI

INSIDE DIAMETER, IN.

PSI	12.	18.	24.	30.	36.	42.	48.	54.	60.	66.	72.	78.	84.	90.	96.	102.	108.	114.	120.	126.	132.
15.	0.008	0.012	0.016	0.020	0.024	0.029	0.033	0.037	0.041	0.045	0.049	0.053	0.057	0.061	0.065	0.069	0.073	0.077	0.082	0.086	0.090
20.	0.011	0.016	0.022	0.027	0.033	0.038	0.043	0.049	0.054	0.060	0.065	0.071	0.076	0.082	0.087	0.092	0.098	0.103	0.109	0.114	0.120
25.	0.014	0.020	0.027	0.034	0.041	0.048	0.054	0.061	0.068	0.075	0.082	0.088	0.095	0.102	0.109	0.116	0.122	0.129	0.136	0.143	0.149
30.	0.016	0.024	0.033	0.041	0.049	0.057	0.065	0.073	0.082	0.090	0.098	0.106	0.114	0.122	0.130	0.138	0.147	0.155	0.163	0.171	0.179
35.	0.019	0.029	0.038	0.048	0.057	0.067	0.076	0.086	0.095	0.105	0.114	0.124	0.133	0.143	0.152	0.162	0.171	0.181	0.190	0.200	0.209
40.	0.022	0.033	0.043	0.054	0.065	0.076	0.087	0.098	0.109	0.120	0.130	0.141	0.152	0.163	0.174	0.185	0.196	0.207	0.217	0.228	0.239
45.	0.024	0.037	0.049	0.061	0.073	0.086	0.098	0.110	0.122	0.135	0.147	0.159	0.171	0.183	0.196	0.208	0.220	0.232	0.245	0.257	0.269
50.	0.027	0.041	0.054	0.068	0.082	0.095	0.109	0.122	0.136	0.150	0.163	0.177	0.190	0.204	0.217	0.231	0.245	0.258	0.272	0.285	0.299
55.	0.030	0.045	0.060	0.075	0.090	0.105	0.120	0.135	0.150	0.164	0.179	0.194	0.209	0.224	0.239	0.254	0.269	0.284	0.299	0.314	0.329
60.	0.033	0.049	0.065	0.082	0.098	0.114	0.131	0.147	0.163	0.179	0.196	0.212	0.228	0.245	0.261	0.277	0.294	0.310	0.326	0.343	0.359
65.	0.035	0.053	0.071	0.088	0.106	0.124	0.141	0.159	0.177	0.194	0.212	0.230	0.247	0.265	0.283	0.300	0.318	0.336	0.353	0.371	0.389
70.	0.038	0.057	0.076	0.095	0.114	0.133	0.152	0.171	0.190	0.209	0.228	0.247	0.266	0.286	0.305	0.324	0.343	0.362	0.381	0.400	0.419
75.	0.041	0.061	0.082	0.102	0.122	0.143	0.163	0.184	0.204	0.224	0.245	0.265	0.286	0.306	0.326	0.347	0.367	0.387	0.408	0.428	0.449
80.	0.044	0.065	0.087	0.109	0.131	0.152	0.174	0.196	0.218	0.239	0.261	0.283	0.305	0.326	0.348	0.370	0.392	0.413	0.435	0.457	0.479
85.	0.046	0.069	0.092	0.116	0.139	0.162	0.185	0.208	0.231	0.254	0.277	0.301	0.324	0.347	0.370	0.393	0.416	0.439	0.462	0.485	0.509
90.	0.049	0.073	0.098	0.122	0.147	0.171	0.196	0.220	0.245	0.269	0.294	0.318	0.343	0.367	0.392	0.416	0.441	0.465	0.490	0.514	0.538
95.	0.052	0.078	0.103	0.129	0.155	0.181	0.207	0.233	0.258	0.284	0.310	0.336	0.362	0.388	0.413	0.439	0.465	0.491	0.517	0.543	0.568
100.	0.054	0.082	0.109	0.136	0.163	0.190	0.218	0.245	0.272	0.299	0.326	0.354	0.381	0.408	0.435	0.462	0.490	0.517	0.544	0.571	0.598
105.	0.057	0.086	0.114	0.143	0.171	0.200	0.228	0.257	0.286	0.314	0.343	0.371	0.400	0.428	0.457	0.486	0.514	0.543	0.571	0.600	0.628
110.	0.060	0.090	0.120	0.150	0.180	0.209	0.239	0.269	0.299	0.329	0.359	0.389	0.419	0.449	0.479	0.509	0.539	0.569	0.598	0.628	0.658
115.	0.063	0.094	0.125	0.156	0.188	0.219	0.250	0.282	0.313	0.344	0.375	0.407	0.438	0.469	0.501	0.532	0.563	0.594	0.626	0.657	0.688
120.	0.065	0.098	0.131	0.163	0.196	0.229	0.261	0.294	0.326	0.359	0.392	0.424	0.457	0.490	0.522	0.555	0.588	0.620	0.653	0.686	0.718
125.	0.068	0.102	0.136	0.171	0.204	0.238	0.272	0.306	0.340	0.374	0.408	0.442	0.476	0.510	0.544	0.578	0.612	0.646	0.680	0.714	0.748
130.	0.071	0.106	0.141	0.177	0.212	0.248	0.283	0.318	0.354	0.389	0.424	0.460	0.495	0.531	0.566	0.601	0.637	0.672	0.707	0.743	0.778
135.	0.073	0.110	0.147	0.184	0.220	0.257	0.294	0.331	0.367	0.404	0.441	0.477	0.514	0.551	0.588	0.624	0.661	0.698	0.735	0.771	0.808
140.	0.076	0.114	0.152	0.190	0.229	0.267	0.305	0.343	0.381	0.419	0.457	0.495	0.533	0.571	0.609	0.648	0.686	0.724	0.762	0.800	0.838
145.	0.079	0.118	0.158	0.197	0.237	0.276	0.316	0.355	0.395	0.434	0.473	0.513	0.552	0.592	0.631	0.671	0.710	0.750	0.789	0.829	0.868
150.	0.082	0.122	0.163	0.204	0.245	0.286	0.327	0.367	0.408	0.449	0.490	0.531	0.571	0.612	0.653	0.694	0.735	0.776	0.816	0.857	0.898
160.	0.087	0.131	0.174	0.218	0.261	0.305	0.348	0.392	0.435	0.479	0.522	0.566	0.610	0.653	0.697	0.740	0.784	0.827	0.871	0.914	0.958
170.	0.093	0.139	0.185	0.231	0.278	0.324	0.370	0.416	0.463	0.509	0.555	0.601	0.648	0.694	0.740	0.787	0.833	0.879	0.925	0.972	1.018
180.	0.098	0.147	0.196	0.245	0.294	0.343	0.392	0.441	0.490	0.539	0.588	0.637	0.686	0.735	0.784	0.833	0.882	0.931	0.980	1.029	1.078
190.	0.103	0.155	0.207	0.259	0.310	0.362	0.414	0.465	0.517	0.569	0.621	0.672	0.724	0.776	0.828	0.879	0.931	0.983	1.034	1.086	1.138
200.	0.109	0.163	0.218	0.272	0.327	0.381	0.436	0.490	0.544	0.599	0.653	0.708	0.762	0.817	0.871	0.926	0.980	1.034	1.089	1.143	1.198
210.	0.114	0.172	0.229	0.286	0.343	0.400	0.457	0.515	0.572	0.629	0.686	0.743	0.800	0.858	0.915	0.972	1.029	1.086	1.143	1.201	1.258
220.	0.120	0.180	0.240	0.300	0.359	0.419	0.479	0.539	0.599	0.659	0.719	0.779	0.839	0.899	0.958	1.018	1.078	1.138	1.198	1.258	1.318
230.	0.125	0.188	0.251	0.313	0.376	0.438	0.501	0.564	0.626	0.689	0.752	0.814	0.877	0.939	1.002	1.065	1.127	1.190	1.253	1.315	1.378
240.	0.131	0.196	0.261	0.327	0.392	0.458	0.523	0.588	0.654	0.719	0.784	0.850	0.915	0.980	1.046	1.111	1.176	1.242	1.307	1.373	1.438
250.	0.136	0.204	0.272	0.340	0.409	0.477	0.545	0.613	0.681	0.749	0.817	0.885	0.953	1.021	1.089	1.158	1.226	1.294	1.362	1.430	1.498
260.	0.142	0.212	0.283	0.354	0.425	0.496	0.567	0.637	0.708	0.779	0.850	0.921	0.991	1.062	1.133	1.204	1.275	1.346	1.416	1.487	1.558
270.	0.147	0.221	0.294	0.368	0.441	0.515	0.588	0.662	0.735	0.809	0.883	0.956	1.030	1.103	1.177	1.250	1.324	1.397	1.471	1.545	1.618
280.	0.153	0.229	0.305	0.381	0.458	0.534	0.610	0.687	0.763	0.839	0.915	0.992	1.068	1.144	1.220	1.297	1.373	1.449	1.526	1.602	1.678
290.	0.158	0.237	0.316	0.394	0.474	0.553	0.632	0.711	0.790	0.869	0.948	1.027	1.106	1.185	1.264	1.343	1.422	1.501	1.580	1.659	1.738
300.	0.163	0.245	0.326	0.409	0.490	0.572	0.654	0.736	0.817	0.899	0.981	1.063	1.144	1.226	1.308	1.390	1.471	1.553	1.635	1.717	1.798
310.	0.169	0.253	0.338	0.422	0.507	0.591	0.676	0.760	0.845	0.929	1.014	1.098	1.183	1.267	1.352	1.436	1.521	1.605	1.690	1.774	1.858
320.	0.174	0.262	0.349	0.436	0.523	0.610	0.698	0.785	0.872	0.959	1.047	1.134	1.221	1.308	1.395	1.483	1.570	1.657	1.744	1.831	1.919
330.	0.180	0.270	0.360	0.450	0.540	0.630	0.720	0.809	0.899	0.989	1.079	1.169	1.259	1.349	1.439	1.529	1.619	1.709	1.799	1.889	1.979
340.	0.185	0.278	0.371	0.463	0.556	0.649	0.741	0.834	0.927	1.019	1.112	1.205	1.297	1.390	1.483	1.576	1.668	1.761	1.854	1.946	2.039
350.	0.191	0.286	0.382	0.477	0.572	0.668	0.763	0.859	0.954	1.050	1.145	1.240	1.336	1.431	1.527	1.622	1.717	1.813	1.908	2.004	2.099
360.	0.196	0.294	0.393	0.491	0.589	0.687	0.785	0.883	0.981	1.080	1.178	1.276	1.374	1.472	1.570	1.668	1.767	1.865	1.963	2.061	2.159
370.	0.202	0.303	0.404	0.504	0.605	0.706	0.807	0.908	1.009	1.110	1.211	1.311	1.412	1.513	1.614	1.715	1.816	1.917	2.018	2.119	2.219

TABLE 2.6k Ellipsoidal head thickness tables in inches: joint efficiency = 0.70

ELLIPSOIDAL HEAD THICKNESS TABLES IN INCHES   MAJOR TO MINOR AXIS RATIO=2:1

JOINT EFF.=0.70   ALLOWABLE STRESS VALUE =13800. PSI

PSI	12.	18.	24.	30.	36.	42.	48.	54.	60.	66.	72.	78.	84.	90.	96.	102.	108.	114.	120.	126.	132.
									INSIDE DIAMETER,IN.												
15.	0.009	0.014	0.019	0.023	0.028	0.033	0.037	0.042	0.047	0.051	0.056	0.061	0.065	0.070	0.075	0.079	0.084	0.089	0.093	0.098	0.103
20.	0.012	0.019	0.025	0.031	0.037	0.043	0.050	0.056	0.062	0.068	0.075	0.081	0.087	0.093	0.099	0.106	0.112	0.118	0.124	0.130	0.137
25.	0.016	0.023	0.031	0.039	0.047	0.054	0.062	0.070	0.078	0.085	0.093	0.101	0.109	0.116	0.124	0.132	0.140	0.148	0.155	0.163	0.171
30.	0.019	0.028	0.037	0.047	0.056	0.065	0.075	0.084	0.093	0.103	0.112	0.121	0.130	0.140	0.149	0.158	0.168	0.177	0.186	0.196	0.205
35.	0.022	0.033	0.043	0.054	0.065	0.076	0.087	0.098	0.109	0.120	0.130	0.141	0.152	0.163	0.174	0.185	0.196	0.207	0.217	0.228	0.239
40.	0.025	0.037	0.050	0.062	0.075	0.087	0.099	0.112	0.124	0.137	0.149	0.162	0.174	0.186	0.199	0.211	0.224	0.236	0.249	0.261	0.273
45.	0.028	0.042	0.056	0.070	0.084	0.098	0.112	0.126	0.140	0.154	0.168	0.182	0.196	0.210	0.224	0.238	0.252	0.266	0.280	0.294	0.308
50.	0.031	0.047	0.062	0.078	0.093	0.109	0.124	0.140	0.155	0.171	0.186	0.202	0.218	0.233	0.249	0.264	0.280	0.295	0.311	0.326	0.342
55.	0.034	0.051	0.068	0.085	0.103	0.120	0.137	0.154	0.171	0.188	0.205	0.222	0.239	0.256	0.273	0.291	0.308	0.325	0.342	0.359	0.376
60.	0.037	0.056	0.075	0.093	0.112	0.131	0.149	0.168	0.186	0.205	0.224	0.242	0.261	0.280	0.298	0.317	0.336	0.354	0.373	0.392	0.410
65.	0.040	0.061	0.081	0.101	0.121	0.141	0.162	0.182	0.202	0.222	0.242	0.263	0.283	0.303	0.323	0.343	0.364	0.384	0.404	0.424	0.444
70.	0.044	0.065	0.087	0.109	0.131	0.152	0.174	0.196	0.218	0.239	0.261	0.283	0.305	0.326	0.348	0.370	0.392	0.413	0.435	0.457	0.479
75.	0.047	0.070	0.093	0.117	0.140	0.163	0.186	0.210	0.233	0.256	0.280	0.303	0.326	0.350	0.373	0.396	0.420	0.443	0.466	0.490	0.513
80.	0.050	0.075	0.099	0.124	0.149	0.174	0.199	0.224	0.249	0.274	0.298	0.323	0.348	0.373	0.398	0.423	0.448	0.472	0.497	0.522	0.547
85.	0.053	0.079	0.106	0.132	0.159	0.185	0.211	0.238	0.264	0.291	0.317	0.343	0.370	0.396	0.423	0.449	0.476	0.502	0.528	0.555	0.581
90.	0.056	0.084	0.112	0.140	0.168	0.196	0.224	0.252	0.280	0.308	0.336	0.364	0.392	0.420	0.448	0.476	0.504	0.532	0.560	0.588	0.615
95.	0.059	0.089	0.118	0.148	0.177	0.207	0.236	0.266	0.295	0.325	0.354	0.384	0.413	0.443	0.473	0.502	0.532	0.561	0.591	0.620	0.650
100.	0.062	0.093	0.124	0.155	0.187	0.218	0.249	0.280	0.311	0.342	0.373	0.404	0.435	0.466	0.497	0.528	0.560	0.591	0.622	0.653	0.684
105.	0.065	0.098	0.131	0.163	0.196	0.229	0.261	0.294	0.326	0.359	0.392	0.424	0.457	0.490	0.522	0.555	0.588	0.620	0.653	0.686	0.718
110.	0.068	0.103	0.137	0.171	0.205	0.239	0.274	0.308	0.342	0.376	0.410	0.445	0.479	0.513	0.547	0.581	0.616	0.650	0.684	0.718	0.752
115.	0.072	0.107	0.143	0.179	0.215	0.250	0.286	0.322	0.358	0.393	0.429	0.465	0.501	0.536	0.572	0.608	0.644	0.679	0.715	0.751	0.787
120.	0.075	0.112	0.149	0.187	0.224	0.261	0.299	0.336	0.373	0.410	0.448	0.485	0.522	0.560	0.597	0.634	0.672	0.709	0.746	0.784	0.821
125.	0.078	0.117	0.155	0.194	0.233	0.272	0.311	0.350	0.389	0.428	0.466	0.505	0.544	0.583	0.622	0.661	0.700	0.739	0.777	0.816	0.855
130.	0.081	0.121	0.162	0.202	0.243	0.283	0.323	0.364	0.404	0.445	0.485	0.526	0.566	0.606	0.647	0.687	0.728	0.768	0.809	0.849	0.889
135.	0.084	0.126	0.168	0.210	0.252	0.294	0.336	0.378	0.420	0.462	0.505	0.547	0.588	0.630	0.672	0.714	0.756	0.798	0.840	0.882	0.924
140.	0.087	0.131	0.174	0.218	0.261	0.305	0.348	0.392	0.435	0.479	0.522	0.566	0.610	0.653	0.697	0.740	0.784	0.827	0.871	0.914	0.958
145.	0.090	0.135	0.180	0.225	0.271	0.316	0.361	0.406	0.451	0.496	0.541	0.586	0.631	0.676	0.722	0.767	0.812	0.857	0.902	0.947	0.992
150.	0.093	0.140	0.187	0.233	0.280	0.327	0.373	0.420	0.467	0.513	0.560	0.607	0.653	0.700	0.747	0.793	0.840	0.886	0.933	0.980	1.026
160.	0.100	0.149	0.199	0.249	0.299	0.348	0.398	0.448	0.498	0.547	0.597	0.647	0.697	0.747	0.796	0.846	0.896	0.946	0.995	1.045	1.095
170.	0.106	0.159	0.212	0.264	0.317	0.370	0.423	0.476	0.529	0.582	0.635	0.688	0.740	0.793	0.846	0.899	0.952	1.005	1.058	1.111	1.164
180.	0.112	0.168	0.224	0.280	0.336	0.392	0.448	0.504	0.560	0.616	0.672	0.728	0.784	0.840	0.896	0.952	1.008	1.064	1.120	1.176	1.232
190.	0.118	0.177	0.236	0.296	0.355	0.414	0.473	0.532	0.591	0.650	0.709	0.769	0.828	0.887	0.946	1.005	1.064	1.123	1.182	1.242	1.301
200.	0.124	0.187	0.249	0.311	0.373	0.436	0.498	0.560	0.622	0.685	0.747	0.809	0.871	0.934	0.996	1.058	1.120	1.183	1.245	1.307	1.369
210.	0.131	0.196	0.261	0.327	0.392	0.458	0.523	0.588	0.654	0.719	0.784	0.850	0.915	0.980	1.046	1.111	1.176	1.242	1.307	1.373	1.438
220.	0.137	0.205	0.274	0.342	0.411	0.479	0.548	0.616	0.685	0.753	0.822	0.890	0.959	1.027	1.096	1.164	1.233	1.301	1.370	1.438	1.507
230.	0.143	0.215	0.286	0.358	0.430	0.501	0.573	0.644	0.716	0.788	0.859	0.931	1.002	1.074	1.146	1.217	1.289	1.360	1.432	1.504	1.575
240.	0.149	0.224	0.299	0.374	0.448	0.523	0.598	0.672	0.747	0.822	0.897	0.971	1.046	1.121	1.196	1.270	1.345	1.420	1.494	1.569	1.644
250.	0.156	0.234	0.311	0.389	0.467	0.545	0.623	0.701	0.778	0.856	0.934	1.012	1.090	1.168	1.245	1.323	1.401	1.479	1.557	1.635	1.713
260.	0.162	0.243	0.324	0.405	0.486	0.567	0.648	0.729	0.810	0.891	0.972	1.053	1.133	1.214	1.295	1.376	1.457	1.538	1.619	1.700	1.781
270.	0.168	0.252	0.336	0.420	0.505	0.589	0.673	0.757	0.841	0.925	1.009	1.093	1.177	1.261	1.345	1.429	1.514	1.598	1.682	1.766	1.850
280.	0.174	0.262	0.349	0.436	0.523	0.610	0.698	0.785	0.872	0.959	1.047	1.134	1.221	1.308	1.395	1.483	1.570	1.657	1.744	1.831	1.919
290.	0.181	0.271	0.361	0.452	0.542	0.632	0.723	0.813	0.903	0.994	1.084	1.174	1.265	1.355	1.445	1.536	1.626	1.716	1.807	1.897	1.987
300.	0.187	0.280	0.374	0.467	0.561	0.654	0.748	0.841	0.935	1.028	1.122	1.215	1.308	1.402	1.495	1.589	1.682	1.776	1.869	1.963	2.056
310.	0.193	0.290	0.386	0.483	0.579	0.676	0.773	0.869	0.966	1.062	1.159	1.255	1.352	1.449	1.545	1.642	1.738	1.835	1.932	2.028	2.125
320.	0.199	0.299	0.399	0.499	0.598	0.698	0.798	0.897	0.997	1.097	1.197	1.296	1.396	1.496	1.595	1.695	1.795	1.894	1.994	2.094	2.194
330.	0.206	0.309	0.411	0.514	0.617	0.720	0.823	0.926	1.028	1.131	1.234	1.337	1.440	1.543	1.645	1.748	1.851	1.954	2.057	2.160	2.262
340.	0.212	0.318	0.424	0.530	0.636	0.742	0.848	0.954	1.060	1.166	1.272	1.378	1.483	1.589	1.695	1.801	1.907	2.013	2.119	2.225	2.331
350.	0.218	0.327	0.436	0.545	0.655	0.764	0.873	0.982	1.091	1.200	1.309	1.418	1.527	1.636	1.745	1.855	1.964	2.073	2.182	2.291	2.400
360.	0.224	0.337	0.449	0.561	0.673	0.786	0.898	1.010	1.122	1.234	1.347	1.459	1.571	1.683	1.796	1.908	2.020	2.132	2.244	2.357	2.469
370.	0.231	0.346	0.461	0.577	0.692	0.807	0.923	1.038	1.153	1.269	1.384	1.500	1.615	1.730	1.846	1.961	2.076	2.192	2.307	2.422	2.538

TABLE 2.6/ Ellipsoidal head thickness tables in inches: joint efficiency = 0.65

ELLIPSOIDAL HEAD THICKNESS TABLES IN INCHES   MAJOR TO MINOR AXIS RATIO=2:1

JOINT EFF.=0.65   ALLOWABLE STRESS VALUE =13800. PSI

INSIDE DIAMETER,IN.

PSI	12.	18.	24.	30.	36.	42.	48.	54.	60.	66.	72.	78.	84.	90.	96.	102.	108.	114.	120.	126.	132.
15.	0.010	0.015	0.020	0.025	0.030	0.035	0.040	0.045	0.050	0.055	0.060	0.065	0.070	0.075	0.080	0.085	0.090	0.095	0.100	0.105	0.110
20.	0.013	0.020	0.027	0.033	0.040	0.046	0.054	0.060	0.067	0.074	0.080	0.087	0.094	0.100	0.107	0.114	0.120	0.127	0.134	0.140	0.147
25.	0.017	0.025	0.033	0.042	0.050	0.059	0.067	0.075	0.084	0.092	0.100	0.109	0.117	0.125	0.134	0.142	0.151	0.159	0.167	0.176	0.184
30.	0.020	0.030	0.040	0.050	0.059	0.070	0.080	0.090	0.100	0.110	0.120	0.130	0.141	0.151	0.161	0.171	0.181	0.191	0.201	0.211	0.221
35.	0.023	0.035	0.047	0.059	0.070	0.082	0.094	0.105	0.117	0.129	0.141	0.152	0.164	0.176	0.187	0.199	0.211	0.222	0.234	0.246	0.258
40.	0.027	0.040	0.054	0.067	0.080	0.094	0.107	0.120	0.134	0.147	0.161	0.174	0.187	0.201	0.214	0.228	0.241	0.254	0.268	0.281	0.294
45.	0.030	0.045	0.060	0.075	0.090	0.105	0.120	0.136	0.151	0.166	0.181	0.196	0.211	0.226	0.241	0.256	0.271	0.286	0.301	0.316	0.331
50.	0.033	0.050	0.067	0.084	0.100	0.117	0.134	0.151	0.167	0.184	0.201	0.218	0.234	0.251	0.268	0.284	0.301	0.318	0.335	0.351	0.368
55.	0.037	0.055	0.074	0.092	0.110	0.129	0.147	0.166	0.184	0.202	0.221	0.239	0.258	0.276	0.294	0.313	0.331	0.350	0.368	0.387	0.405
60.	0.040	0.060	0.080	0.100	0.120	0.141	0.161	0.181	0.201	0.221	0.241	0.261	0.281	0.301	0.321	0.341	0.361	0.382	0.402	0.422	0.442
65.	0.044	0.065	0.087	0.109	0.131	0.152	0.174	0.196	0.218	0.239	0.261	0.283	0.305	0.326	0.348	0.370	0.392	0.413	0.435	0.457	0.479
70.	0.047	0.070	0.094	0.117	0.141	0.164	0.187	0.211	0.234	0.258	0.281	0.305	0.328	0.351	0.375	0.398	0.422	0.445	0.469	0.492	0.515
75.	0.050	0.075	0.100	0.126	0.151	0.176	0.201	0.226	0.251	0.276	0.301	0.326	0.351	0.377	0.402	0.427	0.452	0.477	0.502	0.527	0.552
80.	0.054	0.080	0.107	0.134	0.161	0.187	0.214	0.241	0.268	0.295	0.321	0.348	0.375	0.402	0.428	0.455	0.482	0.509	0.536	0.562	0.589
85.	0.057	0.085	0.114	0.142	0.171	0.199	0.228	0.256	0.285	0.313	0.341	0.370	0.398	0.427	0.455	0.484	0.512	0.541	0.569	0.598	0.626
90.	0.060	0.090	0.121	0.151	0.181	0.211	0.241	0.271	0.301	0.331	0.362	0.392	0.422	0.452	0.482	0.512	0.542	0.572	0.603	0.633	0.663
95.	0.064	0.095	0.127	0.159	0.191	0.223	0.254	0.286	0.318	0.350	0.382	0.413	0.445	0.477	0.509	0.541	0.573	0.604	0.636	0.668	0.700
100.	0.067	0.100	0.134	0.167	0.201	0.234	0.268	0.301	0.335	0.368	0.402	0.435	0.469	0.502	0.536	0.569	0.603	0.636	0.670	0.703	0.737
105.	0.070	0.105	0.141	0.176	0.211	0.246	0.281	0.316	0.352	0.387	0.422	0.457	0.492	0.527	0.563	0.598	0.633	0.668	0.703	0.738	0.773
110.	0.074	0.111	0.147	0.184	0.221	0.258	0.295	0.332	0.368	0.405	0.442	0.479	0.516	0.553	0.589	0.626	0.663	0.700	0.737	0.774	0.810
115.	0.077	0.116	0.154	0.193	0.231	0.270	0.308	0.347	0.385	0.424	0.462	0.501	0.539	0.578	0.616	0.655	0.693	0.732	0.770	0.809	0.847
120.	0.080	0.121	0.161	0.201	0.241	0.281	0.322	0.362	0.402	0.442	0.482	0.522	0.563	0.603	0.643	0.683	0.723	0.764	0.804	0.844	0.884
125.	0.084	0.126	0.167	0.209	0.251	0.293	0.335	0.377	0.419	0.461	0.502	0.544	0.586	0.628	0.670	0.712	0.754	0.795	0.837	0.879	0.921
130.	0.087	0.131	0.174	0.218	0.261	0.305	0.348	0.392	0.435	0.479	0.522	0.566	0.610	0.653	0.697	0.740	0.784	0.827	0.871	0.914	0.958
135.	0.090	0.136	0.181	0.226	0.271	0.316	0.362	0.407	0.452	0.497	0.543	0.588	0.633	0.678	0.723	0.769	0.814	0.859	0.904	0.950	0.995
140.	0.094	0.141	0.188	0.234	0.281	0.328	0.375	0.422	0.469	0.516	0.563	0.610	0.657	0.703	0.750	0.797	0.844	0.891	0.938	0.985	1.032
145.	0.097	0.146	0.194	0.243	0.291	0.340	0.389	0.437	0.486	0.534	0.583	0.631	0.680	0.729	0.777	0.826	0.874	0.923	0.971	1.020	1.069
150.	0.101	0.151	0.201	0.251	0.302	0.352	0.402	0.452	0.503	0.553	0.603	0.653	0.704	0.754	0.804	0.854	0.905	0.955	1.005	1.055	1.106
160.	0.107	0.161	0.214	0.268	0.322	0.375	0.429	0.482	0.536	0.590	0.643	0.697	0.751	0.804	0.858	0.911	0.965	1.019	1.072	1.126	1.179
170.	0.114	0.171	0.228	0.285	0.342	0.399	0.456	0.513	0.570	0.627	0.684	0.741	0.797	0.854	0.911	0.968	1.025	1.082	1.139	1.196	1.253
180.	0.121	0.181	0.241	0.302	0.362	0.422	0.483	0.543	0.603	0.664	0.724	0.784	0.845	0.905	0.965	1.025	1.086	1.146	1.206	1.267	1.327
190.	0.127	0.191	0.255	0.318	0.382	0.446	0.509	0.573	0.637	0.700	0.764	0.828	0.892	0.955	1.019	1.083	1.146	1.210	1.274	1.337	1.401
200.	0.134	0.201	0.268	0.335	0.402	0.469	0.536	0.603	0.670	0.737	0.804	0.872	0.939	1.006	1.073	1.140	1.207	1.274	1.341	1.408	1.475
210.	0.141	0.211	0.282	0.352	0.422	0.493	0.563	0.634	0.704	0.774	0.845	0.915	0.986	1.056	1.126	1.197	1.267	1.338	1.408	1.478	1.549
220.	0.148	0.221	0.295	0.369	0.443	0.516	0.590	0.664	0.738	0.811	0.885	0.959	1.033	1.106	1.180	1.254	1.328	1.401	1.475	1.549	1.623
230.	0.154	0.231	0.308	0.385	0.463	0.540	0.617	0.694	0.771	0.848	0.925	1.003	1.080	1.157	1.234	1.311	1.388	1.465	1.542	1.620	1.697
240.	0.161	0.241	0.322	0.402	0.483	0.563	0.644	0.724	0.805	0.885	0.966	1.046	1.127	1.207	1.288	1.368	1.449	1.529	1.610	1.690	1.771
250.	0.168	0.252	0.335	0.419	0.503	0.587	0.670	0.755	0.838	0.922	1.006	1.090	1.174	1.258	1.342	1.425	1.509	1.593	1.677	1.761	1.845
260.	0.174	0.262	0.349	0.436	0.523	0.610	0.698	0.785	0.872	0.959	1.047	1.134	1.221	1.308	1.395	1.483	1.570	1.657	1.744	1.831	1.919
270.	0.181	0.272	0.362	0.453	0.543	0.634	0.725	0.815	0.906	0.996	1.087	1.177	1.268	1.359	1.449	1.540	1.630	1.721	1.811	1.902	1.993
280.	0.188	0.282	0.376	0.470	0.564	0.658	0.752	0.846	0.939	1.033	1.127	1.221	1.315	1.409	1.503	1.597	1.691	1.785	1.879	1.973	2.067
290.	0.195	0.292	0.389	0.487	0.584	0.681	0.778	0.876	0.973	1.070	1.168	1.265	1.362	1.460	1.557	1.654	1.751	1.849	1.946	2.043	2.141
300.	0.201	0.302	0.403	0.503	0.604	0.705	0.805	0.906	1.007	1.107	1.208	1.309	1.409	1.510	1.611	1.711	1.812	1.913	2.013	2.114	2.215
310.	0.208	0.312	0.416	0.520	0.624	0.728	0.832	0.936	1.040	1.144	1.248	1.353	1.457	1.561	1.665	1.769	1.873	1.977	2.081	2.185	2.289
320.	0.215	0.322	0.430	0.537	0.644	0.752	0.859	0.967	1.074	1.181	1.289	1.396	1.504	1.611	1.719	1.826	1.933	2.041	2.148	2.256	2.363
330.	0.222	0.332	0.443	0.554	0.665	0.775	0.886	0.997	1.108	1.219	1.329	1.440	1.551	1.662	1.772	1.883	1.994	2.105	2.216	2.326	2.437
340.	0.228	0.342	0.457	0.571	0.685	0.799	0.913	1.027	1.141	1.256	1.370	1.484	1.598	1.712	1.826	1.940	2.055	2.169	2.283	2.397	2.511
350.	0.235	0.353	0.470	0.588	0.705	0.823	0.940	1.058	1.175	1.293	1.410	1.528	1.645	1.763	1.880	1.998	2.115	2.233	2.350	2.468	2.585
360.	0.242	0.363	0.484	0.604	0.725	0.846	0.967	1.088	1.209	1.330	1.451	1.572	1.692	1.813	1.934	2.055	2.176	2.297	2.418	2.539	2.660
370.	0.249	0.373	0.497	0.621	0.746	0.870	0.994	1.118	1.243	1.367	1.491	1.615	1.740	1.864	1.988	2.112	2.237	2.361	2.485	2.609	2.734

2. Locate the desired diameter for ellipsoidal heads or inside crown radius for torispherical heads. The diameter or inside crown radius values are printed across the top of the page for each table.

3. Locate the desired pressure. The pressure values are located in the left column.

4. Read the thickness corresponding to the diameter for ellipsoidal heads or inside crown radius for torispherical heads and pressure.

Each table is based on an allowable stress value of 13,800 psi. For other stress values, multiply the thickness read from the table by either the following constant factors or factors read from Fig. 2.2.

Stress	11,300	12,500	13,800	15,000	16,300	17,500	18,800
Constant	1.224	1.105	1.000	0.919	0.846	0.787	0.733

Corrosion allowances are not included in these tables. The thickness equations for ellipsoidal and torispherical (flanged and dished) heads may be found in Table 2.6. (See also Code Par. UW-12.)

ELEMENTS OF JOINT DESIGN FOR HEADS

A number of important factors must be considered in determining the thickness required for heads under internal pressure, such as stress values and joint efficiencies, according to the degree of radiographic examination. Many designers have difficulty in determining these factors.

In an effort to achieve uniformity, *The Transactions of the ASME Series J. Journal of Pressure Vessel Technology,* in their February 1975 issue, and subsequent follow-up article in February 1977, include examples by G. M. Eisenberg, Senior Pressure Vessel Engineering Administrator, clarifying design factors that may be used as a guide. The National Board of Boiler and Pressure Vessel Inspectors also published these examples in their April 1975 *National Board Bulletin.* In this sixth edition we show a bottom head and give examples of calculations for further clarification. See Fig. 2.5, Examples 1 to 4, Fig. 2.6, Examples 5 to 16, and Fig. 2.7, Examples 17 to 28.

It should, however, be remembered that the *ASME Boiler and Pressure Vessel Code* is a progressive, viable code and that when necessary, changes are made. Therefore, the latest edition of the Code should always be used.

GUIDE TO BUTT-JOINT FACTORS BASED ON RADIOGRAPHIC EXAMINATION IN ACCORDANCE WITH UW-11 AND UW-12 OF SECTION VIII, DIVISION 1, JOINT EFFICIENCIES AND ALLOWABLE STRESS VALUES

The requirements based on the degree of radiographic examination of butt welds in pressure vessels constructed to the requirements of Section VIII,

Division 1, of the *ASME Boiler and Pressure Vessel Code* have been the subject of numerous inquiries for interpretation from manufacturers, Authorized Inspectors, and users of these vessels. Most of the questions have been concerned with the examination requirements for circumferential butt welds attaching seamless heads to shell sections.

The Code has always been clear regarding requirements for butt welds attaching shell sections which have fully radiographed, longitudinal, butt welds, and where a joint efficiency of 1.0 is used in the calculation of shell thickness. A revision given in the Summer 1969 Addenda provided, for the first time, a requirement for radiographic examination of the butt weld attaching a seamless head to a seamless section in order to be consistent with the rules for butt-welded sections. However, the unwillingness to accept this new requirement generated further inquiries. Consequently, the Committee reopened the subject for possible further revision.

The degrees of radiographic examination shown in Code Par. UW-11 include full, partial, spot, and no radiography. The factors for allowable stress and joint efficiency to be used in calculations for section and head thicknesses were set forth in Par. UW-12. The use of partial radiography, in lieu of full radiography on Category B or C butt welds, provided a relaxation of the previous requirements. [Refer to Code Par. UW-11(a)(5).]

The examples specified are self-explanatory, giving the joint efficiency and the percentage of the tabulated allowable stress which shall be used in head calculations and shell calculations for both circumferential (hoop) and longitudinal (axial) stresses. Each example has different conditions of weld radiographic examination.

The examples presented show the various factors to be used in head and shell thickness calculations where butt-welded longitudinal (Category A) and butt-welded circumferential (Category B or C) joints are used. Further, they are restricted to the assumptions listed in the preface.

Figure 2.5, Examples 1 to 4, covers vessels with seamless heads and seamless shells; Fig. 2.6, Examples 5 to 16, covers vessels with welded heads and seamless shells; and Fig. 2.7, Examples 17 to 28, covers vessels with seamless heads and welded shells.

PREFACE TO EXAMPLES

The examples shown are based on the following assumptions:

1. The vessel is nonlethal [UW-11(a)(1)].
2. The thicknesses do not exceed those specified in UW-11(a)(2).
3. The vessel is not an unfired steam boiler with a design pressure exceeding 50 psi [UW-11(a)(3)].
4. These examples cover stresses in tension.
5. These examples are based on internal pressure only.
6. All heads are other than hemispherical.
7. All joints are Type 1 or 2 of Table UW-12.

EXAMPLE STATEMENT

A mathematical solution for the required head and shell thickness is shown below each example for the following condition:

Shell inside diameter (D): 24 in

Head 1:   2 : 1 ellipsoidal head; assume butt joints in head are welded with a Type 1 double-welded joint when such a joint is indicated in the figure

Head 2:   Torispherical head with inside crown radius equal to shell inside diameter; assume butt joints in head are welded with a Type 2 single-welded butt joint, when such a joint is indicated in the figure

Maximum working pressure: 250 psi

Material allowable stress value: 13,800 psi

Loadings per Code Par. UG-22: not applicable

Nomenclature: T_s = shell thickness (circumferential stress calculation)*

                 T_{eh} = ellipsoidal head thickness

                 T_{th} = torispherical head thickness

---

* See Table 2.4 for remarks concerning circumferential joint formula.

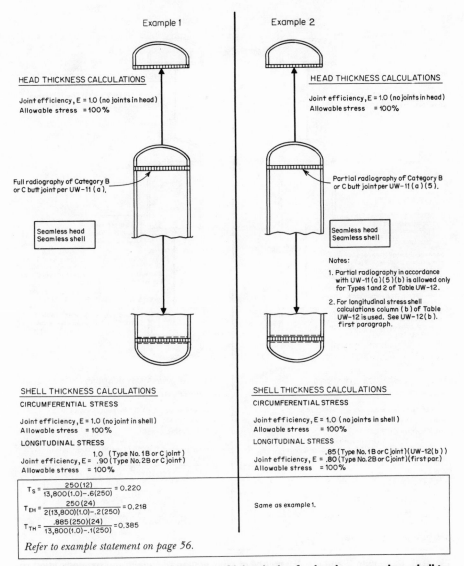

Example 1

Example 2

HEAD THICKNESS CALCULATIONS

Joint efficiency, E = 1.0 (no joints in head)
Allowable stress = 100 %

HEAD THICKNESS CALCULATIONS

Joint efficiency, E = 1.0 (no joints in head)
Allowable stress   = 100%

Full radiography of Category B
or C butt joint per UW-11 (a).

Partial radiography of Category B
or C butt joint per UW-11 (a) (5).

Seamless head
Seamless shell

Seamless head
Seamless shell

Notes:

1. Partial radiography in accordance
   with UW-11 (a) (5) (b) is allowed only
   for Types 1 and 2 of Table UW-12.

2. For longitudinal stress shell
   calculations column (b) of Table
   UW-12 is used. See UW-12(b).
   first paragraph.

SHELL THICKNESS CALCULATIONS

CIRCUMFERENTIAL STRESS

Joint efficiency, E = 1.0 (no joint in shell)
Allowable stress = 100%

LONGITUDINAL STRESS

Joint efficiency, E = $\begin{matrix} 1.0 \ (\text{Type No. 1 B or C joint}) \\ .90 \ (\text{Type No. 2 B or C joint}) \end{matrix}$
Allowable stress = 100%

SHELL THICKNESS CALCULATIONS

CIRCUMFERENTIAL STRESS

Joint efficiency, E = 1.0 (no joints in shell)
Allowable stress   = 100%

LONGITUDINAL STRESS

Joint efficiency, E = $\begin{matrix} .85 \ (\text{Type No. 1 B or C joint}) (\text{UW-12}(b)) \\ .80 \ (\text{Type No. 2 B or C joint}) (\text{first par.}) \end{matrix}$
Allowable stress   = 100%

$$T_S = \frac{250(12)}{13,800(1.0)-.6(250)} = 0.220$$

$$T_{EH} = \frac{250(24)}{2(13,800)(1.0)-.2(250)} = 0.218$$

$$T_{TH} = \frac{.885(250)(24)}{13,800(1.0)-.1(250)} = 0.385$$

Same as example 1.

Refer to example statement on page 56.

Fig. 2.5, Examples 1 and 2.   Elements of joint design for heads — seamless shell to seamless head.

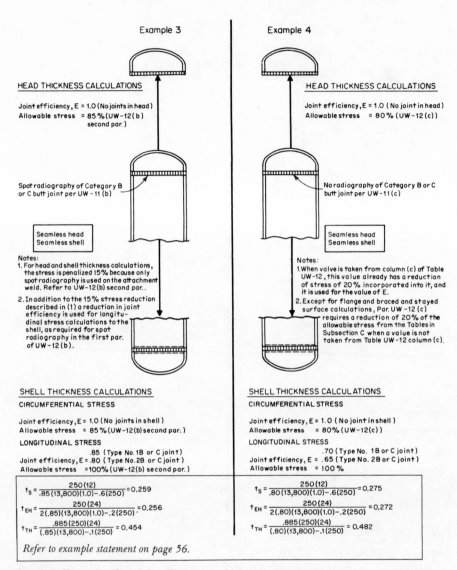

Example 3

HEAD THICKNESS CALCULATIONS

Joint efficiency, E = 1.0 (No joints in head)
Allowable stress = 85 % (UW-12(b)
          second par.)

Spot radiography of Category B
or C butt joint per UW - 11 (b)

Seamless head
Seamless shell

Notes:
1. For head and shell thickness calculations,
   the stress is penalized 15% because only
   spot radiography is used on the attachment
   weld. Refer to UW-12(b) second par..

2. In addition to the 15% stress reduction
   described in (1) a reduction in joint
   efficiency is used for longitudinal stress calculations to the
   shell, as required for spot
   radiography in the first par.
   of UW-12(b).

SHELL THICKNESS CALCULATIONS

CIRCUMFERENTIAL STRESS

Joint efficiency, E = 1.0 (No joints in shell)
Allowable stress = 85 % (UW-12(b) second par.)

LONGITUDINAL STRESS
                    .85 (Type No. 1B or C joint)
Joint efficiency, E = .80 (Type No.2B or C joint)
Allowable stress   =100% (UW-12(b) second par.)

$$t_S = \frac{250\,(12)}{.85\,(13,800)(1.0)-.6(250)} = 0.259$$

$$t_{EH} = \frac{250\,(24)}{2(.85)(13,800)(1.0)-.2(250)} = 0.256$$

$$t_{TH} = \frac{.885\,(250)(24)}{(.85)(13,800)-.1(250)} = 0.454$$

Refer to example statement on page 56.

Example 4

HEAD THICKNESS CALCULATIONS

Joint efficiency, E = 1.0 (No joint in head)
Allowable stress = 80 % (UW-12 (c))

No radiography of Category B or C
butt joint per UW-11 (c)

Seamless head
Seamless shell

Notes:
1. When valve is taken from column (c) of Table
   UW-12 , this value already has a reduction
   of stress of 20% incorporated into it, and
   it is used for the value of E.

2. Except for flange and braced and stayed
   surface calculations , Par. UW -12 (c)
   requires a reduction of 20% of the
   allowable stress from the Tables in
   Subsection C when a value is not
   taken from Table UW-12 column (c).

SHELL THICKNESS CALCULATIONS

CIRCUMFERENTIAL STRESS

Joint efficiency, E = 1.0 (No joint in shell)
Allowable stress = 80% (UW-12(c))

LONGITUDINAL STRESS
                    .70 (Type No. 1B or C joint)
Joint efficiency, E = .65 (Type No. 2B or C joint)
Allowable stress = 100 %

$$t_S = \frac{250\,(12)}{.80(13,800)(1.0)-.6(250)} = 0.275$$

$$t_{EH} = \frac{250\,(24)}{2(.80)(13,800)(1.0)-.2(250)} = 0.272$$

$$t_{TH} = \frac{.885\,(250)(24)}{(.80)(13,800)-.1(250)} = 0.482$$

Fig. 2.5, Examples 3 and 4.   Elements of joint design for heads — seamless shell to seamless head.

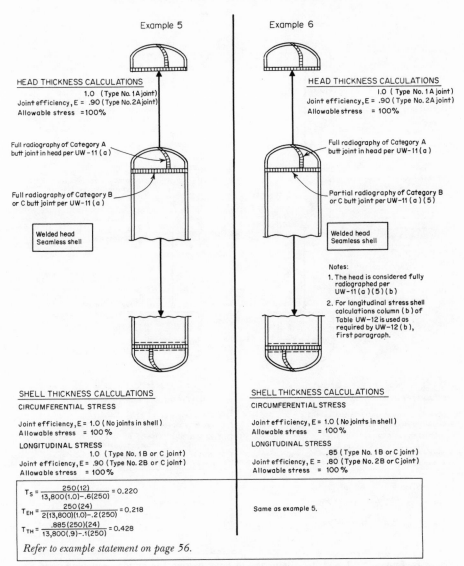

Example 5

Example 6

HEAD THICKNESS CALCULATIONS

1.0   (Type No. 1A joint)
Joint efficiency, E = .90 (Type No.2A joint)
Allowable stress =100%

Full radiography of Category A
butt joint in head per UW-11 (a)

Full radiography of Category B
or C butt joint per UW-11 (a)

Welded head
Seamless shell

HEAD THICKNESS CALCULATIONS

1.0 (Type No. 1 A joint)
Joint efficiency, E = .90 (Type No. 2A joint)
Allowable stress   = 100%

Full radiography of Category A
butt joint in head per UW-11(a)

Partial radiography of Category B
or C butt joint per UW-11 (a) (5)

Welded head
Seamless shell

Notes:

1. The head is considered fully
   radiographed per
   UW-11 (a) (5) (b)

2. For longitudinal stress shell
   calculations column (b) of
   Table UW-12 is used as
   required by UW-12 (b),
   first paragraph.

SHELL THICKNESS CALCULATIONS

CIRCUMFERENTIAL STRESS

Joint efficiency, E = 1.0 (No joints in shell)
Allowable stress = 100 %

LONGITUDINAL STRESS

1.0 (Type No. 1 B or C joint)
Joint efficiency, E = .90 (Type No. 2B or C joint)
Allowable stress = 100 %

SHELL THICKNESS CALCULATIONS

CIRCUMFERENTIAL STRESS

Joint efficiency, E = 1.0 (No joints in shell)
Allowable stress   = 100%

LONGITUDINAL STRESS

.85 (Type No. 1 B or C joint)
Joint efficiency, E = .80 (Type No. 2B or C joint)
Allowable stress   = 100 %

$$T_S = \frac{250(12)}{13,800(1.0)-.6(250)} = 0.220$$

$$T_{EH} = \frac{250(24)}{2(13,800)(1.0)-.2(250)} = 0.218$$

$$T_{TH} = \frac{.885(250)(24)}{13,800(.9)-.1(250)} = 0.428$$

Same as example 5.

Refer to example statement on page 56.

Fig. 2.6, Examples 5 and 6.   Elements of joint design for heads — welded head to seamless shell.

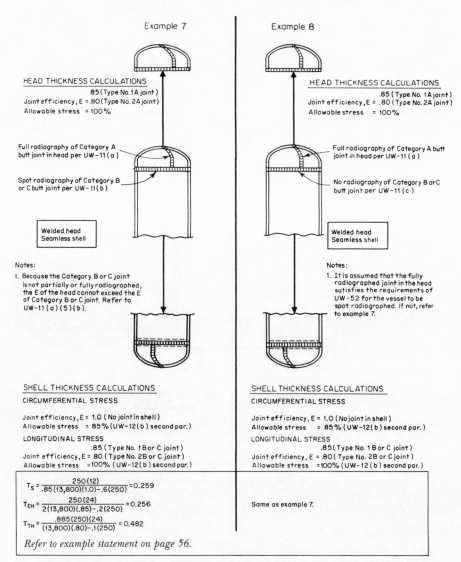

Example 7

Example 8

HEAD THICKNESS CALCULATIONS

.85 (Type No. 1A joint)
Joint efficiency, E = .80 (Type No. 2A joint)
Allowable stress   = 100%

HEAD THICKNESS CALCULATIONS

.85 (Type No. 1A joint)
Joint efficiency, E = .80 (Type No. 2A joint)
Allowable stress   = 100%

Full radiography of Category A
butt joint in head per UW – 11 (a)

Spot radiography of Category B
or C butt joint per UW – 11 (b)

Full radiography of Category A butt
joint in head per UW – 11 (a)

No radiography of Category B or C
butt joint per UW – 11 (c)

Welded head
Seamless shell

Welded head
Seamless shell

Notes:

1. Because the Category B or C joint
   is not partially or fully radiographed,
   the E of the head cannot exceed the E
   of Category B or C joint. Refer to
   UW – 11 (a) (5) (b).

Notes:

1. It is assumed that the fully
   radiographed joint in the head
   satisfies the requirements of
   UW – 52 for the vessel to be
   spot radiographed. If not, refer
   to example 7.

SHELL THICKNESS CALCULATIONS

CIRCUMFERENTIAL STRESS

Joint efficiency, E = 1.0 (No joint in shell)
Allowable stress   = 85% (UW-12 (b) second par.)

LONGITUDINAL STRESS

.85 (Type No. 1 B or C joint)
Joint efficiency, E = .80 (Type No. 2 B or C joint)
Allowable stress   = 100% (UW-12 (b) second par.)

SHELL THICKNESS CALCULATIONS

CIRCUMFERENTIAL STRESS

Joint efficiency, E = 1.0 (No joint in shell)
Allowable stress   = 85% (UW-12 (b) second par.)

LONGITUDINAL STRESS

.85 (Type No. 1 B or C joint)
Joint efficiency, E = .80 (Type No. 2 B or C joint)
Allowable stress   = 100% (UW-12 (b) second par.)

$$T_S = \frac{250\,(12)}{.85\,(13,800)\,(1.0) - .6\,(250)} = 0.259$$

$$T_{EH} = \frac{250\,(24)}{2\,(13,800)\,(.85) - .2\,(250)} = 0.256$$

$$T_{TH} = \frac{.885\,(250)\,(24)}{(13,800)\,(.80) - .1\,(250)} = 0.482$$

Same as example 7.

Refer to example statement on page 56.

Fig. 2.6, Examples 7 and 8.   Elements of joint design for heads — welded head to seamless shell.

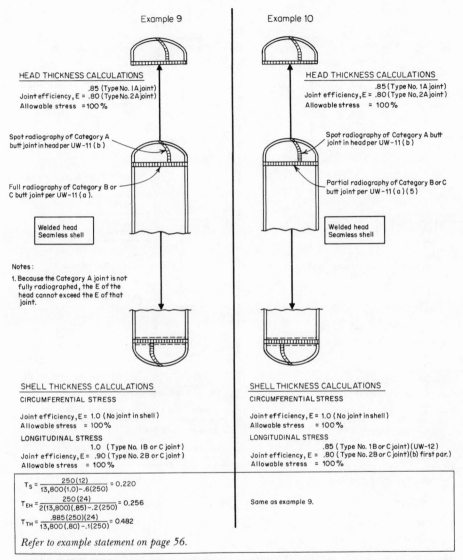

Example 9

Example 10

HEAD THICKNESS CALCULATIONS

.85 (Type No. I A joint)
Joint efficiency, E = .80 (Type No. 2A joint)
Allowable stress = 100%

Spot radiography of Category A
butt joint in head per UW-11 (b)

Full radiography of Category B or
C butt joint per UW-11 (a).

Welded head
Seamless shell

Notes:

1. Because the Category A joint is not
fully radiographed, the E of the
head cannot exceed the E of that
joint.

HEAD THICKNESS CALCULATIONS

.85 (Type No. 1A joint)
Joint efficiency, E = .80 (Type No. 2A joint)
Allowable stress = 100%

Spot radiography of Category A butt
joint in head per UW-11 (b)

Partial radiography of Category B or C
butt joint per UW-11 (a)(5)

Welded head
Seamless shell

SHELL THICKNESS CALCULATIONS

CIRCUMFERENTIAL STRESS

Joint efficiency, E = 1.0 (No joint in shell)
Allowable stress = 100%

LONGITUDINAL STRESS

1.0 (Type No. IB or C joint)
Joint efficiency, E = .90 (Type No. 2B or C joint)
Allowable stress = 100%

SHELL THICKNESS CALCULATIONS

CIRCUMFERENTIAL STRESS

Joint efficiency, E = 1.0 (No joint in shell)
Allowable stress = 100%

LONGITUDINAL STRESS

.85 (Type No. 1B or C joint)(UW-12)
Joint efficiency, E = .80 (Type No. 2B or C joint)(b) first par.)
Allowable stress = 100%

$$T_S = \frac{250(12)}{13,800(1.0)-.6(250)} = 0.220$$

$$T_{EH} = \frac{250(24)}{2(13,800)(.85)-.2(250)} = 0.256$$

$$T_{TH} = \frac{.885(250)(24)}{13,800(.80)-.1(250)} = 0.482$$

Same as example 9.

Refer to example statement on page 56.

**Fig. 2.6, Examples 9 and 10.   Elements of joint design for heads — welded head to
seamless shell.**

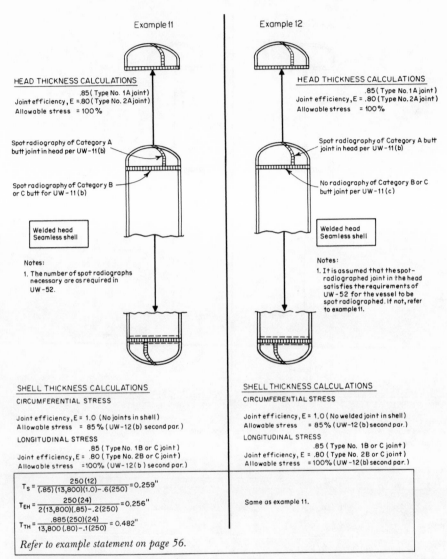

Example 11

Example 12

HEAD THICKNESS CALCULATIONS

.85 (Type No. 1 A joint)
Joint efficiency, E = .80 (Type No. 2 A joint)
Allowable stress   = 100 %

Spot radiography of Category A
butt joint in head per UW-11(b)

Spot radiography of Category B
or C butt for UW-11 (b)

Welded head
Seamless shell

Notes:
1. The number of spot radiographs
   necessary are as required in
   UW-52.

HEAD THICKNESS CALCULATIONS

.85 (Type No. 1 A joint)
Joint efficiency, E = .80 (Type No. 2 A joint)
Allowable stress   = 100 %

Spot radiography of Category A butt
joint in head per UW-11(b)

No radiography of Category B or C
butt joint per UW-11 (c)

Welded head
Seamless shell

Notes:
1. It is assumed that the spot-
   radiographed joint in the head
   satisfies the requirements of
   UW-52 for the vessel to be
   spot radiographed. If not, refer
   to example 11.

SHELL THICKNESS CALCULATIONS

CIRCUMFERENTIAL STRESS

Joint efficiency, E = 1.0 (No joints in shell)
Allowable stress   = 85 % (UW-12 (b) second par.)

LONGITUDINAL STRESS

.85 (Type No. 1B or C joint)
Joint efficiency, E = .80 (Type No. 2B or C joint)
Allowable stress   = 100 % (UW-12 (b) second par.)

SHELL THICKNESS CALCULATIONS

CIRCUMFERENTIAL STRESS

Joint efficiency, E = 1.0 (No welded joint in shell)
Allowable stress   = 85 % (UW-12 (b) second par.)

LONGITUDINAL STRESS

.85 (Type No. 1B or C joint)
Joint efficiency, E = .80 (Type No. 2B or C joint)
Allowable stress   = 100 % (UW-12 (b) second par.)

$$T_S = \frac{250\,(12)}{(.85)\,(13,800)(1.0) - .6(250)} = 0.259''$$

$$T_{EH} = \frac{250\,(24)}{2(13,800)(.85) - .2(250)} = 0.256''$$

$$T_{TH} = \frac{.885(250)(24)}{13,800\,(.80) - .1(250)} = 0.482''$$

Same as example 11.

Refer to example statement on page 56.

Fig. 2.6, Examples 11 and 12.   Elements of joint design for heads — welded head to seamless shell.

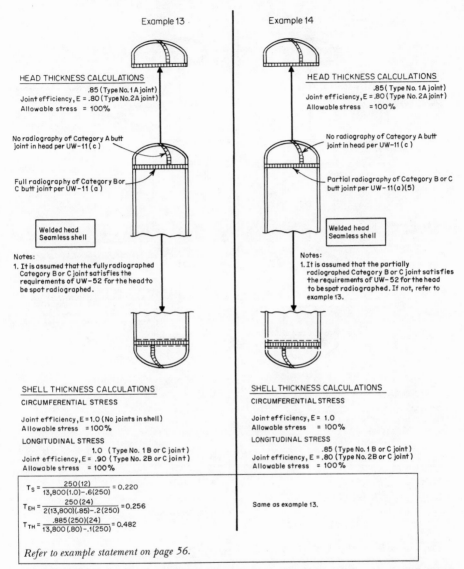

Example 13 · Example 14

HEAD THICKNESS CALCULATIONS

.85 (Type No. 1 A joint)
Joint efficiency, E = .80 (Type No. 2 A joint)
Allowable stress = 100%

No radiography of Category A butt
joint in head per UW-11 (c)

Full radiography of Category B or
C butt joint per UW-11 (a)

Welded head
Seamless shell

Notes:
1. It is assumed that the fully radiographed
Category B or C joint satisfies the
requirements of UW-52 for the head to
be spot radiographed.

HEAD THICKNESS CALCULATIONS

.85 (Type No. 1A joint)
Joint efficiency, E = .80 (Type No. 2A joint)
Allowable stress = 100%

No radiography of Category A butt
joint in head per UW-11 (c)

Partial radiography of Category B or C
butt joint per UW-11 (a)(5)

Welded head
Seamless shell

Notes:
1. It is assumed that the partially
radiographed Category B or C joint satisfies
the requirements of UW-52 for the head
to be spot radiographed. If not, refer to
example 13.

SHELL THICKNESS CALCULATIONS

CIRCUMFERENTIAL STRESS

Joint efficiency, E =1.0 (No joints in shell)
Allowable stress = 100%

LONGITUDINAL STRESS

1.0 (Type No. 1 B or C joint)
Joint efficiency, E = .90 (Type No. 2B or C joint)
Allowable stress = 100%

SHELL THICKNESS CALCULATIONS

CIRCUMFERENTIAL STRESS

Joint efficiency, E = 1.0
Allowable stress = 100%

LONGITUDINAL STRESS

.85 (Type No. 1 B or C joint)
Joint efficiency, E = .80 (Type No. 2B or C joint)
Allowable stress = 100%

$$T_S = \frac{250(12)}{13,800(1.0)-.6(250)} = 0.220$$

$$T_{EH} = \frac{250(24)}{2(13,800)(.85)-.2(250)} = 0.256$$

$$T_{TH} = \frac{.885(250)(24)}{13,800(.80)-.1(250)} = 0.482$$

Same as example 13.

Refer to example statement on page 56.

Fig. 2.6, Examples 13 and 14. Elements of joint design for heads — welded head to seamless shell.

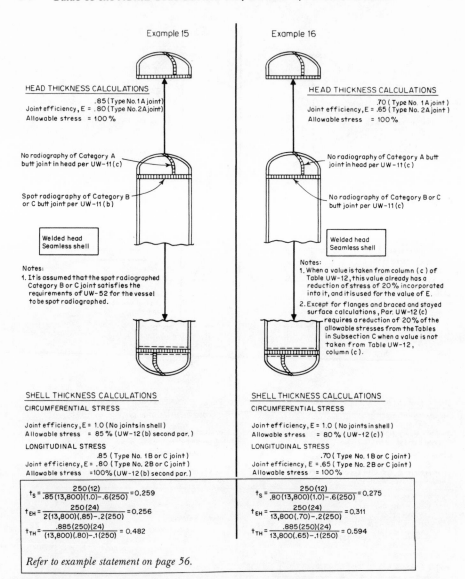

Example 15 · Example 16

HEAD THICKNESS CALCULATIONS

Joint efficiency, E = .85 (Type No. 1 A joint)
        .80 (Type No. 2 A joint)
Allowable stress = 100 %

No radiography of Category A
butt joint in head per UW-11(c)

Spot radiography of Category B
or C butt joint per UW-11(b)

Welded head
Seamless shell

Notes:
1. It is assumed that the spot radiographed
   Category B or C joint satisfies the
   requirements of UW-52 for the vessel
   to be spot radiographed.

HEAD THICKNESS CALCULATIONS

Joint efficiency, E = .70 (Type No. 1 A joint)
        .65 (Type No. 2 A joint)
Allowable stress = 100 %

No radiography of Category A butt
joint in head per UW-11(c)

No radiography of Category B or C
butt joint per UW-11(c)

Welded head
Seamless shell

Notes:
1. When a value is taken from column (c) of
   Table UW-12, this value already has a
   reduction of stress of 20% incorporated
   into it, and it is used for the value of E.
2. Except for flanges and braced and stayed
   surface calculations, Par. UW-12 (c)
   requires a reduction of 20% of the
   allowable stresses from the Tables
   in Subsection C when a value is not
   taken from Table UW-12,
   column (c).

SHELL THICKNESS CALCULATIONS

CIRCUMFERENTIAL STRESS

Joint efficiency, E = 1.0 (No joints in shell)
Allowable stress = 85 % (UW-12 (b) second par.)

LONGITUDINAL STRESS

Joint efficiency, E = .85 (Type No. 1 B or C joint)
        .80 (Type No. 2 B or C joint)
Allowable stress = 100% (UW-12 (b) second par.)

$$t_S = \frac{250(12)}{.85(13,800)(1.0)-.6(250)} = 0.259$$

$$t_{EH} = \frac{250(24)}{2(13,800)(.85)-.2(250)} = 0.256$$

$$t_{TH} = \frac{.885(250)(24)}{(13,800)(.80)-.1(250)} = 0.482$$

SHELL THICKNESS CALCULATIONS

CIRCUMFERENTIAL STRESS

Joint efficiency, E = 1.0 (No joints in shell)
Allowable stress = 80 % (UW-12(c))

LONGITUDINAL STRESS

Joint efficiency, E = .70 (Type No. 1 B or C joint)
        .65 (Type No. 2 B or C joint)
Allowable stress = 100 %

$$t_S = \frac{250(12)}{.80(13,800)(1.0)-.6(250)} = 0.275$$

$$t_{EH} = \frac{250(24)}{13,800(.70)-.2(250)} = 0.311$$

$$t_{TH} = \frac{.885(250)(24)}{13,800(.65)-.1(250)} = 0.594$$

Refer to example statement on page 56.

Fig. 2.6, Examples 15 and 16. Elements of joint design for heads — welded head to seamless shell.

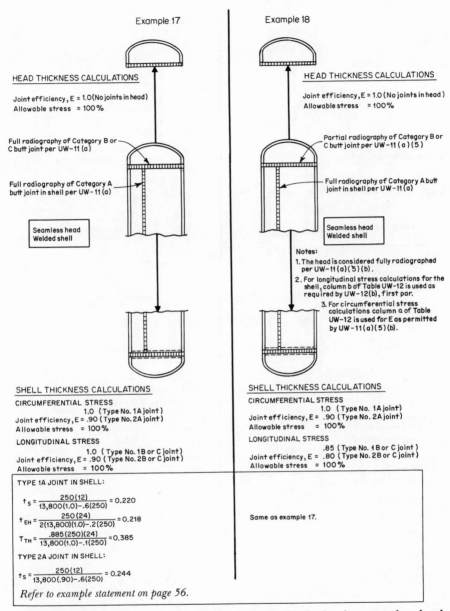

Fig. 2.7, Examples 17 and 18. Elements of joint design for heads — seamless head to welded shell.

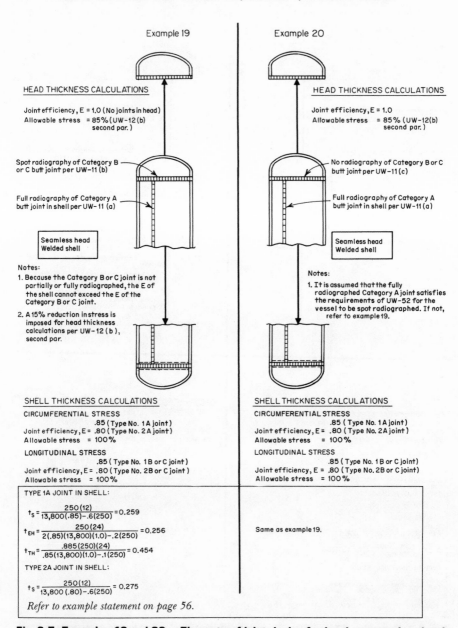

Example 19

Example 20

HEAD THICKNESS CALCULATIONS

Joint efficiency, E = 1.0 (No joints in head)
Allowable stress = 85% (UW-12(b)
second par.)

Spot radiography of Category B
or C butt joint per UW-11(b)

Full radiography of Category A
butt joint in shell per UW-11 (a)

Seamless head
Welded shell

Notes:

1. Because the Category B or C joint is not
   partially or fully radiographed, the E of
   the shell cannot exceed the E of the
   Category B or C joint.

2. A 15% reduction in stress is
   imposed for head thickness
   calculations per UW-12 (b),
   second par.

HEAD THICKNESS CALCULATIONS

Joint efficiency, E = 1.0
Allowable stress = 85% (UW-12(b)
second par.)

No radiography of Category B or C
butt joint per UW-11(c)

Full radiography of Category A
butt joint in shell per UW-11 (a)

Seamless head
Welded shell

Notes:

1. It is assumed that the fully
   radiographed Category A joint satisfies
   the requirements of UW-52 for the
   vessel to be spot radiographed. If not,
   refer to example 19.

SHELL THICKNESS CALCULATIONS

CIRCUMFERENTIAL STRESS
                    .85 (Type No. 1 A joint)
Joint efficiency, E = .80 (Type No. 2 A joint)
Allowable stress = 100%

LONGITUDINAL STRESS
                    .85 (Type No. 1 B or C joint)
Joint efficiency, E = .80 (Type No. 2 B or C joint)
Allowable stress = 100%

SHELL THICKNESS CALCULATIONS

CIRCUMFERENTIAL STRESS
                    .85 (Type No. 1 A joint)
Joint efficiency, E = .80 (Type No. 2 A joint)
Allowable stress = 100%

LONGITUDINAL STRESS
                    .85 (Type No. 1 B or C joint)
Joint efficiency, E = .80 (Type No. 2 B or C joint)
Allowable stress = 100%

TYPE 1A JOINT IN SHELL:

$$t_S = \frac{250(12)}{13,800(.85)-.6(250)} = 0.259$$

$$t_{EH} = \frac{250(24)}{2(.85)(13,800)(1.0)-.2(250)} = 0.256$$

$$t_{TH} = \frac{.885(250)(24)}{.85(13,800)(1.0)-.1(250)} = 0.454$$

TYPE 2A JOINT IN SHELL:

$$t_S = \frac{250(12)}{13,800(.80)-.6(250)} = 0.275$$

Same as example 19.

Refer to example statement on page 56.

**Fig. 2.7, Examples 19 and 20.   Elements of joint design for heads — seamless head
to welded shell.**

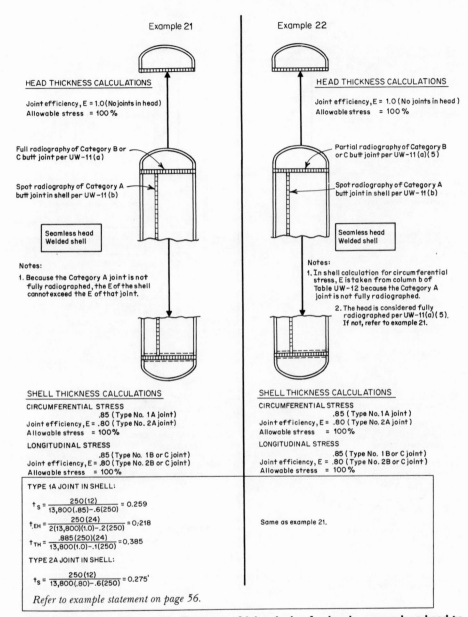

Example 21

HEAD THICKNESS CALCULATIONS

Joint efficiency, E = 1.0 (No joints in head)
Allowable stress = 100%

Full radiography of Category B or
C butt joint per UW-11(a)

Spot radiography of Category A
butt joint in shell per UW-11(b)

Seamless head
Welded shell

Notes:

1. Because the Category A joint is not
   fully radiographed, the E of the shell
   cannot exceed the E of that joint.

SHELL THICKNESS CALCULATIONS

CIRCUMFERENTIAL STRESS
                                    .85 (Type No. 1A joint)
Joint efficiency, E = .80 (Type No. 2A joint)
Allowable stress = 100%

LONGITUDINAL STRESS
                                    .85 (Type No. 1B or C joint)
Joint efficiency, E = .80 (Type No. 2B or C joint)
Allowable stress = 100%

TYPE 1A JOINT IN SHELL:

$$t_s = \frac{250(12)}{13,800(.85)-.6(250)} = 0.259$$

$$t_{EH} = \frac{250(24)}{2(13,800)(1.0)-.2(250)} = 0.218$$

$$t_{TH} = \frac{.885(250)(24)}{13,800(1.0)-.1(250)} = 0.385$$

TYPE 2A JOINT IN SHELL:

$$t_s = \frac{250(12)}{13,800(.80)-.6(250)} = 0.275'$$

Refer to example statement on page 56.

Example 22

HEAD THICKNESS CALCULATIONS

Joint efficiency, E = 1.0 (No joints in head)
Allowable stress = 100%

Partial radiography of Category B
or C butt joint per UW-11(a)(5)

Spot radiography of Category A
butt joint in shell per UW-11(b)

Seamless head
Welded shell

Notes:

1. In shell calculation for circumferential
   stress, E is taken from column b of
   Table UW-12 because the Category A
   joint is not fully radiographed.

2. The head is considered fully
   radiographed per UW-11(a)(5).
   If not, refer to example 21.

SHELL THICKNESS CALCULATIONS

CIRCUMFERENTIAL STRESS
                                    .85 (Type No. 1A joint)
Joint efficiency, E = .80 (Type No. 2A joint)
Allowable stress = 100%

LONGITUDINAL STRESS
                                    .85 (Type No. 1B or C joint)
Joint efficiency, E = .80 (Type No. 2B or C joint)
Allowable stress = 100%

Same as example 21.

Fig. 2.7, Examples 21 and 22. Elements of joint design for heads — seamless head to welded shell.

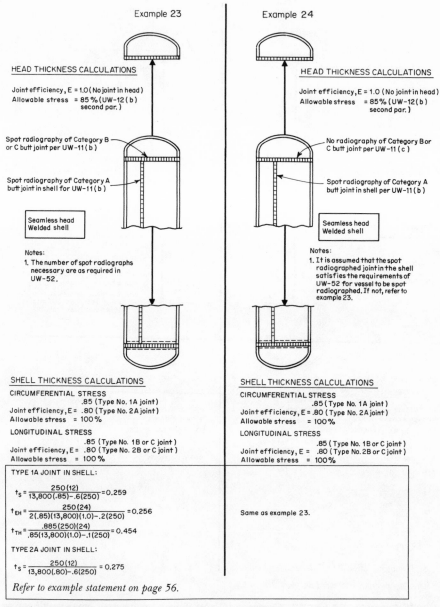

Example 23

HEAD THICKNESS CALCULATIONS

Joint efficiency, E = 1.0 (No joint in head)
Allowable stress = 85% (UW-12(b)
                        second par.)

Spot radiography of Category B
or C butt joint per UW-11(b)

Spot radiography of Category A
butt joint in shell for UW-11(b)

Seamless head
Welded shell

Notes:
1. The number of spot radiographs
   necessary are as required in
   UW-52.

SHELL THICKNESS CALCULATIONS

CIRCUMFERENTIAL STRESS
                    .85 (Type No. 1A joint)
Joint efficiency, E = .80 (Type No. 2A joint)
Allowable stress = 100%

LONGITUDINAL STRESS
                    .85 (Type No. 1B or C joint)
Joint efficiency, E = .80 (Type No. 2B or C joint)
Allowable stress = 100%

TYPE 1A JOINT IN SHELL:

$$t_S = \frac{250(12)}{13,800(.85)-.6(250)} = 0.259$$

$$t_{EH} = \frac{250(24)}{2(.85)(13,800)(1.0)-.2(250)} = 0.256$$

$$t_{TH} = \frac{.885(250)(24)}{.85(13,800)(1.0)-.1(250)} = 0.454$$

TYPE 2A JOINT IN SHELL:

$$t_S = \frac{250(12)}{13,800(.80)-.6(250)} = 0.275$$

Refer to example statement on page 56.

Example 24

HEAD THICKNESS CALCULATIONS

Joint efficiency, E = 1.0 (No joint in head)
Allowable stress = 85% (UW-12(b)
                        second par.)

No radiography of Category B or
C butt joint per UW-11(c)

Spot radiography of Category A
butt joint in shell per UW-11(b)

Seamless head
Welded shell

Notes:
1. It is assumed that the spot
   radiographed joint in the shell
   satisfies the requirements of
   UW-52 for vessel to be spot
   radiographed. If not, refer to
   example 23.

SHELL THICKNESS CALCULATIONS

CIRCUMFERENTIAL STRESS
                    .85 (Type No. 1A joint)
Joint efficiency, E = .80 (Type No. 2A joint)
Allowable stress = 100%

LONGITUDINAL STRESS
                    .85 (Type No. 1B or C joint)
Joint efficiency, E = .80 (Type No. 2B or C joint)
Allowable stress = 100%

Same as example 23.

**Fig. 2.7, Examples 23 and 24.   Elements of joint design for heads — seamless head
to welded shell.**

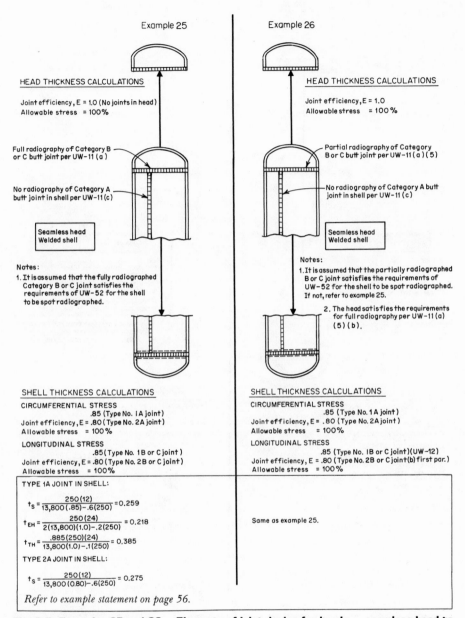

Example 25

Example 26

HEAD THICKNESS CALCULATIONS

Joint efficiency, E = 1.0 (No joints in head)
Allowable stress = 100%

HEAD THICKNESS CALCULATIONS

Joint efficiency, E = 1.0
Allowable stress = 100%

Full radiography of Category B
or C butt joint per UW-11 (a)

Partial radiography of Category
B or C butt joint per UW-11(a)(5)

No radiography of Category A
butt joint in shell per UW-11 (c)

No radiography of Category A butt
joint in shell per UW-11 (c)

Seamless head
Welded shell

Seamless head
Welded shell

Notes:
1. It is assumed that the fully radiographed
Category B or C joint satisfies the
requirements of UW-52 for the shell
to be spot radiographed.

Notes:
1. It is assumed that the partially radiographed
B or C joint satisfies the requirements of
UW-52 for the shell to be spot radiographed.
If not, refer to example 25.

2. The head satisfies the requirements
for full radiography per UW-11 (a)
(5)(b).

SHELL THICKNESS CALCULATIONS

CIRCUMFERENTIAL STRESS
        .85 (Type No. 1A joint)
Joint efficiency, E = .80 (Type No. 2A joint)
Allowable stress = 100%

LONGITUDINAL STRESS
        .85 (Type No. 1B or C joint)
Joint efficiency, E = .80 (Type No. 2B or C joint)
Allowable stress = 100%

SHELL THICKNESS CALCULATIONS

CIRCUMFERENTIAL STRESS
        .85 (Type No. 1 A joint)
Joint efficiency, E = .80 (Type No. 2A joint)
Allowable stress = 100%

LONGITUDINAL STRESS
        .85 (Type No. 1B or C joint)(UW-12)
Joint efficiency, E = .80 (Type No. 2B or C joint(b) first par.)
Allowable stress = 100%

TYPE 1A JOINT IN SHELL:

$$t_S = \frac{250\,(12)}{13,800\,(.85) - .6(250)} = 0.259$$

$$t_{EH} = \frac{250\,(24)}{2(13,800)(1.0) - .2(250)} = 0.218$$

$$t_{TH} = \frac{.885\,(250)(24)}{13,800(1.0) - .1(250)} = 0.385$$

TYPE 2A JOINT IN SHELL:

$$t_S = \frac{250\,(12)}{13,800\,(0.80) - .6(250)} = 0.275$$

Same as example 25.

Refer to example statement on page 56.

Fig. 2.7, Examples 25 and 26. Elements of joint design for heads — seamless head to welded shell.

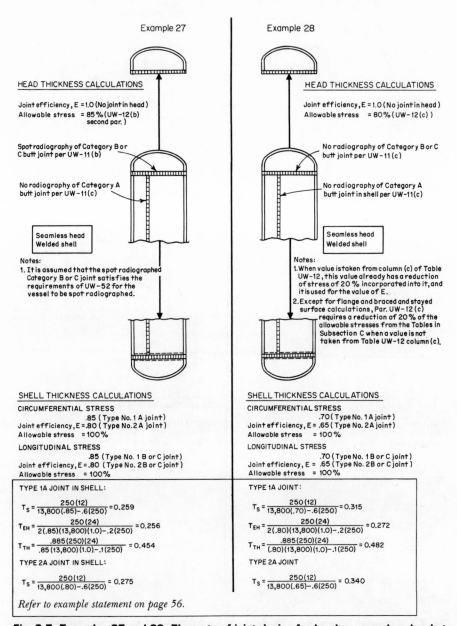

Example 27

HEAD THICKNESS CALCULATIONS

Joint efficiency, E =1.0 (No joint in head)
Allowable stress = 85% (UW-12(b)
second par.)

Spot radiography of Category B or
C butt joint per UW-11 (b)

No radiography of Category A
butt joint per UW-11(c)

Seamless head
Welded shell

Notes:
1. It is assumed that the spot radiographed
Category B or C joint satisfies the
requirements of UW-52 for the
vessel to be spot radiographed.

Example 28

HEAD THICKNESS CALCULATIONS

Joint efficiency,E = 1.0 (No joint in head)
Allowable stress = 80% (UW-12(c))

No radiography of Category B or C
butt joint per UW-11 (c)

No radiography of Category A
butt joint in shell per UW-11(c)

Seamless head
Welded shell

Notes:
1. When value is taken from column (c) of Table
UW-12, this value already has a reduction
of stress of 20 % incorporated into it, and
it is used for the value of E.

2. Except for flange and braced and stayed
surface calculations, Par. UW-12 (c)
requires a reduction of 20 % of the
allowable stresses from the Tables in
Subsection C when a value is not
taken from Table UW-12 column (c).

SHELL THICKNESS CALCULATIONS

CIRCUMFERENTIAL STRESS
          .85 (Type No. 1 A joint)
Joint efficiency, E =.80 (Type No. 2 A joint)
Allowable stress = 100%

LONGITUDINAL STRESS
          .85 (Type No. 1 B or C joint)
Joint efficiency, E =.80 (Type No. 2 B or C joint)
Allowable stress = 100%

SHELL THICKNESS CALCULATIONS

CIRCUMFERENTIAL STRESS
          .70 (Type No. 1 A joint)
Joint efficiency, E = .65 (Type No. 2 A joint)
Allowable stress = 100%

LONGITUDINAL STRESS
          .70 (Type No. 1 B or C joint)
Joint efficiency, E = .65 (Type No. 2 B or C joint)
Allowable stress = 100%

TYPE 1A JOINT IN SHELL:	TYPE 1A JOINT:

TYPE 1A JOINT IN SHELL:

$$T_S = \frac{250(12)}{13,800(.85)-.6(250)} = 0.259$$

$$T_{EH} = \frac{250(24)}{2(.85)(13,800)(1.0)-.2(250)} = 0.256$$

$$T_{TH} = \frac{.885(250)(24)}{.85(13,800)(1.0)-.1(250)} = 0.454$$

TYPE 2A JOINT IN SHELL:

$$T_S = \frac{250(12)}{13,800(.80)-.6(250)} = 0.275$$

TYPE 1A JOINT:

$$T_S = \frac{250(12)}{13,800(.70)-.6(250)} = 0.315$$

$$T_{EH} = \frac{250(24)}{2(.80)(13,800)(1.0)-.2(250)} = 0.272$$

$$T_{TH} = \frac{.885(250)(24)}{(.80)(13,800)(1.0)-.1(250)} = 0.482$$

TYPE 2A JOINT

$$T_S = \frac{250(12)}{13,800(.65)-.6(250)} = 0.340$$

Refer to example statement on page 56.

Fig. 2.7, Examples 27 and 28. Elements of joint design for heads — seamless heads to welded shell.

INSTRUCTIONS FOR JOINT DESIGN CHART

The joint design charts shown as Fig. 2.8*a* and *b* and the following explanation are adopted from the Hartford Steam Boiler Inspection and Insurance Company slide rule chart and are intended to be a quick means of determining the allowable joint efficiency and stress factor to be used in calculating the required thickness for pressure vessel shells and heads. Section VIII, Division 1, has assigned joint efficiency and stress factors based on the degree of radiography and type of joint under consideration. Item numbers have been included in the charts for easy reference of each case.

To determine the joint efficiency and allowable stress value, complete the following steps:

1. Determine the degree of radiography performed on the joints in question. Locate the appropriate column corresponding to the degree of radiography. Note that F indicates full radiography, P indicates partial radiography, S indicates spot radiography, and N indicates no radiography.

2. Determine the type of joint under consideration (if joint type is applicable) and read the allowable joint efficiency or stress factor from the chart.

3. Read all applicable footnotes.

EXAMPLE

A vessel with a Type 1 welded longitudinal joint with seamless torispherical heads is attached by Type 2 joints. The longitudinal joint is fully radiographed. The attachment (circumferential) weld between the heads and shell is partially radiographed. The governing stress is the circumferential stress per Code Par. UG-27.

Shell calculations To determine the joint efficiency and stress factor for the shells, go to the box for "Seamless Head/Welded Shell" in the shell calculations section (Fig. 2.8*b*). Item 2 covers the case when X = Full RT, Y = Partial RT. Read the appropriate column. Footnotes 1 and 5 apply. Since the longitudinal (X) joint is the governing joint per Footnote 1, read the joint efficiency for the Type 1 joint (E $-$ 1 = 1.00). The stress factor is = 100 percent per Footnote 5.

Head calculations To determine the joint efficiency and stress factor for the heads, go to the box for "Seamless Heads/Welded Shell" in the head (except hemispherical) calculation section (Fig. 2.8*a*). Item 2 covers the case when X = Full RT, Y = Partial RT. Read the appropriate column. Footnote 4 applies. The circumferential seam may be either a Type 1 or 2 joint per Code Par. UW-12. Read the stress factor = 100 percent. The joint efficiency = 1.00 per Footnote 4.

HEAD CALCULATIONS
(EXCEPT HEMISPHERICAL)

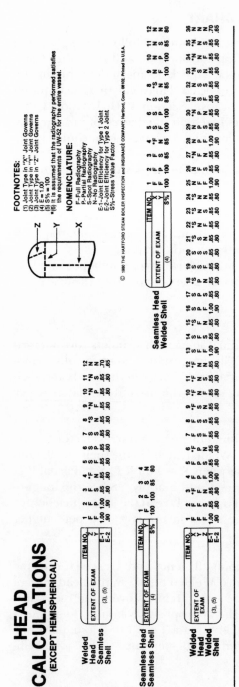

HEMISPHERICAL HEAD

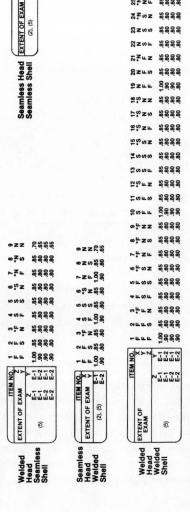

Fig. 2.8a Head calculations.

SHELL CALCULATIONS

FOOTNOTES:
(1) Joint Type in "X" Joint Governs
(2) Joint Type in "Y" Joint Governs
(3) Joint Type in "Z" Joint Governs
(4) E=1.00
(5) S%=100
(6) It is assumed that the radiography performed satisfies the requirements of UW-52 for the entire vessel.

NOMENCLATURE:
F-Full Radiography
P-Partial Radiography
S-Spot Radiography
N-No Radiography
E-1-Joint Efficiency for Type 1 Joint
E-2-Joint Efficiency for Type 2 Joint
S%-Stress Value Factor

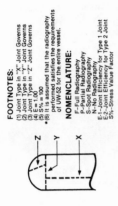

Welded Head / Welded Shell

ITEM NO.		1	2	3	4	5	6	7	8	9	10	11	12	13	14	15	16	17	18	19	20	21	22	23	24	25	26	27	28	29	30	31	32	33	34	35	36
EXTENT OF EXAM	X	F	F	F	F	F	F	F	F	F	F	F	*F	F	F	F	S	S	S	P	S	S	*S	*S	*S	*N	*N	N	N	*N	*N	P	*N	*N	*N	*N	N
	Y	F	F	F	F	P	P	F	F	F	*F	F	F	S	S	S	S	S	P	S	S	S	S	S	S	N	N	N	N	N	N	N	N	N	N	N	N
	Z	F	F	F	S	P	N	S	S	N	N	N	N	F	S	N	P	F	N	N	S	N	N	N	N	F	S	N	N	N	N	N	N	N	N	N	N
CIRCUMFERENTIAL STRESS (1),(5)	E-1	1.00	1.00	1.00	1.00	1.00	1.00	.85	.85	.85	.85	.85	.85	.85	.85	.85	.85	.85	.85	.85	.85	.85	.85	.85	.85	.85	.85	.85	.85	.85	.85	.85	.85	.85	.85	.85	.85
	E-2	.90	.90	.90	.90	.90	.90	.80	.80	.80	.80	.80	.80	.80	.80	.80	.80	.80	.80	.80	.80	.80	.80	.80	.80	.80	.80	.80	.80	.80	.80	.80	.80	.80	.80	.80	.80
LONGITUDINAL STRESS (2),(5)	E-1	1.00	1.00	1.00	.85	.85	.85	.85	.85	.85	.85	.85	.85	.85	.85	.85	.85	.85	.85	.85	.85	.85	.85	.85	.85	.85	.85	.85	.85	.85	.85	.85	.85	.85	.85	.85	.70
	E-2	.90	.90	.90	.80	.80	.80	.80	.80	.80	.80	.80	.80	.80	.80	.80	.80	.80	.80	.80	.80	.80	.80	.80	.80	.80	.80	.80	.80	.80	.80	.80	.80	.80	.80	.80	.65

Seamless Head / Seamless Shell

ITEM NO.		1	2	3	4
EXTENT OF EXAM	Y	F	P	S	N
CIRCUMFERENTIAL STRESS (4)	S%	100	100	85	80
LONGITUDINAL STRESS (2),(6)	E-1	1.00	.85	.85	.70
	E-2	.90	.80	.80	.65

Welded Head / Seamless Shell

ITEM NO.		1	2	3	4	5	6	7	8	9	10	11	12
EXTENT OF EXAM	Z	F	F	F	*F	F	S	S	*S	*N	P	*N	N
	Y	F	P	S	N	F	S	P	S	N	F	P	N
CIRCUMFERENTIAL STRESS (4)	S%	100	100	85	85	85	100	100	85	100	100	85	80
LONGITUDINAL STRESS (2),(5)	E-1	1.00	.85	.85	.85	1.00	.85	.85	.85	1.00	.85	.85	.70
	E-2	.90	.80	.80	.80	.90	.80	.80	.80	.90	.80	.80	.65

Seamless Head / Welded Shell

ITEM NO.		1	2	3	4	5	6	7	8	9	10	11	12
EXTENT OF EXAM	X	F	F	*S	*S	N	S	S	P	*N	*N	*N	N
	Y	F	F	S	S	N	S	S	F	N	S	F	N
CIRCUMFERENTIAL STRESS (1),(5)	E-1	1.00	.90	.90	.90	.80	.80	.80	.80	.80	.80	.85	.70
	E-2	.90	.80	.80	.80	.80	.80	.80	.80	.80	.80	.80	.65
LONGITUDINAL STRESS (2),(5)	E-1	1.00	.90	.80	.80	.85	.85	.85	.85	.85	.85	.85	.70
	E-2	.90	.80	.80	.80	.80	.80	.80	.80	.80	.80	.80	.65

Fig. 2.8b Shell calculations.

TABLE 2.7 **Calculation for ellipsoidal and torispherical flanged and dished heads under external pressure**

Equations	Remarks	Code references
Ellipsoidal and torispherical heads under external pressure		
Compute thicknesses by appropriate procedure in Code. Par. UG-33(a), (d), and (e). Use greater value as indicated	Carbon steel heads	Par. UG-33 Par. UCS-33 Appendix L
	Nonferrous metal heads	Par. UG-33 Par. UNF-33 Appendix L
	High-alloy steel heads	Par. UG-33 Par. UHA-31 Appendix L
Thickness of cast-iron heads shall not be less than that for plus heads nor less than 1 percent of inside diameter of head skirt	Cast-iron heads	Par. UCI-32, -33 Appendix L
	Clad-steel heads	Par. UCL-26 Par. UCS-33
	Cast ductile	Par. UCD-32, -33
	Iron heads	Appendix L
	Cladding may be included in design calculations	Par. UCL-23(b) and UCL-23(c)
	Stress values of materials (use appropriate table)	Table UCS-23 Table UCI-23 Table UNF-23 Table UCN-23 Table UHA-23
	See Code Appendix L for application of Code formulas and rules	Appendix L

FLAT AND BOLTED HEADS

TABLE 2.8 Calculations for flat and bolted heads and bolted flange connections

Equations	Remarks	Code reference
Unstayed flat heads		
Circular: $$t = d \sqrt{\frac{CP}{SE}}$$ or $$P = \frac{St^2E}{d^2C}$$	Values of C and d	Par. UG-34 Fig. UG-34
	Provision if other than full face gasket is used	Par. UG-34,
Noncircular: $$t = d \sqrt{\frac{ZCP}{SE}}$$ Nomenclature: see Code Par. UG-34	Noncircular heads	Par. UG-34
	Modify formulas for unstayed flat heads and covers attached by bolts causing edge moments	Footnote 18
Bolted heads (spherically dished covers)		
Use equations in Code Appendix 1		Appendix 1-6 Fig. 1-6
Bolted flange connections		
Use equations in Code Appendix 2	General recommendations	Par. UG-44 Appendix 2
	Design considerations	Appendix 5
	Required loadings, moments, and stresses	Appendix 2
	Types of flanges	Fig. 2-4
	Optional flanges limited to 300-psi pressure	Appendix 2-4
	Blind flanges conforming to an accepted standard and attached by bolting as shown in Fig. UG-34(J) and (K) may be used for the appropriate pressure-temperature rating	Par. UG-34 Par. UG-11(a)(1) Par. UG-11(a)(2)
	Bolt loads	Appendix 2-5
	Reverse flanges	Appendix 2-13

TABLE 2.9   Maximum allowable working stress per bolt (standard thread and 8-thread series)

Bolt diameter, in	Number of threads per inch	Area at bottom of thread, sq in	Allowable material stress, psi			
			7000*	16,250	18,750	20,000
			Stress per bolt, psi			
Standard thread						
½	13	0.126	882	2047	2362	2520
⅝	11	0.202	1414	3282	3787	4040
¾	10	0.302	2114	4907	5662	6040
⅞	9	0.419	2933	6808	7856	8380
1	8	0.551	3857	8953	10,331	11,020
1⅛	7	0.693	4851	11,261	12,993	13,860
1¼	7	0.890	6230	14,462	16,687	17,800
1⅜	6	1.054	7378	17,127	19,762	21,080
1½	6	1.294	9058	21,027	24,262	25,880
1⅝	5½	1.515	10,605	24,618	28,406	30,300
1¾	5	1.744	12,208	28,340	32,700	34,880
1⅞	5	2.049	14,343	33,296	38,418	40,980
2	4½	2.300	16,100	37,375	43,125	46,000
2¼	4½	3.020	21,140	49,075	56,625	60,400
2½	4	3.715	26,005	60,368	69,656	74,300
2¾	4	4.618	32,326	75,042	86,587	92,360
3	4	5.620	39,340	91,325	105,375	112,400
8-thread series						
1⅛	8	0.728	5,096	11,830	13,650	14,560
1¼	8	0.929	6,503	15,096	17,418	18,580
1⅜	8	1.155	8,085	18,768	21,656	23,100
1½	8	1.405	9,835	22,831	26,343	28,100
1⅝	8	1.680	11,760	27,300	31,500	33,600
1¾	8	1.980	13,860	32,175	37,125	39,600
1⅞	8	2.304	16,128	37,440	43,200	46,080
2	8	2.652	18,564	43,095	49,725	53,040
2¼	8	3.423	23,961	55,623	64,181	68,460
2½	8	4.292	30,044	69,745	80,475	85,840
2¾	8	5.259	36,813	85,458	98,606	105,180
3	8	6.324	44,268	102,765	118,575	126,480

* Maximum temperature, 450°F.

USE OF PRESSURE-THICKNESS CHARTS FOR FLAT HEADS AND BOLTED FLATCOVER PLATES (FIGS. 2.9, 2.10, AND 2.11)

The required head thickness may be found by the following steps: (1) Locate the required inside diameter (for flat heads) or bolt circle (for flatcover plates) at the bottom of the chart, (2) read up vertically to the appropriate pressure line, and (3) read horizontally to the required thickness. Other problems may be similarly solved.

Charts are based on an allowable stress of 13,800 psi. For other stress values, multiply the chart thickness by the appropriate constant in the following table:

Stress (psi)	11,000	11,300	12,500	15,000	16,300	17,500	18,800
Constant	1.120	1.104	1.050	0.959	0.920	0.888	0.856

Corrosion allowances are not included in either chart. Thickness equations for flat heads and bolted flatcover plates may be found in Table 2.8.

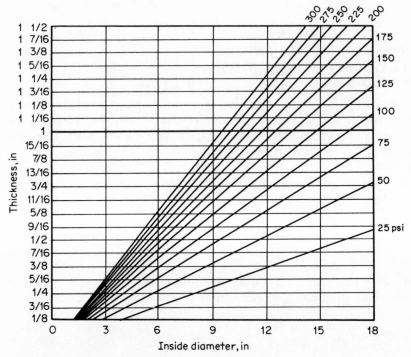

Fig. 2.9   Flat head thickness chart, C = 0.50.

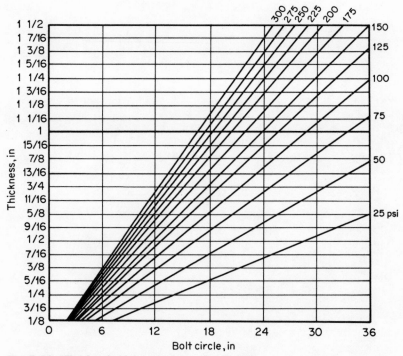

Fig. 2.10   Flat head thickness chart, C = 0.162.

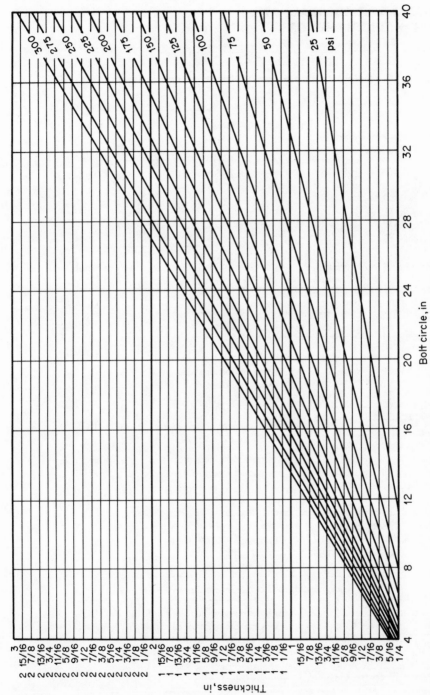

Fig. 2.11 Thickness chart for bolted flat cover plates with full-face gasket, C = 0.25.

CONICAL, TORICONICAL, AND HEMISPHERICAL HEADS

TABLE 2.10   Calculations for conical and toriconical heads

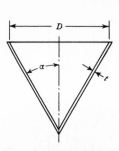

Conical head (no knuckle)

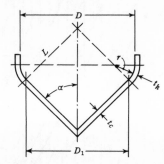

Toriconical head (with knuckle)

Equations	Remarks	Code reference
Conical heads under internal pressure		
$t = \dfrac{PD}{2 \cos \alpha \, (SE - 0.6P)}$	Equations are used when one-half the apex angle (Angle α) does not exceed $30°$	Par. UG-32(g)
$P = \dfrac{2SEt \cos \alpha}{D + 1.2t \cos \alpha}$	Use toriconical head when Angle α exceeds $30°$	Par. UG-32(h)
in which: t = minimum thickness, in P = allowable pressure, psi D = inside diameter, in S = allowable stress, psi E = minimum joint efficiency, percent $\angle \alpha$ = one-half the apex angle, of cone deg	A compression ring shall be provided when required Excess thickness, less corrosion allowance of cones or shells, may be credited to the required area of a compression ring	Appendix 1-8 Table 1-8.1
Toriconical heads under internal pressure		
Cone thickness: $t_c = \dfrac{PD_1}{2 \cos \alpha (SE - 0.6P)}$ $P = \dfrac{2 SEt \cos \alpha}{D_1 + 1.2t \cos \alpha}$	Inside knuckle radius shall not be less than 6 percent of outside diameter of head skirt nor less than 3 times thickness of knuckle	Par. UG-32(h) Appendix 1-4

TABLE 2.10 Calculations for conical and toriconical heads (Cont.)

Equations	Remarks	Code reference
Toriconical heads under internal pressure		
in which: D_1 = inside cone diameter at point of tangency to knuckle, in	Value of factor M Required skirt length	Table 1-4.2 Par. UG-32(l) and UG-32 (m)
Knuckle thickness: $$t_k = \frac{PLM}{2SE - 0.2P}$$ in which: $$L = \frac{D_1}{2 \cos \alpha}$$ M = factor depending on head proportion, L/r		Par. UG-32

TABLE 2.11   Calculations for hemispherical heads

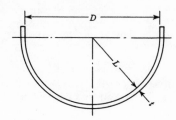

Equations	Remarks	Code reference
Heads under internal pressure		
$t = \dfrac{PL}{2\,SE - 0.2P}$ or	Equations are used when t is less than $0.356L$ or P is less than $0.655\,SE$	Par. UG-32(f)
$P = \dfrac{2\,SEt}{L + 0.2t}$	Thick hemispherical heads (t exceeds $0.356\,L$ or p exceeds $0.665\,SE$)	Appendix 1-3
in which: t = minimum thickness, in P = allowable pressure psi L = inside radius, in S = allowable stress, psi E = minimum joint efficiency, percent	Stress values of materials (use appropriate table)	Table UCS-23 Table UCI-23 Table UHA-23 Table UNF-23 Table UCD-23 Table UHT-23
	Required skirt length and thickness	Table ULT-23 Par. UG-32(L)
Heads under external pressure		
	Thickness determined in same manner as for a spherical shell	Par. UG-33(c) Par. UG-28(d)
	Example of Code calculations	Appendix L-6

BRACED AND STAYED SURFACES

TABLE 2.12 Calculations for braced and stayed surfaces

Equations	Remarks	Code reference
$t = p\sqrt{\dfrac{P}{SC}}$ $P = \dfrac{t^2 SC}{p^2}$	Measurement of pitch p and values for C	Par. UG-47(a)
	Minimum thickness of plate that is stayed is $\frac{5}{16}$ in except in welded construction and cylindrical or spherical outer shell plates	Par. UG-47(b) Par. UW-19
in which: t = minimum plate thickness, in P = allowable pressure psi p = maximum pitch, in S = allowable stress, psi C = a factor depending upon the plate thickness and type of stay	If stayed jackets extend completely around a cylindrical shell or spherical vessel or completely cover a formed head, the required thickness must also meet the requirements for cylindrical shells (Code Par. UG-27) and formed heads (Code Par. UG-32)	Par. UG-47(c)
	If two plates are stayed and only one requires staying, the value of C is governed by the thickness of plate requiring staying	Par. UG-47(e)
	Welded stayed construction	Par. UW-19
	Maximum pitch for welded and screwed stays with the ends riveted over is $8\frac{1}{2}$ in; for welded-in staybolts, the pitch must not exceed 15 times the diameter of the staybolt	Par. UG-47(g)
	End requirements, location, and dimensions of staybolts	Par. UG-48 to -50
	Hole sizes for screw stays	Par. UG-83
	For allowable pressures and thicknesses, see Tables 2.13 and 2.14	
	Dimpled or embossed assemblies	Appendix 14
	When Appendix 14 is used, state appendix identification and first paragraph number on the data report	

ALLOWABLE PRESSURE, PITCH, AND THICKNESS

Table 2.13 gives quick thickness, pressure, and pitch requirements for stayed surfaces of stayed vessels built according to the *ASME Pressure Vessel Code*. To use the table, do the following: (1) Locate the desired pressure block in the table, (2) read up vertically to the appropriate pitch, and (3) read horizontally to the required thicknesses.

Table 2.13 is based on an allowable stress of 16,000 psi. For pressures using other stress values, divide the table pressure by the appropriate constant for the stress desired:

Stress (psi)	11,300	12,500	13,800	15,000	16,300	17,500	18,800
Constant	1.416	1.28	1.159	1.066	0.981	0.914	0.851

Table 2.14 is based on an allowable stress of 13,800 psi. For pressures using other stress values, divide the table pressure by the appropriate constant for the stress desired:

Stress (psi)	11,300	12,500	15,000	16,300	17,500	18,800
Constant	1.221	1.104	0.920	0.846	0.788	0.734

TABLE 2.13 Allowable pressure, pitch, and thickness for braced and stayed surfaces

Table based on allowable stress of 16,000 psi

Thickness in inches	5	5½	6	6½	7	7½	8	8½	9	9½	10	10½	11	11½	12	12½	13	13½	14	14½	15	Thickness in inches
³⁄₁₆	47	39	33	28	24	21	18	16	...	...	...	...	...	...	...	...	...	...	...	...	...	³⁄₁₆
¼	84	69	58	49	43	37	33	29	26	23	21	19	17	15	...	...	...	...	...	...	...	¼
⁵⁄₁₆	131	108	91	78	67	58	51	45	40	36	32	29	27	25	23	21	19	18	17	16	...	⁵⁄₁₆
⅜	189	156	131	112	96	84	74	65	58	52	47	43	39	35	32	30	28	26	24	22	21	⅜
⁷⁄₁₆	257	212	179	152	131	114	100	89	79	71	64	58	53	48	44	41	38	35	32	30	28	⁷⁄₁₆
½	352	290	244	208	180	156	138	122	109	98	88	80	73	66	61	56	52	48	45	41	39	½
⁹⁄₁₆	445	368	309	264	227	198	174	154	137	123	111	101	92	84	77	71	66	61	57	53	49	⁹⁄₁₆
⅝	550	454	382	325	281	244	215	190	170	152	138	125	114	104	95	88	81	75	70	65	61	⅝
¹¹⁄₁₆	666	549	462	394	340	296	260	230	205	180	166	151	138	126	116	106	98	91	85	79	74	¹¹⁄₁₆
¾	792	653	550	469	404	352	309	274	244	219	198	180	164	150	138	127	117	109	101	94	88	¾

Pitch of stays in inches*

Pressure pounds per square inch†

* Maximum pitch shall be 8½ in except for welded-in staybolts; the pitch must not exceed 15 times the diameter of the staybolt. (See Code Par. UG-47.)

† Welded stays as shown on Code Fig. UW-19.2 ASME *Pressure Vessel Code* limited to 300 psi pressure. (See Code Par. UW-19.)

$$t = p\sqrt{\frac{P}{SC}} \qquad P = \frac{t^2SC}{p^2}$$

t = minimum thickness of plate
P = maximum allowable working pressure
p = maximum pitch between centers of stays
S = maximum allowable stress, 16,000 psi
C = 2.1 for plates not over ⁷⁄₁₆ in
C = 2.2 for plates more than ⁷⁄₁₆ in

TABLE 2.14 Allowable pressure, pitch, and thickness for braced and stayed surfaces

Table based on an allowable stress of 13,800 psi; C = 2.1 for plates 7/16 in thick, and less, and 2.2 for thicker plates (see equations in Table 2.12)

Thickness, in	Pitch of stays, in*																				
	5	5½	6	6½	7	7½	8	8½	9	9½	10	10½	11	11½	12	12½	13	13½	14	14½	15
	Allowable pressure, psi†																				
3/16	40	36	28	24	20	18	15														
1/4	72	60	50	42	36	32	28	25	22	20	18	16									
5/16	112	94	79	66	57	50	44	39	34	31	28	25	23	21	19	18	16	15			
3/8	162	133	112	96	83	72	63	56	50	45	40	36	33	30	28	26	24	22	20	19	18
7/16	220	182	153	131	112	98	86	76	68	61	55	50	45	41	38	35	32	30	28	26	24
1/2	302	250	210	179	154	134	118	104	93	83	75	68	62	57	52	48	44	41	38	35	33
9/16	382	313	265	225	194	170	148	131	118	105	95	86	78	72	66	61	56	52	48	45	42
5/8	472	390	327	279	240	210	184	162	145	130	118	106	97	89	82	76	70	65	60	56	52
11/16	572	473	400	340	294	255	225	198	178	159	144	129	118	109	100	92	85	79	73	68	64
3/4	681	568	473	403	347	302	265	235	210	188	171	153	141	128	118	109	100	93	87	81	75

* Maximum pitch shall be 8½ in except for welded-in staybolts, for which the pitch must not exceed 15 times the diameter of the staybolt (Code Par. UG-47).

† Pressures on welded stays shown in Code Fig. UW-19.2 limited to 300 psi (Code Par. UW-19).

TABLE 2.15   Jacketed vessels

Remarks	Code reference
Minimum requirements	Appendix 9
Any combination of pressures and vacuum in vessel and jacket producing a total pressure over 15 psi on the inner wall should be made to Code requirement	Appendix 9.1
Types of jacketed vessels	Appendix 9.2
Partial jackets	Figs. 9.2 to 9.7 Appendix 9.7
Dimpled jackets; indicate Appendix 14 identification and paragraph number on data report.	Appendix 14
Embossed assemblies	Appendix 14 Par. UW-19 Par. UG-101
Proof test of dimpled and embossed assemblies, also partial jackets	Par. UG-101 Appendix 14-10.5
Design of jackets	Appendix 9.4
Types of jackets	Appendix 9.2
Design of openings through jacket space	Appendix 9.6 Fig. 9-6 Par. UG-36 to -45
Braced and stayed surfaces	Appendix 9.4 Par. UG-47 Par. UW-19
Inspection openings need not be over 2 in	Appendix 9.4 Par. UG-46
Test pressure glass-lined vessels	Par UG-99

PLATE THICKNESS

TABLE 2.16   Minimum thickness of plate

The minimum thickness of plate after forming and without allowances for corrosion shall not be less than the minimum thickness allowed for the type of material being used (see Code Par. UG-16, UG-25, and UCS-25).

Material	Minimum thickness	Code reference
Carbon and low-alloy steel	$\frac{3}{32}$ in for shells and heads used for compressed air, steam, and water service	Par. UG-16 Par. UCS-25
	Minimum thickness of shells and heads after forming shall be $\frac{1}{16}$ in plus corrosion allowance	Par. UG-16
	$\frac{1}{4}$ in for unfired steam boilers	

TABLE 2.16   Minimum thickness of plate (Cont.)

Material	Minimum thickness	Code reference
Heat-treated steel	¼ in for heat-treated steel	Par. UHT-16
Clad vessels	Same as for carbon and low-alloy steel based on total thickness for clad construction and the base plate thickness for applied-lining construction	Par. UCL-20
High-alloy steel	$\frac{3}{32}$ in for corrosive service $\frac{1}{16}$ in for noncorrosive service	Par. UHA-20
Nonferrous materials	$\frac{1}{16}$ in the welded construction in noncorrosive service $\frac{3}{32}$ in for welded construction in corrosive service	Par. UNF-16
	Single-welded butt joint without use of backing strip may be used only for circumferential joints not over 24 in outside diameter and material not over $\frac{5}{8}$ in thick	Table UW-12
Low-temperature vessels: 5, 8, 9 percent nickel steel	$\frac{3}{16}$ in	Par. ULT-16
Cast-iron dual metal cylinders	5 in (Max. diameter: 36 in)	Par. UCL-29

PIPING AND THREADED CONNECTIONS

TABLE 2.17   Code jurisdiction over piping

Remarks	Code reference
Code jurisdiction over external piping connected to the vessel terminates at the following junctures: 1. The first circumferential joint for welding end connections 2. The face of the first flange in bolted flange connections 3. The first threaded joint 4. The first sealings surface for proprietary connections or fittings	Par. U-1(c)

TABLE 2.18   Threaded connections

Remarks	Code reference
Maximum size is 3 in if pressure is over 125 psi, except for plug closures used for inspection openings, end closures, and the like	Par. UG-43(e) Table UG-43 Par UW-16(g) Fig. UW-16.1 and UW-16.2

OPENINGS AND REINFORCED OPENINGS

TABLE 2.19   Openings and reinforced openings

Remarks	Code reference
For vessels under internal pressure, reinforced area required, A, equals $d \times t_r$ (see Fig. 2.12 for simplified calculations)	Par. UG-37(b)
For vessels under external pressure, reinforced area required need be only one-half that for internal pressure if shell thickness satisfies external-pressure requirements	Par. UG-37(c)
Limits of reinforcement	Par. UG-40
Strength of reinforcement	Par. UG-41
Reinforcement of multiple openings	Par. UG-42
Welded connections	Par. UW-16(c) and UW-16(e) Appendix L-7
Openings in or adjacent to welds	Par. UW-14
Large openings in cylindrical shells	Par. UA-7
Openings in flat heads	Par. UG-39
Single openings and not requiring reinforcement 3-in pipe size, vessel thickness ⅜ in or less	Par. UG-36
2-in pipe size, vessel thickness over ⅜ in	Par. UG-36
Openings in head or girth seam requiring radiography	Par. UW-14 and UG-37
Flued openings in formed heads	Par. UG-38 and Fig. UG-38

SIMPLIFIED CALCULATIONS FOR REINFORCEMENT OF OPENINGS (FIG. 2.12)

To determine whether an opening is adequately reinforced, it is first necessary to determine whether the areas of reinforcement available will be sufficient without the use of a pad. Figure 2.12 includes a schematic diagram of an opening showing the involved areas and also all the required calculations. The total cross-sectional area of reinforcement required (in square inches) is indicated by the letter A, which is equal to the diameter (plus corrosion allowance) times the required thickness. The area of reinforcement available without a pad includes:

1. The area of excess thickness in the shell or head, A_1
2. The area of excess thickness in the nozzle wall, A_2
3. The area available in the nozzle projecting inward, A_3
4. The cross-sectional area of welds, A_4

If $A_1 + A_2 + A_3 + A_4 \geqslant A$, the opening is adequately reinforced.
If $A_1 + A_2 + A_3 + A_4 < A$, a pad is needed.

If the reinforcement is found to be inadequate, then the area of pad needed (A_5) may be calculated as follows:

$$A_5 = A - (A_1 + A_2 + A_3 + A_4)$$

If a pad is used, the factor $(2.5t_n)$ in the equation for A_2 in Fig. 2.12 is measured from the top surface of the pad and therefore becomes $(2.5t_n + T_p)$. The area A_2 must be recalculated on this basis and the smaller value again used. Then:

If $A_1 + A_2 + A_3 + A_4 + A_5 = A$, the opening is adequately reinforced.

Other symbols for the area equations in Fig. 2.12 (all values except E_1, and F are in inches) are as follows:

d = diameter in the plane under consideration of the finished opening in its corroded condition

t = nominal thickness of shell or head, less corrosion allowance

t_r = required thickness of shell or head as defined in Code Par. UG-37

t_{rn} = required thickness of a seamless nozzle wall

T_p = thickness of reinforcement pad

W_p = width of reinforcement pad

t_n = nominal thickness of nozzle wall, less corrosion allowance

W_1 = cross-sectional area of weld

W_2 = cross-sectional area of weld

E_1 = 1, when an opening is in the plate or when the opening passes through a circumferential point in a shell or cone (exclusive of head-to-shell joints); or

E_1 = the joint efficiency obtained from Code Table UW-12 when any part of the opening passes through any other welded joint

$F =$ a correction factor which compensates for the variation in pressure stresses on different planes with respect to the axis of a vessel. A value of 1.00 is used for F in all cases except when the opening is integrally reinforced. If integrally reinforced, see Code Fig. UG-37.*

$h =$ distance nozzle projects beyond the inner surface of the vessel wall, before corrosion allowance is added.

To correct for corrosion, deduct the specified allowance from shell thickness, t, and nozzle thickness, t_n, but add twice its value to diameter of opening, d. [*Note:* Total load to be carried by attachment welds may be calculated from the equation, $W = S(A - A_1)$, in which S equals the allowable stress specified by Code Subsection C and Code Par. UW-15(b). For large openings in cylindrical shells, see Code Appendix 1-7. For openings in flat heads, see Code Par. UG-39. For weld sizes required, see Code Par. UW-16 and Code Fig. UW-16.1.] On all welded vessels built under column C of Code Table UW-12, 80 percent of the allowable stress value must be used in design formulas and calculations.

---

* Reinforcement is considered integral when it is inherent in shell plate or nozzle. Reinforcement built up by welding is also considered integral. The installation of a reinforcement pad is not considered integral.

Shell or head data	Nozzle data†
	P = pressure, psi
	S = allowable stress, psi
	R_n = inside radius of nozzle, in
	E_1 = joint efficiency, percent
	$t_m = \dfrac{PR_n}{SE_1 - 0.6P}$
t = actual thickness of shell or head (minus corrosion)	t_n = actual thickness of nozzle (minus corrosion)
t = calculated thickness of shell or head	t_m = calculated thickness of nozzle
$E_1 t - F t_r$ = excess thickness in shell or head	$t_n - t_m$ = excess thickness in nozzle

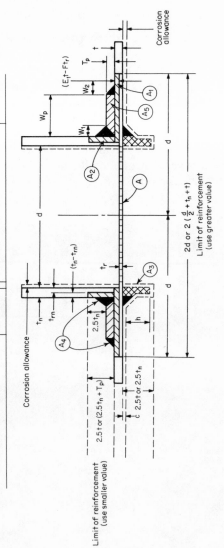

| Area of reinforcement required† | $A = d t_r F$ | = _____ |

| Area of excess thickness in shell or head (use greater value) | $A_1 = (E_1 t - F t_r)d$ or $A_1 = 2(E_1 t - F t_r)(t + t_n)$ | = _____ |

| Area available in nozzle projecting outward; use smaller value | $A_2 = 2(2.5 t_n)^*(t_n - t_{rn})f_r$ or $A_2 = 2(2.5t)(t_n - t_{rn})f_r$ | = _____ |

| Area available in nozzle projecting inward | $A_3 = 2(t_n - c)f_r \times h$ | = _____ |

| Cross-sectional area of welds | $A_4 = 2\left[\dfrac{(W_1)^2 + (W_2)^2}{2}\right] \times f_r$ | = _____ |

| Area of reinforcement available without pad | $(A_1 + A_2 + A_3 + A_4)$ | = _____ |

If $A_1 + A_2 + A_3 + A_4 \geq A$ Opening is adequately reinforced

If $A_1 + A_2 + A_3 + A_4 < A$ Opening is not adequately reinforced so reinforcing element must be added or thicknesses increased

| Area in pad | $A_5 = 2W_p T_p$ | = _____ |
| | Total area available | _____ |

* If reinforcing pad is used, the factor $2.5t_n$ becomes $(2.5t_n + T_p)$.

† For cases when the allowable stress of the nozzle of reinforcing element is less than the allowable stress of the vessel, refer to Code Par. UG-41(a) and Appendix L, latest Addenda, for consideration of this effect.

Fig. 2.12 Calculation sheet for reinforcement of openings.

CORROSION ALLOWANCES

TABLE 2.20   Corrosion allowances and requirements

Remarks	Code reference
General requirements	Par. UG-25
Suggested good practice	Appendix E
For carbon and low-alloy steel vessels with a required minimum thickness of less than ¼ in used for air, steam, and water, add one-sixth of calculated thickness. No corrosion allowance necessary for vessels built according to Code Table UW-12 (column C)	Par. UCS-25
For brazed vessels	Par. UB-13
For clad vessels	Par. UCL-25, UG-26 Appendix F
Vessels subject to corrosion must have suitable drain opening	Par. UG-25
Vessels subject to internal corrosion, erosion, or mechanical abrasion must have inspection openings	Par. UG-46

TABLE 2.21 Tolerances

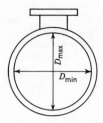

Tolerance	Remarks	Code reference
Out-of-roundness in cylindrical shells under internal pressure	$D_{max} - D_{min}$ must not exceed 1 percent of nominal diameter	Par. UG-80(a) Fig. UG-80.2
	At nozzle openings, increase above tolerance by 2 percent of inside diameter of opening	
	Tolerance must be met at all points on shell	
Out-of-roundness in cylindrical shells under external pressure	Maximum plus-or-minus deviation from true circle using chord length equal to twice the arc length obtained from Code Fig. UG-29.2 is not to exceed the value e determined by Code Fig. UG-80.1	Par. UG-80(b) Fig. UG-80.1 Fig. UG-29.2 Appendix L-4
	Take measurements on unwelded surface of plate	
	For shells with lap joint, increase tolerance by value t	Par. UG-80(b)
	Do not include corrosion allowance in value t	
Permissible variation in plate thickness	No plate shall have an under-tolerance exceeding 0.01 in or 6 percent of the ordered thickness, whichever is less	Par. UG-16
Formed heads	Inside surface must not deviate from specified shape by more than 1.25 percent of inside skirt diameter	Par. UG-81 Par. UW-13
	Minimum machined thickness is 90 percent of that required for a blank head	

TABLE 2.21   Tolerances (Cont.)

Tolerance	Remarks	Code reference
Pipe under tolerance	Manufacturing undertolerance must be taken into account	Par. UW-16
	Follow pipe specifications	Subsection C and Code Material Section II

	Remarks		
Tolerance	Plate thickness, in	Offset limits	Code reference
Longitudinal joints (Category A)	Up to ½ in inclusive Over ½ in to ¾ in Over ¾ in to 1½ in Over 1½ in to 2 in Over 2 in	¼t ⅛ in ⅛ in ⅛ in Lesser of $\frac{1}{16}t$ or ⅜ in	Par. UW-33 Table UW-33
Circumferential joints (Category B)	¾ in or less Over ¾ in to 1½ in Over 1½ in to 2 in Over 2 in	¼t $\frac{3}{16}$ in ⅛t Lesser of ⅛t or ¾ in	Par. UW-33 Table UW-33
Offset type joint	⅜ in maximun; longitudinal weld within offset area shall be ground flush with parent metal prior to offset and examined by magnetic particle method after offset		Par. UW-13.(c) Fig. UW-13.1(k) plus note.
Welding of plates of unequal thickness			Par. UW-9, UW-42 Fig. UW-9
Spherical vessels and hemispherical heads	Offset must meet longitudinal joint requirements		Par. UW-33
	Offset must be faired at a 3 to 1 taper over width of finished weld		Par. UW-42 Par. UW-9 Fig. UW-9
	Additional weld metal may be added to edge of weld if radiographed or examined by liquid penetrant or magnetic particle		Par. UW-42 Par. UW-9

TABLE 2.22 Welded joint efficiency values

Type of joint and radiography	Efficiency allowed, percent	Code reference
Double-welded butt joints (Type 1)		Par. UW-11
Fully radiographed	100	Par. UW-51, UW-35
Spot-radiographed	85	Par. UW-12, UW-52
No radiograph	70	Table UW-12
Single-welded butt joints (backing strip		Par. UW-52
left in place) (Type 2)		Par. UCS-25
Fully radiographed	90	Par. UW-51
Spot-radiographed	80	Par. UW-52
No radiograph	65	
Single-welded butt joints no backing strip (Type 3) limited to circumferential joints only, not over ⅝-in thick and not over 24-in outside diameter	60	Table UW-12
Fillet weld lap joints and single-welded butt circumferential joints		Table UW-12

RADIOGRAPHIC EXAMINATION

TABLE 2.23 Radiographic requirements

Remarks	Code reference
Full radiography	
Full radiograph all joints over 1 ½ in [required radiographing on lesser thickness]	Par. UW-11(a)2 Par. UCS-57 Table UCS-57
Circumferential butt joints of nozzles and sumps not exceeding 10 in in diameter or 1⅛ in in thickness do not require radiography	Par. UW-11(a)5
Vessels containing lethal substances must be fully radiographed	Par. UW-11(a)1 Par. UW-2
Efficiency of fully radiographed joints	Par. UW-12 Table UW-12
Radiographic acceptance standards	Par. UW-51
Carbon and low-alloy steel materials	Par. UCS-57
High-alloy steel materials	Par. UHA-33
Clad-plate materials	Par. UCL-35
Nonferrous materials	Par. UNF-57
Ferritic steels whose tensile properties have been enhanced by heat treatment	Par. UHT-57
Spot radiography	
General recommendations	Par. UW-11(b)
Spot radiographing required for higher weld joint efficiencies	Par. UW-12 Par. UW-52 Table UW-12
Minimum length of spot radiograph, 6 in	Par. UW-52(c)
If cladding is included in computing required thickness, spot radiograph is mandatory	Par. UCL-23(c)
Vessels having 50 ft or less of main seams require one spot examination; larger vessels require one spot for each 50 ft of welding	Par. UW-52(b)1
Additional spots may be selected to examine welding of each welder or welding operator	Par. UW-52(b)2
Joints welded with austenitic chromium-nickel steel	Par. UCL-36

TABLE 2.24 Hydrostatic, pneumatic, magnetic particle, and liquid penetrant tests

Remarks	Code reference
Hydrostatic tests	
Test must be at at least 1½ times the maximum design pressure multiplied by the lowest ratio of the stress value for the test temperature to that for the design temperature	Par. UG-99, footnote 28 Par. UG-21, footnote 8 Par. UA-60(e)
If the allowable stress at design temperature is less than the allowable stress at test temperature the hydrostatic test pressure must be increased proportionally. Thus	

$$\text{Test pressure} = 1.5 \times \frac{\text{allowable stress at test temperature} \times \text{design pressure}}{\text{allowable stress at design temperature}}$$

Remarks	Code reference
Combination units must be so tested that hydrostatic pressure is on one chamber without pressure in the other parts	Par. UG-21 and UG-98 Par. UA-60(b)
Include corrosion allowance in calculating test pressure	
Inspection must be made at a pressure not less than ⅔ of test pressure	
Test cast-iron vessels at 2 times the design working pressure; for design pressure under 30 psi, test at 2½ times design pressure but not to exceed 60 psi	Par. UCI-99
Test for clad-plate vessels	Par. UCL-52
Pneumatic tests	
Pneumatic test may be used instead of hydrostatic test when:	Par. UG-100
1. Vessels are so designed and supported that they cannot safely be filled with water 2. Vessels for service in which traces of testing liquid cannot be tolerated are not easily dried	
Test pressure must not be less than 1¼ times maximum allowable pressure	
All attachment welds of a throat thickness greater than ¼ in must be given a magnetic particle or penetrating oil test before pneumatic test	Par. UW-50
Special precautions should be taken when usng air or gas for testing	Par. UG-100, footnote 30

TABLE 2.24   Hydrostatic, pneumatic, magnetic particle, and liquid penetrant tests (Cont.)

Remarks	Code reference
Magnetic particle and penetrating oil tests	
When vessel is to be pneumatically tested, welds around openings and attachment welds of throat thickness greater than ¼ in must be given a magnetic particle or liquid penetrant examination for ferromagnetic materials (liquid penetrant for nonmagnetic materials)	Par. UW-50
Liquid penetrant examination must be made on all austenitic chromium-nickel alloy-steel welds in which shell thickness is over ¾ in and on all thicknesses of 36 percent nickel steel welds. Examinations shall be made after heat treatment if heat treatment is given	Par. UHA-34
Pressure part welded to a plate thicker than ½ in to form a corner joint must be examined by magnetic particle or liquid penetrant method	Par. UG-93 Par. UW-13 Fig. UW-13.2
Layered vessels	Par. ULW-56

TABLE 2.25 Postweld heat treatment

Remarks	Code reference
All carbon and low-alloy steel seams must be postweld heat-treated if nominal thickness exceeds 1 ¼ in or 1 ½ in if preheated to 200°F before welding	Par. UCS-56
Some materials must be postweld heat-treated at lower thickness	Table UCS-56
Vessels containing lethal substances must be postweld heat-treated	Par. UW-2
Unfired steam boilers	Par. UW-2
Carbon-steel vessels for service at temperature below −20°F must be postweld heat-treated unless exempted from impact test	Par. UCS-67
Welded vessels	Par. UW-10 Par. UW-40
Carbon- and low-alloy steel vessels	Par. UCS-56 Par. UCS-66 Par. UCS-67 Par. UCS-79 Par. UCS-85 Par. U-1(e)
High-alloy steel vessels	Par. UHA-32
Cast-iron vessels	Par. UCI-3
Clad-plate vessels	Par. UCL-34
Bolted flange connections	Par. UA-46
Castings	Par. UG-24
Forgings	Par. UF-31
When only part of vessel is postweld heat-treated, extent of postweld heat treatment must be stated on manufacturer's data report Form U-1	Par. UG-116
The letters HT must be stamped under the Code symbol when the entire vessel has been postweld heat-treated	Par. UG-116
The letters PHT must be stamped under the Code symbol when only part of the vessel has been postweld heat-treated	Par. UG-116
Low-temperature vessels	Par. ULT-56
Ferritic steel vessels	Par. UHT-80, -81
Nonferrous vessels	Par. UNF-56
Layered vessels	Par. ULW-26
Vessels or parts subject to direct firing, when thickness exceeds ⅝ in	Par. UW-2

TABLE 2.26 Requirements for low-temperature operation (below − 20°F)

Remarks	Code reference
Some materials and welds do not require impact test; consult paragraphs in Code Subsection C to which material applies (see Table 2.1 of this chapter)	Par. UG-84 Par. UCS-66(c) Par. UHA-51
When impact tests are required weld metal and heat-affected zone must have impact tests.	Par. UCL-27 Par. UNF-65 Appendix NF(6)
Low-temperature vessels	Part ULT
Welded carbon-steel vessels must be postweld heat-treated unless exempted from impact tests	Par. UCS-66
All longitudinal seams and girth seams must be double-welded butt joints or the equivalent	Par. UCS-67
Remove backing strip on long seams if possible	Par. UW-2(b)
Vessels must be stamped according to Code requirements (see Fig. 2.15)	Par. UG-116(b)
Austenitic stainless steels may not require impact tests, providing weld procedure specifications also included qualified impact test plates of the welding procedure	Par. UG-84
All joints of Category A must be Type No. (1) of Code Table UW-12	Par.UW-2 Par. UW-3
All joints of Categories B must be Type No. (1) or No. (2) of Code Table UW-12	Par. UW-2
All joints of Category C and D must be full penetration welds	Par. UW-2
Brazed vessels	Par. UB-22
High-alloy vessels	Par. UHA-105
Nonferrous vessels	Par. UNF-65, NF-6

CRYOGENIC AND LOW-TEMPERATURE VESSELS

Use of cryogenic and low-temperature pressure vessels is increasing rapidly, but there have been many defects caused by improper materials and poorly designed vessels, particularly those meant for −20°F and lower. Many materials exposed to low temperatures lose their toughness and ductility. As the temperature goes down, the metal has a tendency to become brittle. This may cause brittle fracture at full stress.

Most materials used for service at normal temperatures have a marked decrease in impact resistance at temperatures below −20°F. These materials must have the proven ability to safely resist high-stress changes and shock loads. Only notch tough materials able to withstand unavoidable stress concentrations should be used for cryogenic vessels. All sharp changes of sections, corners, and notches should be eliminated in design of the vessel and fabrication of its parts.

Manufacturers, purchasers, and users of low-temperature pressure vessels often misunderstand certain paragraphs in the *ASME Pressure Vessel Code*. Perhaps it might be helpful to clarify various paragraphs applying to low-temperature vessels and to indicate when impact tests are required on cryogenic vessels.

If such tests are necessary, each part of the vessel must be considered: shells, heads, nozzles, and other parts. Therefore, to impact test or not can be a large

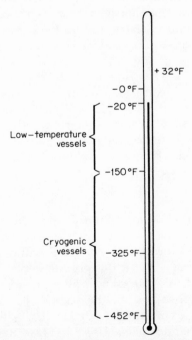

Fig. 2.13   Temperature range for low-temperature and cryogenic vessels.

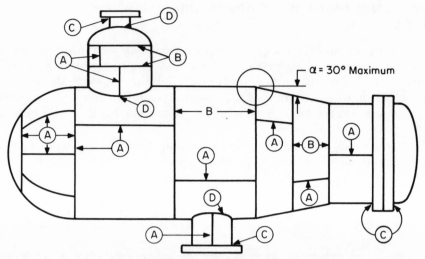

Fig. 2.14 Illustration defining locations of welded joint categories. *A. Longitudinal joints; circumferential joints connecting hemispherical heads to main shell; any welded joint in a sphere. B. Circumferential joints; circumferential joints of torispherical ellipsoidal heads; angle joints not greater than 30 degrees. C. Flange joints; Van Stone laps; tube sheets; flat heads to main shell. D. Nozzle joints to main shell; heads; spheres; flat-sided vessels.*

cost factor.   The important thing to remember is that vessels operating below −20°F demand special attention in design and materials.

There is one major difference between these and other vessels: Low-temperature vessels have been impact-tested and postweld heat-treated or have been designed with 2½ times a given coincident pressure-temperature service condition below −20°F.

Specific paragraphs (Code Par.) in the *ASME Pressure Vessel Code* prescribe the type of welded joints involved in low-temperature operations.   These joints must meet the requirements of Code Par. UW-2, "Service Restrictions."   Code Par. UW-3, "Welded Joint Category," determines the location of the joint.

If the material is carbon steel, Code Par. UCS-66 and UCS-67 state whether impact tests are required.   If tests are required, Code Par. UG-84 describes how they must be made.   Code Par. UG-84 also instructs on size of specimens and on location of areas in the specimens, taking impacts, test values, and properties.

However, impact testing is not a "must" for welds or materials if certain design requirements are met.   For example, a vessel designed for 100 psi at 100°F could not be used for 100 psi at −50°F without impact testing of the vessel's welds and material.   Yet if this product were designed for 250 psi at 100°F, it could be used for 100 psi at −50°F without impact testing.   The reason is that of these two vessels, the second, designed for pressure and temperature conditions according to Code Par. UCS-66, has a design pressure planned for 2½ times that needed at 100°F.   This corresponds to using ¹⁄₁₀ of the tensile strength, or 40 percent of allowable stress.

Other materials exempt from impact tests are austenitic stainless steel, with no impact tests mandatory to −325°F, and various stainless steels, such as SA-240, Type 304, Type 304L, and Type 347, with no impact tests necessary to −425°F. Code Par. UHA-51 permits these exemptions, basing them on the fact that these materials do not show a transition from ductile to brittle fracture. However, welding procedure specifications require additional testing to include impact test plates of the procedure. (See Code Par. UG-84.)

Wrought aluminum alloys do not undergo a marked drop in impact resistance below −20°F. As indicated in Code Par. UNF-65, no impact tests are required to −452°F. Other nonferrous metals for copper, copper alloys, nickel, and nickel alloys can be used to −325°F without impact tests. Code Par. UNF-65 will indicate the need for impact testing.

The major concern for metals at low temperature is brittle fracture, which can cause vessel failure. At normal temperatures, metals usually give warning by bulging, stretching, leaking, or other failure. Under low-temperature conditions, some normally ductile metals may fail at low-stress levels without warning of plastic deformation.

Four conditions must be controlled to avoid brittle fractures in low-temperature vessels:

1. Material should not show a transition from ductile to brittle fracture.
2. Welding materials for low-temperature service must be selected carefully, and a proven welding procedure must be used.
3. Stress raisers from design or fabrication should not be permitted.
4. Localized yielding in the stress raiser area must not be allowed.

Pressure vessel users should use a material that behaves in a ductile manner at all operating temperatures. This will allow for the possible stress raiser that may be overlooked in fabrication or design. Impact tests could determine the correct materials and welding procedures for these vessels.

SERVICE RESTRICTIONS AND WELDED JOINT CATEGORY

Many designing engineers and pressure vessel manufacturers have been using the wrong type of welded joints in vessels which must meet the requirements in Code Par. UW-2, "Service Restrictions," and Par. UW-3, "Welded Joint Category," of the *ASME Pressure Vessel Code,* Section VIII, Division 1.

When the paragraph on welded joint category was added to the Code, it was thought that the intent of Code Par. UW-2, "Service Restriction," was clarified, but evidently it has not been clear enough to some pressure vessel fabricators. The requirements of Code Par. UW-2 have often been misinterpreted and at times not referred to until too late. For example, if a vessel designed for −140°F and requiring impact tests had flanges, they would have to be Welded Joint Category C and would require full penetration welds extending through the entire section of the joint. Also, all longitudinal seams would have to be Category A, without backing strips.

We have reviewed drawings of vessels designed for cryogenic service requiring impact tests, where the design on the flanges did not call for full penetration welds and the longitudinal seams had metal backing strips which were to remain in place. This, of course, does not meet the intent of the Code. It is important that designing engineers and pressure vessel manufacturers have a clear understanding of the paragraphs in the Code on service restrictions and welded joint category and that the vessels, where required, receive the proper attention during design and in handling in the shop.

What are the vessels which demand special requirements listed in Code Par. UW-2, "Service Restrictions"? They are vessels intended for lethal service; vessels that are to operate below $-20°F$; unfired steam boilers with design pressures exceeding 50 psi; pressure vessels subject to direct firing; and vessels with special requirements based on material, thickness, and service.

What is it you must remember about the welded joint category? Only those joints to which special requirements apply are included in the categories. The term *category* defines only the location of the welded joint in the vessel and never implies the type of welded joint that is required. The type of joint is prescribed by other paragraphs in the Code suitable to the design, which may specify special requirements regarding joint type and degree of inspection for certain welded joints.

The purpose of this review is to spell out the requirements of the Code with regard to service restrictions and welded joint category. Designers and manufacturers should examine their practices to ensure that the service restrictions paragraph of the Code is receiving appropriate attention in the engineering department and the shop, for they must accept full responsibility for the design and construction of a Code pressure vessel when the vessel is completed and the data sheets are to be signed.

The ASME Code Form U-1 "Manufacturers Data Report," in the section for remarks, requires a brief description of the purpose of the vessel and also requires that the contents of each vessel be indicated. It will be the Code inspector's duty to determine whether this item is accurately completed and whether service restrictions were taken into consideration.

Therefore, in designing Code vessels, be sure you have taken the paragraphs on service restrictions and welded joint category into consideration. Also be sure the Authorized Code Inspector has reviewed the design before fabrication starts so that no difficulties will arise when the vessel is completed.

TABLE 2.27   Vessels exempt from inspection

Types and sizes exempted	Remarks	Code reference
Vessels not within scope of the Code		
Vessels with a nominal capacity of 120 gal or less containing water (or water cushioned with air)	Even though exempted from Code, any vessel meeting all Code requirements may be stamped with the Code symbol	Par. U-1
Vessels having an internal or external operating pressure of 15 psi or less with no limitation on size		
Vessels having inside diameter, width, height, or cross section diagonal of 6 in or less with no vessel length or pressure limitation		
Vessels within scope of the Code		
Vessels not exceeding 5-cu ft volume and 250-psi design pressure	Exempted vessels must comply in all other respects with Code requirements, including stamping with UM symbol by manufacturer	Par. U-1 Par. UG-91 Par. UG-116 Fig. UG-116(b)
Vessels not exceeding 1.5-cu ft volume and 600-psi design pressure	A certificate for UM vessels must be furnished by manufacturer on Code Form U-3 when requested	Par. UG-120

PRESSURE VESSELS STAMPED WITH UM SYMBOL

Vessels stamped with the UM symbol are limited to size and design pressures in accordance with the ASME Code Section VIII, Division 1, Code Par. U-1. The manufacturer should understand that the authority given to the company by the ASME Committee to use the UM symbol is dependent on strict conformance to the ASME Code requirements and is subject to cancellation. It is given when requested following a shop review by the inspection agency and jurisdiction to holders of a Certificate of Authorization for construction of pressure vessels.

It is necessary that the UM stamp with the required data be completed as requested by Code Par. UG-116 and Code Fig. UG-116(b). A certificate for UM-stamped vessels, exempted from inspection by an Authorized Inspector, should be furnished on Code Form U-3 when required. (Some jurisdictions will not recognize the UM stamp and will require inspection by an Authorized

TABLE 2.28   Welding procedure qualification, welders' identifying symbols, and welding records

Remarks	Code reference
Manufacturer must assign an identifying number, letter, or symbol to permit identification of welds made by welder or welding operator	Par. UW-29
Manufacturer must maintain a record of welders and welding operators, showing date, test results, and identification mark assigned to each	
Records must be certified by manufacturer and made accessible to Inspector	
Welder must stamp symbol on plate at intervals of 3 ft or less on work done on steel plate ¼ in thick or more; for steel plate less than ¼ in thick or nonferrous plate less than ½ in thick, suitable stencil or other surface marking must be used	Par. UW-37
A record must be kept by manufacturer of all welders and welding operators working on a vessel and the welds made by each so that this data will be available for the Inspector	Par. UW-48
No production work can be done until both the welding procedure and the welders or welding operators have been qualified (see the section in Chap. 6 "Examination of Procedure and Operator")	Par. UW-5 Par. UW-26(c) Par. UW-28 Par. UW-29 Par. UW-47 Section IX
Procedure for clad vessels	Par. UCL-31 Par. UCL-40 to -45
Lowest permissible temperature for welding	Par. UW-30
No welding should be done when temperature of base metal is lower than 0°F	
Heat surface area to above 60°F where a weld is to be started when temperature is between 0 and 32°F	

Inspector.)   The responsibility of the manufacturer is to design, fabricate, inspect, and test all UM vessels to see that they meet Code requirements, with the exception of inspection by an Authorized Inspector.

STAMPING REQUIREMENTS (FIG. 2.15)

If the markings required are stamped directly on the vessel, the minimum letter size is ⁵/₁₆ in.   If they appear on a nameplate, the minimum letter size is ⁵/₃₂ in.   If a nameplate is used, only the Code symbol and the manufacturer's serial

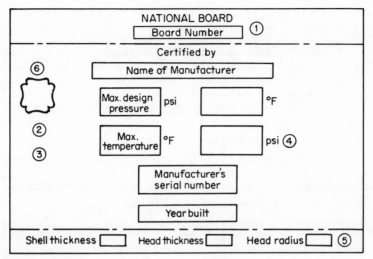

Fig. 2.15 Stamping form
Key: *1. Stamp* National Board *and the National Board number at top of form if National Board inspection is required.*

2. Stamp Part *under Code symbol on those parts of a vessel for which partial data sheets are required.*

3. Stamp type of construction and radiographic and postweld heat treatment status of main longitudinal and circumferential joints under Code symbol by means of the following letters:

W — arc or gas welded

S — seamless

RES — resistance welded

F — forged welded

B — brazed

RT1 — completely radiographed

RT2 — partially radiograhed

RT3 — spot-radiographed

RT4 — part of vessel radiographed where none of RT1, 2 or 3 apply

HT — completely postweld heat-treated

PHT — partially postweld heat-treated

Vessels intended for special service must also be appropriately stamped separated by a hyphen after lettering of 3. above.

L — Lethal Service

UB — Unfired Boiler

D — Direct Firing

4. Maximum temperature seldom required if less than 650°F (Code Par. UG-116, footnote 31). For vessels operating below − 20°F, stamping must also give minimum allowable temperature and the minimum and maximum temperatures supplied. If the material has passed the required impact tests, the letters IT *should be placed after the figures for minimum temperature.*

When in accordance with Code Par. UCS-66 or UHA-51 no impact test is made, stamp Maximum Allowable Pressure *for coincident temperature of − 20°F and above. ' For the operating pressure at coincident temperature below − 20°F, such marking is followed by stamping UCS-66 or UHA-51.*

5. Pennsylvania requires shell and head thickness and head radius.

6. Stamp User *above Code symbol if vessel is inspected by the user's inspector.*

number are required to be stamped; the other information may be applied by another method, for example, etching.

The stamping should be located in a conspicuous place, preferably near a manhole or handhole opening.

Removable pressure parts should be marked to identify them with the vessel. For vessels having two or more pressure chambers, see Code Par. UG-116.

Calculation Sheet

CUSTOMER'S NAME						ORDER NO.		
Design pressure		Material specification	Stress	Radiography		Postweld heat treat	Joint efficiency	
Design temperature		Shell		None		Yes	percent	
Corrosion allowance		Heads		Spot		No	percent	
Inside diameter		Bolts		Complete				
Head radius		Nuts						

Shell thickness

$$t = \frac{PR}{SE - 0.6P} =$$

SE	
0.6P	
SE − 0.6P	Code Par. UG-27

+ _____ Corrosion allowance
_____ Thickness required
_____ Thickness used

Torispherical head thickness

$$t = \frac{0.885PL}{SE - 0.1P} =$$

SE	
0.1P	
SE − 0.1P	Code Par. UG-32

+ _____ Corrosion allowance
_____ Thickness required after forming
_____ Thickness used

Ellipsoidal head thickness

$$t = \frac{PD}{2SE - 0.2P} =$$

2SE	
0.2P	
2SE − 0.2P	Code Par. UG-32

+ _____ Corrosion allowance
_____ Thickness required after forming
_____ Thickness used

Flat head thickness

$$t = d\sqrt{\frac{CP}{S}} = \sqrt{}$$

Code Par. UG-34

+ _____ Corrosion allowance
_____ Thickness required
_____ Thickness used

Maximum Allowable Working Pressure for New and Cold Vessels (see Code Par. UG-99)

Shell

R	
0.6t	
R + 0.6t	

$$P = \frac{SEt}{R + 0.6t} = = \text{ psi}$$

Torispherical head

0.885L	
0.1t	
0.885L + 0.1t	

$$P = \frac{SEt}{0.885L + 0.1t} = = \text{ psi}$$

Ellipsoidal head

D	
0.2t	
D + 0.2t	

$$P = \frac{2SEt}{D + 0.2t} = = \text{ psi}$$

Flat head

$$P = \frac{St^2}{d^2C} = = \text{ psi}$$

Maximum allowable working pressure = _____ psi (limited by_____)

Hydrostatic test pressure = _____ X _____ psi = _____ psi

Fig. 2.16 Calculation sheet for ASME pressure vessels.

TABLE 2.29   Vessels of noncircular cross section

Remarks	Code reference
1. Scope:	Appendix 13-1
Types of vessels, descriptions, and figures	Appendix 13-2
2. Design requirements and stress acceptance criterion	Appendix 13-4
3. Thermal expansion reactions and loads due to attachments	Appendix 13-4
4. Nomenclature, and terms used in design formulas	Appendix 13-5 Par. UG-47 Par. UG-34 Par. UG-101
5. End closures	Appendix 13-4
a. If ends are unstayed, flat plate design closure per UG-34 except a C factor of no less than 0.20 shall be used in all cases	
b. All other end closures are special design per U-2, or they may be subject to proof test per UG-101	Par. U-2 Par. UG-101
Braced and stayed surfaces may be designed per Code Par. UG-47	Par. UG-47 Par. UG-48 Par. UG-49 Par. UW-19 Appendix 13-4
6. Special consideration needs to be given to vessels of aspect ratio less than 4	Appendix 13-4
7. Vessels are designed for membrane and bending stress throughout the structure (for both the short-sided and long-sided plates). Membrane, bending, and total stresses are considered. Formulas for the stresses are given as follows:	Appendix 13-4
a. Vessels of rectangular cross section (nonreinforced)	Appendix 13-7
b. Reinforced vessels of rectangular cross section	Appendix 13-8
c. Stayed vessels of rectangular cross section	Appendix 13-9
d. Vessels having an obround cross section	Appendix 13-10
e. Reinforced vessels of obround cross section	Appendix 13-11
f. Stayed vessels of obround cross section	Appendix 13-12
g. Vessels of circular cross section having a single diametral staying member	Appendix 13-13
h. Vessels of noncircular cross section subject to external pressure	Appendix 13-14
Vessel fabrication	Appendix 13-15
8. Examples illustrating application of formulas	Appendix 13-17
For corner joints	Par. UW-13

THREE

Design for Safety

BASIC CAUSES OF PRESSURE VESSEL ACCIDENTS

Two new vessels built for equal pressures and temperatures and the same type of service can fare very differently in different locations. In one plant, capable maintenance personnel will keep the vessel in top shape by means of a systematic schedule of inspection and maintenance. In another, an equal number of personnel, through ignorance, incompetence, or neglect, can bring about the failure of the vessel in a comparatively short time. Competent personnel are obviously a key factor in the longevity and safety of a pressure vessel.

CORROSION FAILURES

Most pressure vessels are subject to deterioration by corrosion or erosion, or both. The effect of corrosion may be pitting or grooving over either localized or large areas. Corrosion over large areas can bring about a general reduction of the plate thickness.

Some of the many causes of corrosion are inherent in the process being used, but others are preventable. Vessels containing corrosive substances, for example, deteriorate more rapidly if the contents are kept in motion by stirring paddles or by the circulation of steam or air. If the corrosion is local, it may not appear serious at first; as the metal thickness is reduced, however, the metal becomes more highly stressed, even under normal pressures. The more stressed the metal becomes, the faster it corrodes, at last reaching the point of sudden rupture. Corrosion in unstressed metal is less likely to cause such sudden failure.

STRESS FAILURES

Many plate materials are not at all suitable for the service required of them. Other materials may be serviceable in an unstressed condition but will fail rapidly in a stressed condition. To give an example, a large chemical company once used stainless steel corrosion buttons to measure corrosion in a large fired cast-steel vessel. When the cast-steel vessel had to be replaced, it ordered a new vessel made of the same stainless steel used for the buttons because of its proved resistance. After a few days of service, however, the operator noticed leakage through the 1-in shell, immediately reduced the pressure, and shut down the vessel. Examination showed many cracks in the stainless steel. The substance being processed was highly corrosive to the grain boundaries of the metal when under stress. (See Code Par. UHA-103.) This fact was verified when the metal was tested in both stressed and unstressed conditions in the same solution. The stressed pieces failed within 100 hours.

When metals are subjected to repeated stresses exceeding their limits, they ultimately fail from "fatigue." It may take thousands of stress applications to produce failure, but a vessel such as an air tank in which there are rapid pressure fluctuations may fail at the knuckle radius. In a large laundry, the writers once saw grooving that penetrated halfway into a ⅝-in head in the knuckle radius of a hot water tank.

Excessive stresses can develop in vessels that are not free to expand and contract with temperature changes. If these stresses are allowed to recur too frequently, eventual failure from fatigue can be expected.

OTHER FAILURES

Vessels are often installed in positions that prevent entry through the inspection openings for examination of interior surfaces. Any owner or user of a vessel who is interested in its actual condition would never let this happen.

If a vessel is supported on concrete saddles, the difference in the coefficients of expansion of the concrete and steel will sometimes cause the concrete to crack just beneath the tank, allowing accumulations of moisture and dust that will cause corrosion.

Safety valves or devices that become plugged or otherwise inoperative can lead to accidents. Safety valves may also be adjusted too high for the allowable pressure so that the vessel fails if overpressure occurs.

Defective material not visible at the time of fabrication, like laminations in the plate or castings with internal defects, may bring on failure. Inferior workmanship at the time of fabrication — such as poor joint fit-up or bad welding that may appear satisfactory on the surface but suffers from lack of penetration, slag, excessive porosity, or lack of ductility — can also cause serious trouble.

It requires consistent effort on the part of plant personnel as well as a well-organized system of maintenance and inspection to safeguard pressure vessel equipment properly.

DESIGN PRECAUTIONS

The reliable functioning of a pressure vessel depends upon the foresight of the user and the information the user gives to the designer and fabricator. If the process involved is corrosive, for example, these data must include the chemicals to be used and, if possible, the expected rate of corrosion based on a history of vessels in similar service. (Corrosion resistance data are available for most materials and chemicals, but actual operating conditions can seldom be duplicated in the laboratory and must be determined by frequent, close examination after the vessel is placed in service.)

Usually no one but the user is fully familiar with the operating conditions and the causes of defects and deterioration in the type of vessel being ordered. If the designer is presented with all the facts, however, the materials best suited for the job will be used and all openings placed in the most desirable locations.

Although the Code permits many types of construction, some of them may not be suitable for the service required. Therefore, merely to prescribe an ASME vessel is not always sufficient. For example, the Code permits small openings to be welded from one side only and also provides for offset construction of girth seams and heads. [See Code Par. UW-13 also Fig. UW-13.1(K).] This type of construction may not be suitable for some vessels, however, because it leaves some inside seams and fittings unwelded and thus subject to corrosion. The purchaser aware of such facts will be specific in ordering and state the type of welded joints desired. The drawings will also be checked before fabrication to make sure that the openings are where desired and that the inspection openings permit easy access to the interior.

The Code Inspector must also check the design of the vessel to be assured that the shell, head, and nozzle thickness are correct, the knuckle radius sufficient, the openings properly reinforced, the flange thickness sufficient, and the like, but there is no concern with the position of the outlets so long as they comply with the Code.

Code Par. UG-22 and Appendix G state that loadings must be considered when designing a vessel, but do not give clear information as to how to take them into account. When vessels and piping systems operate out-of-doors, wind must be considered. Also to be taken into account are external loadings such as snow; the minimum ambient temperature; the reaction of piping systems; the restricted thermal expansion on their attachment to the vessel and on its hangers or anchors; the weight of valves; and pumps, blowers, and rotating machinery. It is imperative for the designer to see that all machinery, piping, and equipment is properly balanced and supported. Nonmandatory Code Appendixes D and G also give more useful information to be taken into consideration.

Internal corrosion usually develops on surfaces susceptible to moisture. For this reason, it is important to make sure that the vessel can be properly drained.

In vertical vessels equipped with agitating devices, the greatest wear usually occurs on the bottom head. In horizontal vessels with agitators, the most wear will be between the six o'clock and nine o'clock positions for clockwise rotation and between the six o'clock and three o'clock positions for counterclockwise rotation. Horizontal vessels with an agitator should therefore be designed for turning, if possible, so that the thinned-out area can be rotated to the top half. The useful life of the vessel will thus be extended.

Vessels to be subjected to heavy corrosion, erosion, or mechanical abrasion require an extra corrosion allowance and telltale holes. Leakage from these holes will give a warning that the vessel has corroded to its minimum allowable thickness. For example, if the minimum thickness required is ⅜ in but the plate thickness used (to increase the length of service) is ⅞ in, the corrosion allowance would thus be ½ in. Telltale holes placed on about 18-in centers would then be drilled ⅜ in in depth from the surface not affected by corrosion.

Code Par. UG-25(e) states: "When telltale holes are drilled, they shall be ¹⁄₁₆ to ³⁄₁₆ in in diameter and shall be drilled to a depth not less than 80 percent of the thickness required for a seamless shell of like dimensions. These holes shall be drilled into the surface opposite to that at which deterioration is expected." For clad or lined vessels, see Code Par. UCL-25(b).

INSPECTION OPENINGS

Accessibility to the inside of a vessel is essential. Carrying out regular internal inspections is of importance to the safety and extended life of a vessel. Code Par. UG-46 requires inspection openings in all vessels subject to internal corrosion or having internal surfaces subject to erosion or mechanical abrasion so that these surfaces may be examined for defects. Actually, almost all pressure vessels are subject to some kind of interior corrosion, and its serious-ness can be determined only by examining the inside of the vessel. The alternative is possible failure and consequent injury to or even loss of human life.

Minimum sizes for the various types of openings have been established by Code. An elliptical or obround manhole opening must not be less than 11×15 or 10×16 in and a circular manhole not less than 15 in in inside diameter. Handholes must not be less than 2×3 in. However, for vessels exceeding 36 in in inside diameter, the minimum size handhole used in place of a manhole is 4×6 in.

On small vessels it is sometimes difficult to provide extra openings. The Code states that vessels more than 12 but less than 18 in in inside diameter must have not less than two handholes or two threaded plug openings of at least 1½-in pipe size. Vessels 18 to 36 in, inclusive, in inside diameter must have one manhole or two handholes or two threaded plugs of not less than 2-in pipe size. Vessels more than 36 in in inside diameter must have a manhole. Those of a shape or use that make a manhole impracticable must have at least two 4×6 in hand-holes or two openings of equal area.

Removable heads or cover plates may be used in place of inspection openings provided they are large enough to permit an equivalent view of the interior.

If vessels not more than 16 in in inside diameter are so installed that they must be disconnected from an assembly to permit inspection, removable pipe connections of not less than 1½-in pipe size can be used for inspection openings.

Only vessels 12 in or less in inside diameter are not required to have openings, but to omit them is poor practice. The exact internal condition of a vessel should be known at all times whatever its size.

Openings in small tanks are required to be located in each head, or near each head, so that a light placed in one end will make the interior surfaces visible from the other.

When a plant orders a vessel, it should be sure that the design provides for inspection openings that will be accessible *after* the vessel has been installed. It should also have the vessel examined periodically by competent Inspectors to make sure the internal surfaces remain free from deterioration.

QUICK-OPENING CLOSURES

Code Par. UG-35 cautions users about removable closures on the ends of pressure vessels, manways, and similar openings that are not of the multibolt type (charge openings, dump openings, quick-opening doors, cover plates). It states that these openings must be so arranged that the failure of one holding device will not release all the others.

Because of the serious failures incurred by such closures, fabricators and Inspectors must give them special attention and see that they are equipped with a protective device that will ensure their being completely locked before any pressure can enter the vessel and that will prevent their being opened until all pressure has been released.

As there are many types of quick-opening closures, it is not possible to give detailed requirements for protective devices. One example will have to suffice. The most common of these closures is the breech-lock door, which is supported on a hinge or davit and is locked or unlocked by being revolved a few degrees. When locked, the lugs in the door are in a position behind the lugs in the door frame or locking ring. An interlocking device can be easily installed to make sure these lugs are properly engaged before steam is fed to the vessel. If a positive locking device is not possible, some type of warning device should be installed, such as a locking pin that will prevent the door or locking device from turning until an exhaust valve opens. The steam escaping from the opened valve will warn the operator to release all the pressure in the vessel before the door is opened. Likewise, the valve cannot be closed until the door is in the proper closed position. All quick-opening closure vessels must have a pressure gage visible from the operating area.

The total force acting on the head of a quick-opening closure when it is under maximum pressure is very large; see also page 129 for an example.

Figures 3.1 and 3.2 show the result from the loosening of a set screw in the collar on a valve stem which allowed the valve in a safety locking device to be closed when the door was not fully locked. Vibration when the vessel was

Fig. 3.1 Drying cylinder explodes when an almost foolproof safety device failed. *(Courtesy of The Locomotive of the Hartford Steam Boiler Inspection and Insurance Company)*

under pressure caused the locking device to become fully disengaged, allowing the door to be blown away.　The 8 ft 8 in diameter door traveled about 500 ft through the air before shearing off the roof of a car on a highway.　The door continued down the highway about 400 ft, then veered, tore down a chain-link fence, and cut down a 10-in-diameter tree.

In reaction, the 129-ft-long cylinder vaulted from the 2-ft-deep pit in which it

Fig. 3.2 Door of cylinder. *(Courtesy of the Locomotive of the Hartford Steam Boiler Inspection and Insurance Company)*

was located, traveled 200 ft before ramming through a building, and then struck, overturned, and demolished a railroad freight car half loaded with cement.

Safety interlocking devices will help prevent many failures, but they are not infallible. The condition of the operating parts of a door should be carefully examined and the door itself observed to be in its intended position before pressure is applied. Precautions should be taken to ensure that the closed door cannot be moved while the pressure in on and that the pressure will be entirely released before the door is opened.

PRESSURE RELIEF DEVICES

The importance of pressure relief devices in the safe operation of pressure vessels cannot be overemphasized. Improper functioning can result in disaster both to life and equipment. The *ASME Pressure Vessel Code,* in Par. UG-125, states that all pressure vessels must be provided with a means of overpressure protection and that the number, size, and location are to be listed on the manufacturer's data report. There are times when the vessels are designed for a system and properly protected against overpressure by one safety valve. If this is the case, the vessel manufacturer may add the statement, "Safety valves elsewhere in the system." The item on the data sheets for safety valves should never be left blank, nor should it contain a statement that no overpressure protection is provided.

It is the manufacturer's responsibility to see that vessels be fabricated according to the ASME Code and meet the minimum design requirements. Simply because the customer did not indicate on the purchase order or drawing the size and location of the safety valve opening, does not relieve the vessel manufacturer of this responsibility.

The safety devices need not be provided by the vessel manufacturer, but overpressure protection must be installed before the vessel is placed in service or tested for service, and a statement to this effect should be made in the remarks section on the manufacturer's data report.

The user of a pressure vessel should assure that the requirements of the pressure relief devices, be they pressure relief valves (Code Par. UG-126), nonreclosing pressure relief devices such as rupture disk devices, or liquid relief valves, are designed in accordance with the ASME Code requirements and so stamped in accordance with Code Par. UG-129 as follows:

1. ASME Standard Symbols on the valve or nameplate (see Code Fig. UG-129)
2. Manufacturer's name or identifying trademark
3. Manufacturer's design or type number
4. Size in inches (mm) of the pipe size of the valve inlet
5. Pressure at which the valve is set to blow, in psi (kPa)
6. Capacity, in cu ft/min (m³/min) of air; or capacity, in lb/hr of saturated steam (See Code Par. UG-131.)
7. Year built

Rupture disk devices shall be marked by the manufacturer as required in Code Par. UG-129 in such a way that the marking will not be obliterated in service.

Safety valves must be installed on the pressure vessel or in the system without intervening valves between the vessel and protective device or devices and point of discharge. (See also Code Appendix M.)

To prevent excessive pressure from building up in the pressure vessel, a safety valve is set at or below the maximum allowable working pressure for the vessel it protects.

Safety valve construction is important in the selection of safety relief valves. By specifying that the construction must conform to ASME requirements, such construction may be ensured.

Before a manufacturer of relief devices is allowed to use the Code stamp on a relief device, the manufacturer must apply to the Boiler and Pressure Vessel Committee in writing for authorization to use the stamp. A manufacturer or assembler of safety valves must demonstrate to the satisfaction of a designated representative of the National Board of Boiler and Pressure Vessel Inspectors that the manufacturing, production, testing facilities, and quality control procedures will ensure close agreement between performance of random production samples and the capacity performance of valves submitted to the National Board for capacity certification. (For capacity conversion formulas, see Code Appendix 11.)

Escape pipe discharge should be so located as to prevent injury to anyone where it is discharging. It is important to have the discharge pipe of the relief device at least equal to the size of the relief device. It must be adequately supported and provisions must be made for expansion of the piping system.

No pressure vessel unit can be considered completed until it is provided with a properly designed safety valve or relief valve of adequate capacity and in good operating condition.

All safety devices should be tested regularly by a person responsible for the safe operation of the pressure vessel.

Guide to a Quality Control System for ASME Code Vessels

To assure that the manufacturer has the ability and integrity to build vessels according to the Code, the *ASME Boiler and Pressure Vessel Code* now requires each manufacturer of Code power boilers, heating boilers, pressure vessels, and fiberglass-reinforced plastic pressure vessels to have and to demonstrate a quality control system. (For pressure vessels, see Code Par. U-2, UG-90 and Code Appendix 10.) This system must include a written description or check list explaining in detail the quality-controlled manufacturing process. The written description, which for clarification we will call a manual, must explain the company's organized and systematic way of specifying procedures involving each level of design, material, fabrication, testing, and inspection. An effective quality program will find errors before fabrication and will reduce or prevent defects during fabrication. To be effective, this system must have managerial direction and technical support. Proper administration is essential in order to provide the quality control and quality assurance necessary for fabrication of ASME Code pressure vessels.

Before issuance or renewal of an ASME Certificate of Authorization to manufacture code vessels or piping, the manufacturer's facilities and organization are subject to a joint review by an inspection agency and the legal jurisdiction concerned. In areas where a jurisdiction does not review a manufacturer's facility, that function may be carried out by a representative of the National Board of Boiler and Pressure Vessel Inspectors.

The quality control manual, therefore, will form the basis for understanding the manufacturer's controlled manufacturing system. It will be reviewed by the inspection agency, the legal jurisdiction, or the representative of the

National Board.   The manual will serve as the basis for continuous auditing of the system by the Authorized Inspectors and by later survey review when the manufacturer's Certificate of Authorization is to be renewed.

The scope and details of the system will depend upon the size, type of vessel being manufactured, and work to be performed.   All welded pressure vessels require certain basic provisions which must be included in the manual.   An outline of the procedures to be used by the manufacturer in describing a quality control system should include the following items:

1. STATEMENT OF AUTHORITY AND RESPONSIBILITY

Management's statement of authority, signed by the president of the company or someone else in top management, must specify that the quality control manager shall have the authority and responsibility to establish and maintain a quality assurance program and the organizational freedom to recognize quality control and quality assurance problems and to provide solutions to these problems.

2. ORGANIZATION

Manufacturer's organizational chart giving titles showing the relationship between management, quality control, purchasing, engineering, fabrication, testing, and inspection must be presented.   In an approved Code shop an Authorized Inspector's first step in the shop inspection of Code vessels is to review the design.   It is the manufacturer's responsibility to submit drawings and calculations; therefore, in the quality control manual under design specifications the following items should be listed.

3. DESIGN SPECIFICATIONS

   a. The fabricator must verify that specifications have been reviewed and are available to the Authorized Inspector.   (See Code Par. U-2, UG-90.)
   b. Design calculations for the pressure vessel must be signed by a responsible engineering department person and sent to the Authorized Inspector for review.
   c. State the method of handling specification and drawing changes so that any changes in design specifications and calculations can be verified with the Authorized Inspector.
   d. Show the table of revisions, revision number, and date that must be applied.

An important factor when designing a vessel is designing for accessibility of welding and for the position the joint is to be welded in.   For example, if it is difficult to weld a joint because the welder cannot reach it, see it, and be comfortable at the same time, it is impossible to expect and get good quality work as a result.   This is especially important in field erection.   Start quality control by reviewing the drawings, for it is here that difficult-to-weld joints may be discovered and possibly corrected before fabrication is begun.

In reviewing a drawing, the Inspector will be looking for the following design information: dimensions, plate material and thickness, weld details, corrosion allowance, design pressure and temperatures, hydrostatic testing pressure, calculations for reinforcement of openings, and the size and type of flange. It is especially important to indicate whether the vessel is to be postweld heat-treated, radiographed, or spot-radiographed because the Code allows 100 percent efficiency for double-welded butt joints that are fully radiographed, 85 percent for joints that are spot-radiographed, and 70 percent efficiency for joints not radiographed.

Any comments by the Inspector on the design should be brought to the attention of the manufacturer's engineering department immediately. It is easy to make corrections before construction. To make them afterwards is costly and sometimes not possible.

Another important function that the engineer has in the control of quality is the selection of materials. The method of material procurement control must be shown in the manufacturer's quality control manual and should list the following information.

4. MATERIAL PROCUREMENT CONTROL

a. Engineering should have control of all materials ordered for an ASME Code pressure vessel, clearly stating the approrpiate SA or SB and SFA specification of the material ordered. (See Code Par. UG-5.)

b. The system of material control used in purchasing, to assure identification of all material used in Code vessels and to check compliance with the Code material specification, should be explained. (See Code Par. UG-93.)

c. The engineering department must supply all quality assurance requirements on their requisition before it is given to the purchasing department for procurement.

d. In the forming of shell sections and heads, the purchase order should state that the requirements of Code Par. UCS-79 and UHT-79 must be satisfied.

Suppliers can sometimes cause problems. For example, while examining the material and fit-up on a flange on a carbon-steel vessel, the Inspector noticed the flange was 2.25 percent chrome – 1 percent molybdenum material. The vendor had substituted 2.25 percent chrome material because there was no carbon-steel flange this size in stock. The vendor thought a flange of better material was being shipped. But the 2.25 percent chrome – 1 percent molybdenum material required heat treatment and the carbon steel did not, plus the fact that this shop did not have welding qualifications for chrome-molybdenum welding or the consideration of the special welding procedure for the difference of welding the 2.25 percent chrome-molybdenum material and the carbon-steel metal.

At another plant, a welder was attempting to weld a nozzle to a stainless steel type 321 vessel, but the metal was not flowing as it should. The welder notified the Inspector who always carried a small magnet as part of the equipment. As the Inspector touched the magnet to the nozzle, there was a magnetic attraction. This easy test proved that the nozzle material was not type 321 stainless steel, since all 300 series stainless steels are nonmagnetic. Somehow a type 400 series stainless steel material, which is definitely magnetic, was shipped and not properly checked when received. These are examples of why material inspection is so important. The method of material inspection must be included in a quality control manual and should include the following items.

5. MATERIAL CONTROL AND IDENTIFICATION

    a. The quality control department must verify the mill test reports with the material received by checking the heat numbers, material specifications, and tensile strength with the mill test reports. (See Code Par. UG-77, UG-85.)

    b. Procedures must be established for examination of plates, forgings, casting, pipe, nozzles, etc., for laminations, scars, and surface defects to assure that all materials meet the quality assurance requirements of the Code. (See Code Par. UG-93, UG-84.)

    c. Nonconforming materials, components, and parts that do not meet the Code requirements shall be corrected in accordance with Code requirements or shall not be used for Code vessels.

    d. Partial data reports must be requested for parts requiring Code inspection that are furnished by another shop. (See Code Par. UG-120.)

The Inspector's first check on the job itself is making sure that the plate complies with the mill test reports and Code specifications. All material should be identifiable and carry the proper stamping. The thickness of the plates can be gaged at this time to see if it meets Code tolerances, and a thorough examination can be made for defects like laminations, pit marks, and surface scars.

If the plates are to be sheared or cut, the shop is required to transfer the original mill stamping to each section. (See Code Par. UG-77.) The Inspector need not witness the stamping but must check the accuracy of the transfer. To guard against cracks from stamping in steel plate less than ¼ in thick, and in nonferrous plate less than ½ in thick, the method of transfer must be acceptable to the Inspector.

The ASME and National Board have always pressed for better management procedures and require an orderly pervasive system of quality assurance in all phases of Code vessel engineering and construction. A quality control manual must include the manufacturer's system of process control.

6. PROCESS CONTROL, EXAMINATION, AND INSPECTION

    a. The scope of this section is to detail the system established to control quality in the manufacturer's operational sequence.

    b. Detailed process flowcharts should be prepared so that each operation can be examined and signed off by the manufacturer's quality control representatives and by the Authorized Inspector.

c. The chief engineer must take the responsibility for correct design instructions to the shop, specifying on the drawings any special Code requirements which would affect the fabrication or quality control operations.

d. The quality control manager must prepare and maintain procedures for controlling fabrication and testing processes—such as welding, nondestructive examinations, heat treating, hydrostatic or pneumatic testing, and inspection—in addition to providing an operations process sheet with sign-off space for use by the manufacturer's representatives and the Authorized Inspector, since even the best check-off list may not indicate the hold points an Authorized Inspector will want to see. The list should include a remarks space for the Inspector to indicate hold points of parts that are to be examined. (See Fig. 4.1, Inspection Checklist, also Code Par. UG-90 to UG-99.)

e. Explain procedures for covering processes such as preheat, postweld heat treatment, hot forming, etc.

INSPECTION CHECKLIST

Shop Order No._____

Quality Control Inspector (QCI) Hold Point — 0
Authorized Inspector (AI) Hold Point — X

Hold point QCI	AI	Description	Checked by & date QCI	AI	Hold point QCI	AI	Description	Checked by & date QCI	AI
		Drawings					Dye penetrant test		
		Calculations					Magnetic particle test		
		Material insp.					Radiograpic insp.		
		Heat numbers					Ultrasonic insp.		
		Material test reports					Postweld heat treat.		
		Weld procedures					Hydrostatic test		
		Welders qualification					Air test		
		Layout plates & heads					Final check		
		Weld edge preparation					Cleanup		
		Forming					Heat number sketch		
		Layout fittings					Code stamp		
		Fit-up long seam					ASME data report		
		Fit-up girth seam					Paint		
		Fit-up heads					Shipping		
		Fit-up nozzles							
		Preheat check							
		Welding inside							
		Welding outside							

Remarks
_____
_____

Fig. 4.1   Inspection Checklist.

f. Material indentification record should be made on an as-built sketch. (See Code Par. UG-77.)

7. Correction of Nonconformities

a. This is an important part of the quality control manual. A *nonconformity* is any condition in material or quality of the finished product that does not comply with the applicable Code rules. Nonconforming materials, components, and parts that do not meet the Code requirements shall be corrected in accordance with Code requirements or shall not be used for Code vessels.

b. When a nonconforming situation is found, a hold tag or other sticker should be placed on the part or material by the company inspector describing the nature of the nonconformity and also giving all information possible on the tag and on the nonconformity report. (See Figs. 4.2 and 4.3.)

c. In order to determine a proper disposition of the nonconformity, it should be brought to the attention of the appropriate company person or nonconformity review board as well as the Authorized Inspector. The method of repair or corrective action should be documented on the nonconformity report and hold tag.

d. When the corrective action has been completed, the report should be

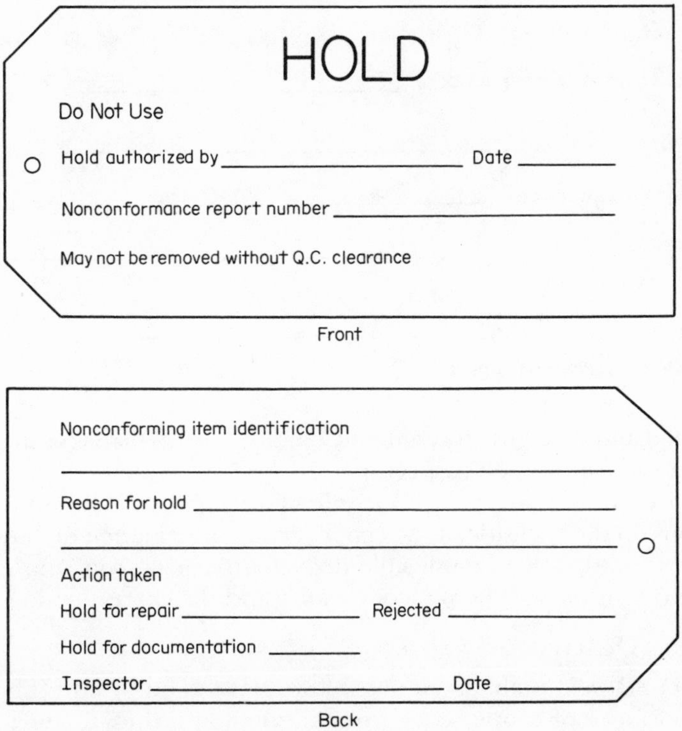
HOLD

Do Not Use

Hold authorized by ____________ Date ________

Nonconformance report number ____________

May not be removed without Q.C. clearance

Front

Nonconforming item identification

____

Reason for hold ____________

____

Action taken

Hold for repair ________ Rejected ________

Hold for documentation ________

Inspector ____________ Date ________

Back

Fig. 4.2 Nonconforming hold tag.

Nonconformance Report

Nonconformance report number: _____

Date: _____

Identification and description: _____

_____

_____

_____

_____

Corrective action: _____

_____

_____

_____

_____

Corrective action completed date: _____

Hold tags removed by: _____ Date: _____

Quality control manager: _____ Date: _____

Authorized inspector: _____ Date: _____

Fig. 4.3 Nonconformance report.

signed and dated by the company quality control manager in conjunction
with the Authorized Inspector.

To assure correct use of a nonconformity report, some companies print
instructions on the backside of the report giving information on how the report
should be routed to the responsible department or person whose job it is to
verify the disposition of the nonconformity and the correct action to take.

8. WELDING QUALITY ASSURANCE

a. A sound weld is almost impossible without good fit-up. Therefore, the
   importance of proper edge preparation and fitting of edges on longitu-

dinal seams, girth seams, heads, nozzles, and pipe should be clearly emphasized in your manual. (See Code Par. UW-31.)

b. State procedure for fit-up and inspection before welding. (See Code Par. UG-90.)

c. Sign-off spaces should be initialed by the Inspectors for fit-up inspection before welding. The Inspectors must assure themselves that the Code tolerances for the alignment of the abutting plates of the shell courses, heads, and openings are met. A multiple-course vessel may also require staggering of the longitudinal joints. To avoid a flat section along the longitudinal seams, the edges of the rolled plates should be prepared with the proper curvature. The out-of-roundness must be within the amount specified by the Code for the diameter of the vessel and the type of pressure (either internal or external) used. (See Code Par. UG-96.) Following the fit-up of a part for welding, the quality of the weld is controlled by providing qualified welding procedures and welders' qualifications. Therefore, in your quality control manual show your system for welding quality assurance. (See Code Par. UW-26, UW-28, UW-29, UW-31; see also Code Section IX, Par. QW-103, QW-322.)

d. Describe the manufacturer's system for assuring that all welding procedures and welders or welding operators are properly qualified in accordance with the ASME Code Section IX, "Welding Qualifications."

e. The engineering department, with the concurrence of the manager of quality control, must determine which welds require new procedure qualification and must provide for qualifying the welding procedure, welding operators, and welders.

f. Specify the method of surveillance of the welding during fabrication to assure that only qualified procedures are being used.

g. Explain the method of qualification of personnel and assigning welders' identification symbols. (See Code Par. UW-28, UW-29.)

h. Describe the method of identifying the person making welds on a Code vessel. (See Code Par. UW-37.)

i. Describe the method of welding electrode control, i.e., electrode ordering, to an SFA specification, inspection when received, storage of electrodes, how they are issued, and the in-process protection of the electrode. (See Code Section II, Part C, Table A-1.)

j. Give the name of the person responsible for seeing that item i is carried out.

k. Describe the method used on the drawing to tell what type and size of weld and electrode to use, i.e., whether the drawing shows weld detail in full cross section or uses welding symbols.

l. Describe the method of tack welding. Do the tack welds become part of the weld or are they removed? (See Code Par. UW-31.) The control of tack welding must be part of the welding quality control system.

Tack welds to secure alignment are often under high stress when fit-up is made. Also, the rapid cooling of the tack weld to the parent metal can form a brittle crack-sensitive microstructure and cause cracks in the base metal as well as in the tack weld.

All tack welds must be made with a qualified welding procedure specification and welded by qualified welders.

A most important control in welding is actual surveillance on the job. This cannot be minimized. The welding of the vessel must be done according to a qualified procedure. The back-chip of the welded seams must be inspected to see that all defects have been removed and that the chipped or ground groove is smooth and has a proper bevel to allow welding without entrapping slag. Each piece of welding must be cleaned of all slag before the next bead is applied; many defects can be caused by improper cleaning.

When a backing strip is used, it is important to see that the strip is closely fitted and that the plates have sufficient clearance to permit good penetration on the first pass. The first pass is more susceptible to cracking because of the large amount of parent metal involved and the fact that the plate may be cold. As the root pass solidifies quickly, it tends to trap slag or set up stresses that can induce cracking if the fit is not good. The importance of good fit-ups to a sound weld cannot be overemphasized.

The finished butt welds should be free of undercuts and valleys and also should avoid too high a crown or too heavy an overlay. Fillet welds should be free of undercuts and overlap. The throat and leg of the welds should be the size specified by the Code.

9. NONDESTRUCTIVE EXAMINATION

A weld may look perfect on the outside and yet reveal cracks, lack of penetration, heavy porosity, or undercutting at the backing strip when the weld is given a nondestructive examination by radiography or ultrasonic tests. As with welding, nondestructive examinations may require a procedure and like the welding procedure it must satisfy the requirements of the Code and be acceptable to the Authorized Inspector.

Various sections of the ASME Code now refer to Section V, "Nondestructive Examinations," for nondestructive examinations. These examinations should be made to a detailed written procedure according to Section V and the referencing Code section. The manufacturer's quality control manual must include the following:

a. A description of the manufacturer's system for assuring that all nondestructive examination personnel are qualified in accordance with the applicable code.

b. Written procedures to assure that nondestructive examination requirements are maintained in accordance with the Code.

c. If nondestructive examination is subcontracted, a statement of who is responsible for checking the subcontractor's site, equipment, and qualification of procedures to assure they are in conformance with the applicable Code.

In a nondestructive examination certification program, the manufacturer has the responsibility for qualification and certification of its NDE personnel. To obtain these goals, personnel may be trained and certified by the manufacturer, or an independent laboratory may be hired. If an independent laboratory is hired, the manufacturer must state in the quality control manual what department and person (by name and title) is responsible for assuring that the necessary quality control requirements are met, because the manufacturer's responsibility is the same as if the parent company had done the work.

For the pressure vessel fabricator, nondestructive testing provides a means of spotting faulty welds, plates, castings, and forgings before, during, and after fabrication. For the user, it offers a maintenance technique that more or less assures the safety of a vessel until the next scheduled inspection period.

10. HEAT TREATMENT

   a. Explain when preheat is necessary, the welding preheat, and interpass temperature control methods.
   b. For postweld heat treatment, describe how the furnace is loaded, how and where the thermocouples are placed, the time-temperature recording system, marking on charts, and how these details are recorded for each thermocouple. (See Code Par. UW-40, UW-49.)
   c. Describe the methods of heating and cooling, metal temperature, metal temperature uniformity, and temperature control.
   d. Describe methods of vessel or parts support during postweld heat treatment.

Many fabricators take testing equipment for granted and seldom check the accuracy of gages and measuring equipment. We have often had gages checked that proved to be inaccurate. For example, in testing a vessel 12 ft in diameter with a quick-opening door, the pressure gage was found to be 25 lbs off on the low side. On a head 12 ft in diameter you have 16,350 sq in. For each pound over the test pressure the head would carry 16,350 lbs. If this gage had been used, the extra load on the head would be $25 \times 16,350$, or 410,000 lbs, above the test requirement. This extra pressure would have been enough to cause considerable damage to this type of opening. The quality control manual must state how the measuring and test equipment is controlled.

11. MEASURING AND TEST EQUIPMENT CONTROL

   a. Describe the system used for assuring that the gage measuring and testing devices used are calibrated and controlled.
   b. Show how records and identification are kept to assure proper calibration. These records shall be available to Authorized Inspectors.
   c. All gages should be calibrated with a calibrated master gage or a standard dead-weight tester every 6 months or at any time there is reason to believe that they are in error. State in detail the method for testing gages and instruments with references to the standards of the U.S.

Bureau of Standards or other recognized standard. Having accurate gages will assure the correct testing pressure when the vessel is undergoing a hydrostatic or pneumatic test.

Before the hydrostatic test, the vessel should be blocked properly to permit examination of all parts during the test. Large, thin-walled vessels may require extra blocking to guard against undue strains caused by the water load. Adequate venting must be provided to ensure that all parts of the vessel can be filled with water.

The water should be approximately room temperature but not less than 60°F. If the temperature is lower, moisture in the air may condense on the surfaces, making proper testing difficult. Two gages should be used to provide a double check on the pressure. If only one gage is used and it is incorrect, the vessel can be damaged by overpressure.

The hydrostatic test must be performed at pressure equal to at least 1½ times the maximum design pressure multiplied by the lowest ratio of the stress value for the test temperature to the stress value for the design temperature.

$$1.5 \times DP = \frac{\text{allowable stress at test temperature} \times \text{design pressure}}{\text{allowable stress at design temperature}}$$

Examinations of the vessel must be made at not less than two-thirds of the test pressure. After the vessel has been drained, a final examination should be made of the internal and external surface, the alignment, and the circularity. (See Chap. 5, "Hydrostatic Testing, Case 3.")

The Authorized Inspector, after witnessing the final hydrostatic tests or pneumatic tests and carrying out the examinations that are necessary, will check the vessel data report sheets to make sure they are properly filled out and signed by the manufacturer and that the material and quality of the work in the vessel are accurately identified. The Authorized Inspector will then allow the Code symbol to be applied in accordance with the Code and any local or state requirements. If everything is in order, the data sheets are signed by the Inspector. Many Code vessels, especially boilers, are fabricated partly in the shop and partly in the field. When construction of a Code vessel is completed in the field, some organization must take responsibility for the entire vessel, including the issuing of data sheets on the parts fabricated in shop and field and the application of Code stamping. Possible arrangements include the following:

(1) The vessel manufacturer assumes full responsibility, supplying the personnel to complete the job.

(2) The vessel manufacturer assumes full responsibility, completing the job with personally supervised local labor.

(3) The vessel manufacturer assumes full responsibility but engages an outside erection company to assemble the vessel by Code procedures specified by the manufacturer. With this arrangement the manufacturer must make inspections to be sure these procedures are being followed.

(4) The vessel may be assembled by an assembler with the required ASME

certification and Code symbol stamp, which indicates that the organization is familiar with the Code and that the welding procedures and operators are Code-qualified. Such an assembler, if maintaining an approved quality control system, may assume full responsibility for the completed assembly.

Other arrangements are also possible. If more than one organization is involved, the Authorized Inspector should find out on the first visit which one will be responsible for the completed vessel. If a company other than the shop fabricator or certified assembler is called onto the job, the extent of association must be determined. This is especially important if the vessel is a power boiler and an outside contractor is hired to install the external piping. This piping, often entirely overlooked, lies within the jurisdiction of the Code and must be fabricated in accordance with the applicable rules of the ANSI B31-1 Code for pressure piping by a certified manufacturer or contractor and must be inspected by the Inspector who conducts the final tests before the master data sheets are signed.

Once the responsible company has been identified, the Inspector must contact the person directly responsible for the erection to agree upon a schedule that will not interfere with the erection but will still permit a proper examination of the important fit-ups, chip-outs, and the like. The first inspection will include a review of the company's quality control manual, welding procedures, and test results to make sure they cover all positions and materials used on the job. Performance qualifications records of all welders working on parts of the vessel under jurisdiction of the Code must also be examined.

It is the responsibility of the manufacturer, contractor, or assembler to submit drawings, calculations, and all pertinent design and test data so that the Inspector may make comments on the design or inspection procedures before the field work is started and thereby expedite the inspection schedule. New material received in the field for field fabrication must be acceptable Code material and carry proper identification markings for checking against the original manufacturer's certified mill test reports. Material may be examined at this time for thicknesses, tolerances, and possible defects.

The manufacturer who completes a vessel or part of a vessel has the responsibility for the structural integrity of the whole or the part fabricated. Included in the responsibilities of the manufacturer are documentation of the quality assurance program to the extent specified in the Code. The type of documentation needed varies in the different sections of the Code. The quality control manual should include the following.

12. PROGRAM DOCUMENTATION

    a. List the type of check-off form used, including those for welding procedures and welding performance qualification, traveler sheets, welder's log records and other charts used in the completion of records for material purchase, identification, examinations, and tests that were taken before, during, and on completion of the vessel.

    b. Show examples of all forms and describe their use, such as the check-off list and data reports which will be presented to the Authorized Inspector for review. The Inspector will carefully check these documents to make sure they are properly listed in the data reports, the vessel complies with the Code, and the data reports are signed by the manufacturer before being signed by the Inspector. Documentation requires good procedure and organizational review. Experience has proved that with good documentation a manufacturer is able to assess the operation accurately and to improve efficiency of production.

The manufacturer should be carefully acquainted with the provisions in the Code that establish the duties of the Authorized Inspector. This third party in the manufacturer's plant, by virtue of authorization by the state to do Code inspections, is the legal representative. Having this party there permits the manufacturer to fabricate under state laws. Under the ASME Code rules, the Authorized Inspector has certain duties as specified in the Code.

It is the obligation of the manufacturer to see that the Inspector has the opportunity to perform as required by law. The details of an authorized inspection at the plant or site are to be worked out between the Inspector and the manufacturer. The manufacturer's quality control manual should state provisions for assisting the Authorized Inspector in performing duties. (See Code Par. UG-92.)

13. AUTHORIZED INSPECTOR

    a. The manual should state that the Authorized Inspector is the Code inspector, and it is the obligation of the company to assist in the performance of these duties.

    b. The Inspector can indicate "hold points" and discuss with the quality control manager arrangements for additional inspections and audits of personal choosing to establish the company's compliance with the ASME Code.

    c. The manual should state that the check list of records, drawings, calculations, process sheets, and any other quality control records shall be available for the Authorized Inspector's review.

    d. The final hydrostatic or pneumatic tests required by the Code shall be witnessed by the Authorized Inspector, and any examinations deemed necessary shall be made before and after these tests.

    e. The manufacturer's data reports shall be carefully reviewed by a responsible representative of the manufacturer or assembler. If this representative finds them properly completed, they shall be signed for the company. The Authorized Inspector, after gaining assurance that the requirements of the Code have been met and the data sheets properly completed, may sign these data sheets.

Audits are not required by all sections of the Code, but in order to make certain that the quality control system is being properly used during manufacture, a comprehensive system of planned and periodic audits should be carried out by the manufacturer to assure compliance with all aspects of the quality control program. The manufacturer's quality control manual should describe the system of self-auditing.

14. RECORD RETENTION

    a. The manufacturer must have a system for the maintenance of manufacturer's data reports, radiographs, impact test results, and the like.

    b. The manager of quality control shall assemble all the manufacturing and inspection documents to be filed for a minimum of 5 years (see Code Par. UW-51) and shall include, but not be limited to, the following: design specifications and drawings, design calculations, certified material test reports, nondestructive examination reports and radiographic films, heat treatment records, hydrostatic and pneumatic test records, impact test results, copy of signed ASME data reports, nonconformity records, inspection check-off list, and signoffs.

    c. All records shall be kept in a safe environment.

15. AUDITS

    a. Explain manufacturer's system of planned periodic audits to determine the effectiveness of the quality assurance program.

    b. Audits should be performed in accordance with manufacturer's procedure, using the check list developed for each audit, to ensure that a consistent approach is used by audit personnel.

    c. Reports of audit results should be given to top management. It should also be indicated what positive corrective action will be taken, if necessary.

    d. Corrective action must be taken by the quality control manager to resolve deficiencies, if any, revealed by the audits.

    e. Reaudit should be made within 30 days to confirm results when corrective action is taken.

    f. Results of audit must be made available to the Authorized Inspector.

    g. Give name and title of person responsible for the audit system. Good management will have a self-auditing system that regularly reviews the adequacy of its quality control program and provides for immediate and effective corrective measures when required.

The maintenance of records and revisions in accordance with the Code should be the quality control manager's responsibility. A system should be maintained for controlling the issuance of the latest quality control manual revisions. Notify the Authorized Inspector of any contemplated revisions for a nonnuclear quality control manual. Verify any revisions with the Authorized Inspector. Revisions and record forms could be listed in the appendix. For nuclear construction, the manual revisions must be submitted to the authorized inspection agency for its review. No revisions are to be issued without prior review and acceptance by the authorized inspection agency.

16. APPENDIX

    a. Show the table of revisions, the revision number, and the date of revision.

    b. Exhibit all forms and describe the form usage.

In our comments, we have given an outline for inspection and quality control systems that the average-sized manufacturer of boilers and pressure vessels might use in establishing a quality control program and manual as required by the ASME Code Section I, "Power Boilers," Section IV, "Heating Boilers," Section VIII, Divisions 1 and 2, "Pressure Vessels." A quality assurance program for nuclear vessels would require a more comprehensive program and manual. (See Chap. 10.) We have also not included Section X, "Fiberglass Reinforced Plastic Pressure Vessels." For quality control systems required by these Code vessels, the applicable Code sections must be followed.

It is important to remember that before the issuance or renewal of an ASME Certificate of Authorization to manufacture Code vessels, the manufacturer's facilities and organization will be reviewed by an inspection agency and/or legal jurisdiction, or a representative of the National Board of Boilers and Pressure Vessel Inspectors.

In reviewing the various operations listed in the manufacturer's quality control manual, their objective will be to see in operation the controlled manufacturing system described by the manufacturer's quality control manual. This survey must be jointly made before the survey team can write a report of the manufacturer's system for the American Society of Mechanical Engineers.

The responsibility for establishing the required quality control system is that of the manufacturer. In writing a quality control manual, effective means of planning during all phases of design, fabrication, and inspection should be considered in order to assure a properly constructed Code vessel. Rear Admiral Charles Curtze, in a report to the congressional committee investigating the loss of the *USS Thresher,* talked about the need for a higher IQ (in this case, integrity and quality) in industry. If your plant has the ability to build Code vessels, and a good IQ, you should have no trouble acquiring and maintaining the ASME Code stamp.

Inspection and Quality Control of ASME Code Vessels

SHOP INSPECTION

An Authorized Inspector's first step in the shop inspection of Code vessels is to review the design. It is the manufacturer's responsibility to submit drawings, calculations, and the following design information: dimensions, plate material and thickness, weld details, corrosion allowance, design pressures and temperatures, hydrostatic testing pressure (based on the maximum allowable working pressure) and the part of the vessel to which the maximum pressure is limited, calculations for reinforcement of openings, and the size and type of flange. It is especially important to indicate whether the vessel is to be postweld heat-treated and the degree of radiography because the Code allows 100 percent efficiency for double-welded butt joints that are fully radiographed, 85 percent efficiency for joints that are spot-radiographed, and 70 percent efficiency for joints not radiographed.

Any comment by the Inspector on the design should be brought to the attention of the manufacturer's engineering department immediately. It is easy to make corrections before construction; to make them afterwards is costly and sometimes not possible.

The Inspector's first check on the job itself is making sure that the materials of construction comply with the mill test reports and Code specifications. All material should be identifiable and carry the proper stamping. The thickness of the plates can be gaged at this time to see if it meets Code tolerances, and a thorough examination can be made for defects like laminations, pit marks, and surface scars.

If the plates are to be sheared or cut, the shop is required to transfer the original mill stamping to each section. The Inspector need not witness the stamping but must check the accuracy of the transfer. To guard against cracks from stamping in steel plate less than ¼ in thick and in nonferrous plate less than ½ in thick, the method of transfer must be acceptable to the Inspector.

Code tolerances for the alignment of the abutting plates of the shell courses, heads, and openings must be met. A multiple-course vessel may also require staggering of the longitudinal joints. To avoid a flat section along the longitudinal seams, the edges of the rolled plates should be prepared with the proper curvature. The out-of-roundness must be within the amount specified by the Code for the diameter of the vessel and the type of pressure (internal or external) used. (See Code Par. UG-80 and Code Fig. UG-80.)

The welding of the vessel must be done according to a qualified procedure (see Chap. 6). The back-chip of the welded main seams must be inspected to see that all defects have been removed and that the chipped or ground groove is smooth and has a proper bevel to allow welding without entrapping slag. Each bead of welding must be cleaned of all slag before the next bead is applied; many defects can be caused by improper cleaning.

When a backing strip is used, it is important to see that the ring is closely fitted and that the plates have sufficient clearance to permit good penetration on the first pass. The first pass is more susceptible to cracking because of the large amount of parent metal involved and the fact that the plate may be cold. As the root pass solidifies quickly, it tends to trap slag or set up stresses that can induce cracking if the fit is not good. The importance of good fit-up to a sound weld cannot be overemphasized.

The finished butt welds should be free of undercuts and valleys and also should avoid too high a crown or too heavy an overlay. Fillet welds should be free of undercuts and overlap. The throat and leg of the weld should be of the size specified by the drawing and in compliance with the Code.

Automatic welding properly performed is superior to hand welding, but it requires rigid supervision and inspection. The best check is provided by spot-radiographing. If radiography is unavailable, a test plate can be welded and sectioned.

TESTS

A weld may look perfect on the outside and yet reveal deep cracks, lack of penetration, heavy porosity, or undercutting at the backing strip when it is radiographed.

If the vessel is to be radiographed, all irregularities on the weld plate surfaces must be removed so as not to interfere with the interpretation of subsurface defects. If a vessel is to have only a pneumatic test, all attachment welds of a throat thickness greater than ¼ in must be given a magnetic particle test or penetrating oil test before the pneumatic test.

If a vessel is constructed of austenitic chromium-nickel alloy steels over ¾ in thick, all welds must be given a penetrant oil test (after heat treatment if such treatment is performed).

If a vessel is to be fully postweld heat-treated, the furnace should be checked to make sure that the heat will be applied uniformly and without flame impingement on the vessel.

Vessels with thin walls or large diameters should be protected against deformation by internal bracing. The vessel should be evenly blocked in the furnace to prevent deformation at the blocks and sagging during the stress-relieving operation. Furnace charts should be reviewed to determine if the rate of heating, holding time, and cooling time are correct.

All work should be completed before a vessel is presented for final testing, especially if it is to be postweld heat-treated. A minor repair can assume major proportions because of a lack of time to make the repair and repeat postweld heat treatment and testing prior to shipping.

A vessel undergoing a hydrostatic test should be blocked properly to permit examination of all parts during the test. Large, thin-walled vessels may require extra blocking to guard against undue strains caused by the water load. Adequate venting must be provided to insure that all parts of the vessel can be filled with water. The water should be at approximately room temperature. If the temperature is lower, moisture in the air may condense on the surfaces, making proper testing difficult. Two gages should be used to provide a double check on the pressure. If only one gage is used and it is incorrect, the vessel can be damaged by overpressure.

The hydrostatic test must be performed at pressures equal to at least 1½ times the maximum design pressure multiplied by the lowest ratio of the stress value for the test temperature to the stress value for the design temperature.

Examinations of the vessel must be made at not less than two-thirds of the test pressure. After the vessel has been drained, a final examination should be made of the internal and external surfaces, the alignment, and the circularity.

If everything is satisfactory, the vessel may be stamped in accordance with Code and any local or state requirements. The data report sheets are then checked to make sure they are properly filled out and signed by the manufacturer and that the material and workmanship in the vessel are accurately stated. If everything is in order, the data sheets are signed by the inspector.

FIELD ASSEMBLY INSPECTION

Many code vessels are fabricated partly in the shop and partly in the field, especially boilers. Although this book deals primarily with pressure vessels, many plants will want to know about the field assembly of boilers as well.

RESPONSIBILITY FOR FIELD ASSEMBLY

When construction of a Code vessel is completed in the field, some organization must take responsibility for the entire vessel, including the issuing of data sheets on the parts fabricated in shop and field and the application of Code stamping. Possible arrangements include the following:

1. The vessel manufacturer assumes full responsibility, supplying the personnel to complete the job.

2. The vessel manufacturer assumes full responsibility, completing the job with local labor under the manufacturer's supervision.

3. The vessel manufacturer assumes full responsibility but engages an outside erection company to assemble the vessel by Code procedures specified by the manufacturer. With this arrangement, the manufacturer must make inspections to be sure these procedures are being followed.

4. The vessel may be assembled by an assembler with the required ASME Certificate and Code symbol stamp, which indicates that the assembler's organization is familiar with the Code and that the welding procedures and operators are Code qualified. Such an assembler may assume full responsibility for the complete assembly.

Other arrangements are also possible. If more than one organization is involved, the Authorized Inspector should find out on the first visit which one will be responsible for the completed vessel. If a company other than the shop fabricator or an authorized assembler is called onto the job, the extent of association must be determined. This is especially important if the vessel is a power boiler and an outside contractor is hired to install the external piping. This piping, often entirely overlooked, lies within the jurisdiction of the Code and must be fabricated in accordance with the applicable rules of ANSI B31-1 Code for pressure piping by a certified manufacturer or contractor and be inspected by the inspector who conducts the final tests before signing the master data sheets.

INSPECTION PROCEDURE

Once the responsible company has been identified, the inspector must contact the person directly responsible for the erection to agree upon a schedule that will not interfere with the erection but will still permit a proper examination of the important fit-ups, chip-outs, and the like. The first inspection will include a review of welding procedures (and test results) to make sure they cover all positions and materials used on the job. Performance qualification records of all welders working on parts of the vessel under the jurisdiction of the Code must also be examined.

It is the responsibility of the manufacturer, contractor, or assembler to submit drawings, calculations, and all pertinent design and test data (see "Shop Inspection") so that the Inspector may make comments on the design or inspection procedures before the field work is started and thereby expedite the inspection schedule.

New material received in the field for field fabrication must be acceptable Code material and carry proper identification markings for checking against the mill test reports. It may be examined at this time for thicknesses, tolerances, and defects like laminations, pit marks, and surface scars.

Field examinations of the vessel for fit-up, for finished welds, and during and after hydrostatic or pneumatic testing should be conducted in the manner already prescribed for shop inspections.

If all is satisfactory, the Inspector will check the data sheets to see that they are properly filled out and signed by the manufacturers and shop inspectors and that the portion of the vessel that was field-constructed is accurately described and the material and quality of the work accurately stated. The inspector then signs the certificate of field assembly.

RADIOGRAPHY

Radiography is a tool permitting the examination of the invisible parts of a welded joint. Radiographic equipment requires skilled personnel for safe and efficient handling. Special provisions must be made for their protection, moreover, and unauthorized persons must be kept from entering the surrounding area while the equipment is in use.

The use of radiography is increasing rapidly in the welded pressure vessel industry, especially because of the higher weld-joint efficiencies allowed by the Code for vessels that are radiographically examined. Such vessels can be built for the same pressure and diameter with thinner shells and heads and consequently with less weld metal and welding time.

CONDITIONS REQUIRING RADIOGRAPHY

The ASME Code requires radiographing of pressure vessels for certain design and service conditions, the type of radiography varying with materials, thicknesses, joint efficiency, type of construction, and the like. Conditions for which full or spot radiography is specified are the following:

1. To increase the basic joint efficiency, either full or spot radiography may be required (Code Par. UW-11 and UW-12, Code Table UW-12).

2. Vessels using lethal substances require full radiography of all longitudinal and circumferential joints (Code Par. UW-2 and UW-11).

3. If the plate or vessel wall thickness at the weld joint exceeds the thicknesses given in Code Par. UW-11, Par. UCS-57 and Table UCS-57, Par. UNF-57, UHA-33, UCL-35 or UCL-36, and UHT-57, full radiography is required. (See also Table 5.1.)

4. To increase allowable stress and joint efficiency to be used in the calculations for shell and heat thicknesses (Code Par. UW-11 and UW-12).

5. Cast ductile iron repair of defects (Code Par. UCD-78).

6. Vessels conforming to type 405 materials that are welded with straight chromium electrodes or to types 410 and 430 materials welded with any electrode require radiography for all thicknesses. If types 405 or 410 material with carbon content less than 0.08 percent are welded with electrodes that produce an austenitic chromium-nickel weld deposit, however, radiography is required only if the plate thickness at the welded joint exceeds 1½ in (Code Par. UHA-33).

7. Clad and lined vessels may require either full or spot radiography (Code par. UCL-35 and UCL-36).

TABLE 5.1   Full radiographic thickness requirements*

Carbon and Low-Alloy Steels

P number and group number† — metals		When thickness exceeds
P = 1 Group 1 1 2 1 3 Carbon steels		1.25 in
P = 3 Group 1 3 2 3 3 Alloy steels with 0.75 maximum chromium and those with 2.00 maximum total alloy.		0.75 in
P = 4 Group 1 4 2 Alloy steels with 0.75 to 2.00 chromium and those with 2.75 maximum total alloy		0.625 in
P = 5 Group 1 5 2 Alloy steels with 10.00 maximum total alloy		0.0 in
P = 9A Group 1 9B 1 Nickel alloy steels		0.625 in
P = 10A Group 1 10F 6		0.75 in
P = 10B Group 2 10C 3 Other alloy steels		0.625 in

 * These requirements are in addition to those of full or spot radiography needed to increase weld efficiency. See Code Par. UW-11, UW-12, and Table UW-12.

 † P numbers and group numbers as given in Code Table UCS-23, Pressure Vessel Code Section VIII, Division 1, and Code Table QW-422, ASME Code Section IX, "Welding and Brazing Qualifications."

    8. Welded repairs of clad vessels damaged by seepage in the applied lining may require radiography if the Inspector so requests (Code Par. UCL-51).

    9. If longitudinal joints of adjacent courses are not staggered or separated by a distance at least 5 times the plate thickness, each side of the welded intersection must be radiographed for a distance of 4 in [Code Par. UW-9(d)].

    10. Unreinforced openings located in a circumferential seam must be radiographed for a distance 3 times greater than the diameter of the opening (Code Par. UW-14).

    11. Castings for which a higher quality factor is desired require radiography at all critical sections (Code Par. UG-24).

12. Leaking chaplets of cast-iron vessels to be repaired by plugging require radiography of metal around defect (Code Par. UCI-78).

13. Forgings repaired by welding may require radiographing (Code Par. UF-37).

14. Unfired steam boilers with a design pressure exceeding 50 psi must be fully radiographed [Code Par. U-1(e)].

15. Inserted-nozzle butt welds joining the flange or saddle to a vessel requiring radiography must be fully radiographed (Code Par. UW-11 and Code Fig. UW-16.1, f-1, and f-2).

INTERPRETATION OF WELDING RADIOGRAPHS

Some people falsely assume that because a weld had been radiographed, no defect which has not been revealed by the radiographing process can exist in the weld.   On the other hand, some persons interpret each visible mark on the film as a defect in the weld and request chip-out and rewelding.   In each of the above cases the radiograph is subjected to false or poor interpretation.

In order for the viewer to interpret the film correctly to find all possible defects in the weld, a proper understanding of the nature and appearance of weld defects, of common processing defects, of the degree of contrast and detail possible in radiography, of the film and film sensitivity, and of the technique of making a radiograph is necessary.

Radiography uses either x-rays or gamma rays.   X-ray machines operate through a range of 10,000 to 2 million volts.   Gamma rays, measured in curies, are produced from radioactive materials of various strengths.   (The most common sources of gamma rays in the welding industry are cobalt 60, cesium 136, iridium 192, and radium.)

Both x-rays and gamma rays are a form of electromagnetic waves of short wavelength, which allows them to penetrate metals impervious to ordinary light.   Some of the radiation is absorbed after penetration, the amount depending on the wavelength and the thickness and density of the material.   If a defect such as a crack, slag inclusion, or porosity exists in a material of uniform thickness and density, the thickness and density are less at the point of defect and more radiation will pass through.   When this radiation is recorded on sensitive film, the defect will show up as an x-ray shadow.   The shadow picture is called a radiograph.

X-ray radiographs show defects more sharply than gamma ray radiographs, a fact that must be taken into account in the interpretation of the film.   When persons experienced with gamma rays are shifted to x-ray work, they tend to see any defect as a cause for rejection, whereas those experienced with x-ray film may tend to underestimate the seriousness of a defect seen on gamma ray film.

The Code states: "The weld shall be radiographed with a technique that will indicate the size of defects having a thickness equal to, or greater than, 2 percent of the thickness of the base metal."   As a check on the radiographic technique, penetrameters or thickness gages are used. The thickness of the penetrameters is given in *ASME Boiler and Pressure Vessel Code,* Section V, "Nondestructive Examination," Table T-262 or Table T-272. This is not more than 2 percent of the thickness of the plate being radiographed, but for thin plates, it need not be less than 0.005 in. As measured by the penetrameter, sensitivity is understood to be the smallest fractional difference in the thickness of the material that can be detected visually on a radiograph. The appearance of the penetrameter image on the film defines a vertical sensitivity of 1 to 2 percent. The required appearance of the 2 T-hole or slit in the penetrameter proves a lateral sensitivity of 2 to 4 percent of the thickness being radiographed.

If the weld reinforcement or backing strip of a welded seam is not removed, a shim of the same material as the backing strip must be placed under the penetrameter so that the thickness being radiographed under the penetrameter is the same as the total thickness of the weld, including the reinforcement and backing strip.

Each penetrameter must have three holes with diameters that are usually 2, 3, and 4 times the penetrameter thickness, one of which must be of diameter equal to 2 times the penetrameter thickness, but not less than ¹⁄₁₆ in. However, smaller holes are permitted. The smallest hole must be distinguishable on the film.

The penetrameters should be placed on the side nearest the source of the radiation. Where it is not possible to do this, they may be placed on the film side of the joint provided that the radiographic technique can be proved adequate and a lead letter F ½ in minimum is placed adjacent to the penetrameter; at least one penetrameter must be used for each exposure. The penetrameter should be placed parallel and adjacent to the welded seam except when the weld metal is not similar to the base material, in which case it may be placed over the weld metal with the plane of the penetrameter normal (perpendicular) to the radiation beam. This provides for penetrameter image sharpness and minimum image distortion and allows proper evaluation of radiographic quality.

To assure getting good contrast and seeing a weld defect, the ASME Code gives specific requirements for film density, based on the Hurter and Driffield (H & D) curve. * Density through acceptable weld metal shall be 1.8 minimum for single film viewing for radiographs made with an x-ray source and 2.0 for radiographs made with a gamma ray source for composite viewing of double film exposures. The minimum density shall be 2.6. Each radiograph of a composite set shall have a density of 1.3. The maximum density shall be 4.0 for either case. In the upper limits of density, the viewing equipment is important for correct interpretation of the film. It is essential to have available a strong source of light with flexibility both as to light intensity and area.

If the source of radiation is placed at the axis of the joint and the complete

---

* The Hurter-Driffield method of defining quantitative blackening of the film expresses the relationship between the exposure applied to a photographic material and the resulting photographic density.

circumference is radiographed with a single exposure, three uniformly spaced penetrameters are to be used.

Many persons in the welding industry who know how to interpret radiographic film sometimes err in their judgment. Moreover, radiographs often require retakes because of improper techniques or differences of opinion in interpretation. Only the retake can prove whether the technique or the interpretation of the original film was wrong. Basically, the important decisions in a radiographic inspection are these:

1. Has the radiograph been properly taken?
2. Has complete penetration of the weld been obtained?
3. Are the revealed defects acceptable or should they be repaired?

Proper radiography will show any change in metal thickness greater than 2 percent, the contrast sensitivity required by the Code. Any defect surpassing the allowable minimum — whether slag entrapment, porosity, lack of penetration, undercutting, or cracks — will be recorded on the film as a dark area compared to the normal density of the sound metal. Weld reinforcements, backing strips, splatter, and icicles will show on the film as light areas because they represent metal sections thicker than the average plate. Burn-throughs show up as dark, irregular areas, but the drip-through surrounding the burn-through presents a lighter image. In highly absorbent material like the burn-off from an electrode used in heliarc welding, the burn-off shows up as a light area that is usually spherical in shape.

In heavy wall welds, knowing how deep a defect is from each side is helpful. We have seen chip-outs 2 in deep made in a 3-in weld that could have been only 1 in deep from the opposite side if stereoscopic principles of radiography had been used to make proper depth interpretation. (See Fig. 5.1 for an example.) When available, ultrasonic techniques can and often are utilized to determine the depth of many types of defects. Not every discontinuity in film density is objectionable if vessels are fully radiographed. (See Code Par. UW-51 and Appendix 4.) The Code allows for a maximum length of slag inclusion according to the plate thickness, as follows: ¼ in for plates less than ¾ in thick; ⅓ the plate thickness for plates ¾ to 2¼ in thick; and ¾ in for plates more than 2¼ in thick.

Any group of slag inclusions in a line with a total length greater than the thickness of plate in a length of 12 times the plate thickness is unacceptable, except when the distance between the successive imperfections exceeds 6 times the length of the largest imperfection. On full radiography, moreover, any type of crack or lack of penetration is not acceptable and must be repaired. And porosity is judged by comparison with the standards given in the Code.

If only spot radiography is required, the standards are less demanding. (See Code Par. UW-52.) The length of film need be only 6 in. Porosity, moreover, ceases to be a factor in the acceptability of welds. There is a difference in the other defects allowed as well. Any weld that shows cavities or elongated slag inclusions is unacceptable if these have a length greater than two-thirds of the thickness of the thinner plate of the welded joint.

D = Shift of defect image

M = Shift of marker image

t = Thickness of weld

L = Depth of defect from film side

$$L = \frac{D \; t}{M}$$

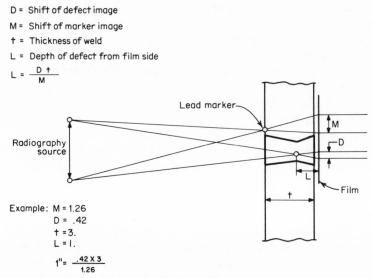

Example: M = 1.26
         D = .42
         t = 3.
         L = 1.

$$1'' = \frac{.42 \times 3}{1.26}$$

Fig. 5.1   Stereoscopic principle in radiography.

Any group of imperfections in a line with a total length less than the plate thickness in a length of 6 times the plate thickness is acceptable only if the longest defects are separated by acceptable weld metal equal to 3 times the length of the longest defect. (As the minimum length required for a spot radiograph is 6 in, the total length of imperfections in a line shall be proportional for radiographs shorter than 6 times the plate thickness.) Any defect shorter than ¼ in is acceptable in spot radiography, with the exception of any type of crack or lack of penetration, which is never acceptable and must be repaired. Any defect longer than ¾ in is not acceptable.

If a spot radiograph shows welding that violates Code requirements, two additional spots must be taken in the same seam at locations chosen by the Inspector away from the original spot, preferably one on each side of the defective section. If these additional spots do not meet the acceptable standard, the rejected welding must be removed and the joint rewelded, or the fabricator may choose to reradiograph the entire unit of weld represented, and the defective welding need only be repaired. The rewelded areas must then be spot-radiographed to prove that they meet Code requirements.

For the proper interpretation of films, the view box should be lighted with sufficient intensity to illuminate the darkest area of the film. The view box should be located in a room that can be darkened. Film handling and processing at times will cause marks that give the impression of defects. A fine scratch on the film surface can look like a crack; water marks, stains, and streaks can resemble other defects. If the film surface is viewed in reflected light, these defects can be properly detected. Other apparent but nonexistent flaws in the weld can be caused by a weld overlay that contrasts sharply at the weld edge, by scratches on intensifying screens, or by lines of weld penetration into the backing ring if a backing strip is used. Slag under the backing strip and tack welds on the strip will produce a lighter image and should not be interpreted as defects.

Such conditions may be obvious to the experienced technician, if not to the novice. There are often real defects in a weld that are difficult to determine, however, and it is these one must look out for. Some critical materials are more prone to crack than others, and their film images must be evaluated with extreme care.

A crack that runs perpendicular to the plate surface can be seen easily, but cracks along the fusion line are often at an angle, and the defect image may not be sharp. Defects that are parallel to the plate also may not show up clearly. If a laminated plate is radiographed with the radiographic beam perpendicular to the plate surface, for example, the lamination very probably will not be seen. However, if the plate is cut and radiographed with the beam parallel to the plate as rolled, these laminations will show clearly. (See Figs. 7.1 and 7.2 for examples.)

Most cracks show up longitudinally to the weld, in the weld itself, in the fusion line of the weld, at craters, and occasionally in the plate adjacent to the weld. Transverse cracks are not seen as often but do appear in the weld and plate edges. Slag inclusions are usually sharp and irregular in shape. As stated before, certain small slag inclusions are acceptable to the Code, but elongated slag or too many inclusions in a line are not permitted.

Incomplete penetration, which is not acceptable, is usually sharply defined. Undercuts are usually easily seen at the edge of a weld. Porosity shows up as small, dark spots. These spots may be defined as either fine or coarse porosity; if full radiography is required, their acceptability is measured by the standards of the Code.

The important thing in viewing radiographs is to be sure that the penetrameters are properly placed so that the penetrameter holes show and that the H & D density requirements are given in the negative. This will assure better contrast to show any possible defects. The required appearance of the essential hole in the penetrameter is the indication that defects in the weld will be seen in the film. (See also Table 7.1.)

SECTIONING

The Code permits the examination of welded joints by sectioning if this method is agreed to by both the user and the manufacturer, but sectioning of a welded joint is not considered a substitute for spot-radiographic examination and permits no increase in joint efficiency; therefore, it is seldom used.

Sectioning is accomplished with a trepanning saw, which cuts a round specimen, or with a probe saw, which cuts a boat-shaped specimen. The specimen removed must be large enough to provide a full cross section of the welded joint.

If sections are oxygen-cut from the vessel wall, the opening must not exceed 1½ in or the width of the weld, whichever is greater. A flame-cut section must be sawed across the weld. A saw-cut surface must show the full width of the weld. The specimens must be ground or filed smooth and then etched in a solution that will reveal the defects. Care must be taken to avoid deep etching.

The etching process should be continued just long enough to reveal all defects and clearly define the cross-sectional surfaces of the weld. Several acid solutions are used to etch carbon and low-alloy steels, one of which is hydrochloric acid (commonly known as muriatic acid) mixed with an equal part of water. This solution should be kept at or near its boiling point during the etching process.

Many shops prefer to use ammonium persulfate because the method of its application allows better observation of the specimen to prevent a deep etch. The solution, one part ammonium persulfate to nine parts water, is used at room temperature. A piece of cotton saturated with the solution is rubbed hard on the surface to be etched. The etching process is continued until it shows a clear definition of the cross-sectional surfaces of the weld and any defects present.

Other solutions are used to etch metal specimens, such as iodine, potassium iodide, and nitric acid, but hydrochloric acid and ammonium persulfate are the most common ones.

To be acceptable, the specimens should not show any type of crack or lack of penetration. Some slag or porosity is permitted if (1) the porosity is not greater than $\frac{1}{16}$ in and there are no more than six porosity holes of this maximum size per square inch of weld metal, or (2) the combined area of a greater number of porosity holes does not exceed 0.02 sq in per square inch of weld.

The width of a single slag inclusion lying parallel to the plate must not be greater than one-half the width of the sound metal at that point. The total thickness of all slag inclusions in any plane at approximate right angles to the plate surface must not be greater than 10 percent of the thinner plate thickness.

The plugs or segments should be suitably marked or tagged after removal with a record of their original location, the job number, and the welder or welding operator who did the welding. After etching, they should be oiled and kept in an identifiable container. A record should be made on a drawing of the vessel of all these specimens (with identification marks) after the vessel has been completed and stamped. The plugs may be retained by the purchaser, or they may be discarded.

Holes left by the removal of the trepanned plugs may be closed by any welding procedure approved by the Inspector. The best method is to cut a plug from the same plate material and then bevel the plate around the hole and the plug, insert the plug, weld from one side, back-chip the other side, and reweld.

If it is possible to weld from only one side, the bottom of the plug should be fitted to permit full penetration of the weld closure. Each layer of weld should be peened to reduce the stresses set up by welding.

If the plate thickness is less than one-third the diameter of the hole, a backing strip may be placed inside the shell over the hole and the hole filled completely with weld metal. If the plate thickness is not less than one-third or greater than two-thirds the diameter of the hole, the hole can be completely filled with weld metal from one side and then chipped clean on the other side.

Other methods exist (see Code Appendix K), but the ones described here are the most common.

HYDROSTATIC TESTING

Most fabricating shops take pride in their work and do everything possible to turn out a good job. All this effort can be spoiled by carelessness in preparing for or conducting a hydrostatic test. Several actual incidents may serve as illustrations.

CASE 1: A large fabricating shop, one of the best, put all its care and know-how into a critically needed vessel. During hydrostatic testing, however, a 3-in shell plate bulged because of overpressure. This near accident resulted in a costly delay in shipping that should never have occurred. The company had been cautioned by an Authorized Inspector on the first day in its shop, and several times thereafter, that the water pressure on the line to the test floor was 2200 lb and that a reducing valve or some other means of protection was required. The shop's chief inspector replied that dependable people worked on the test floor and that the two shut-off valves provided for each vessel were sufficient. The Authorized Inspector told the shop's inspector that, as a preacautionary measure, the presence of an Authorized Inspector would be necessary at all hydrostatic tests of Code vessels. The general manager was then informed of the possibility of a vessel's being damaged during these tests. The general manager said that the matter would be taken up with the chief inspector, but the precautionary safety devices recommended were never installed. A few months later, an urgently needed non-Code vessel was badly damaged.

The Code does not specify an upper limit for the hydrostatic test pressure. However, if this pressure exceeds the required test pressure to the degree that a vessel is visibly distorted, the Inspector shall reserve the right to reject the vessel.

CASE 2: A thin-shelled vessel was hydrostatically tested at 30 lb. It was filled late in the afternoon and left overnight with the pressure on and all valves closed. During the night the water in the vessel warmed up, expanded, and bulged out a flat stayed section. This accident occurred despite the cautionary footnote to Code Par. UG-99, which states: "A small liquid relief valve set to $1\frac{1}{3}$ times the test pressure is recommended for the pressure test system, in case a vessel, while under test, is likely to be warmed up materially with personnel absent." (For a graph showing the increase in water pressure caused by increase of water temperature, see Fig. 5.2.)

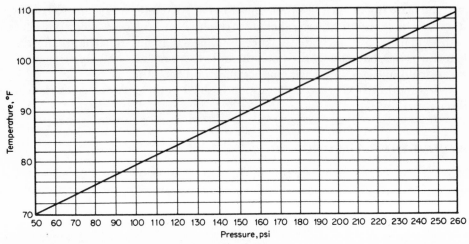

Fig. 5.2   Water pressure and temperature graph.

CASE 3: A vessel 6 ft in diameter and 15 ft long was damaged while being drained.   A tester opened a large opening at the bottom of the tank without first opening a valve or other inlet on top of the vessel to permit air to enter. The large opening at the bottom drained the water rapidly, creating a partial vacuum in the tank.   As the vessel was not designed for external pressure, this vacuum caused the shell to crack between two nozzle outlets.

All three of these mishaps occurred in shops whose work is far above average; they prove that similar acidents are possible in any shop if care is not taken.

TEST GAGE REQUIREMENTS

Code Par. UG-102 states: "An indicating gage shall be connected directly to the vessel.   If it is not readily visible to the operator controlling the pressure applied, an additional gage shall be provided where it will be visible throughout the duration of the test."   It also recommends that a recording gage be used in addition to indicating gages on larger vessels.   Not all shops, especially small ones, have recording gages, but good practice indicates the use of two gages on all tests because they are sensitive instruments and can easily be damaged.

The Code requires vents at all the high points of a vessel (in the position in which it is being tested) to purge air pockets while the vessel is filling.   During a recent hydrostatic test of a large vessel, the frame of a large, quick-opening door failed, and the door dropped to the floor, thereby flooding the shop with several inches of water.   Had there been any air in the vessel, serious damage to the building and possible injury to personnel might have occurred.

The Code requires test equipment to be inspected and all low-pressure filling lines and other appurtenances that should not be subjected to test pressure to be disconnected before test pressure is applied.   The reason can be seen in the recent test of a vessel for 1500-lb pressure with a 125-lb valve in the line.   As this valve was leaking at the bonnet, the Inspector immediately had the pressure reduced and the valve replaced. The Inspector had seen the damage in another shop when a bonnet, valve, and stem had been blown through the roof, leaving a hole a foot wide (fortunately, no one was over the valve when it failed).

Not so long ago, a tester was seen placing a cast-iron plug in a vessel that was to be tested to 1400-lb pressure. The tester was stopped by the Inspector, who requested that a steel plug be used. Replying, "Is that why the plug blew out?" the tester then showed the Inspector a dent in the concrete wall. Ignorance of correct procedures can be fatal in testing vessels.

SEVEN SIMPLIFIED STEPS TO EFFICIENT WELD INSPECTION

A manufacturer's preparation of pressure vessels for examination by a Code Inspector or a customer can be routine and easy or complicated and time-consuming. The speediest and most efficient inspection takes place in the shop that prepares for inspection in an orderly manner. And such preparation will save not only time but very possibly money.

A job should have advanced to the point at which specifications called for an inspection before one is undertaken; if there is no such specification, the job should have advanced to the point requested by the Inspector. Any vessel failing to meet minor requirements should be corrected before being shown to the Inspector. Major defects must be shown to the Authorized Inspector before repairs are made. The Inspector makes an inspection on the basis of a written code or customer specification only. It is the responsibility of the manufacturer, not of the Inspector, to see that these standards are met.

A good Inspector tries to find all defects before the final or hydrostatic tests, especially if the vessel is to be postweld heat-treated. The Inspector knows — and certainly the manufacturer knows — that a minor repair can assume major proportions because of a lack of time to make the repair and to repeat postweld heat treatment and testing operations before the scheduled date of delivery.

A careful observation of the following items will help expedite inspection:

1. Inspectors should be provided with drawings that clearly show weld details and dimension, types of material, corrosion allowances, and design pressures and temperatures. If the vessel is to be fully radiographed or spot-radiographed, this fact should be clearly stated because the Code allows different efficiencies for fully radiographed, spot-radiographed, and nonradiographed joints. Hydrostatic test pressure should be based on the maximum working pressure allowed for a new and cold vessel, and the part of the vessel to which the maximum pressure is limited should be defined.

2. A Code Inspector must check and verify all calculations. If they are neatly arranged and include all pertinent design information, the job will be much easier. A well-managed engineering department will have the calculation sheets printed or mimeographed and will fill in pertinent design information.

3. If the vessel is a Code vessel, the data sheet should be checked and all corrections made before it is typed. Data sheets prepared at the start of a job after drawings are made keep the engineering department and the Inspector from having to review the design and calculations a second time.

4. Mill test reports of material should be filed with the Inspector's copies of drawings to save the manufacturer's personnel from being called from what they are doing to produce these reports.

5. Stamping information, including the location of stamps, should be readily available. Corrections should be made beforehand.

6. A written copy of the welding procedure (or a note stating that both the procedure ad operator are qualified) and of the test results sould be available on request. These items should be clipped or filed together.

7. A place for the Inspector to sit and prepare the written part of the inspection should be provided to avoid the feeling of being an intruder which could arise if someone else's desk has to be used. This could even cause the Inspector to do the rest of the work at home or in a hotel room. This inconvenience could conceivably cause important data to be overlooked that might not have been missed had the report been written in the shop.

Welding, Welding Procedure, and Operator Qualification

EXAMINATION OF PROCEDURE AND OPERATOR

One of the most important functions of an Authorized Inspector is examining welding procedures and operators for their compliance with the intent of the Code. The Code does not dictate methods to the manufacturer, but it does require proof that the welding is sound (Code Par. QW-201) and that operators are qualified in accordance with the ASME Code Section IX, "Welding and Brazing Qualifications" (see Code Par. QW-301). *To avoid confusion, it should be noted that all Code references in this chapter, unlike the rest of this book, refer to ASME Code Section IX, "Welding and Brazing Qualifications."*

An Inspector must be familiar with the physical properties of materials and electrodes and should have experience in the evaluation of the written procedures and test plates required by Section IX. An Inspector also must know how to find out the cause of test plate failures. The manufacturer should actually be more interested than the Inspector in finding the cause, for it is far less expensive to prevent failures than to make expensive repairs. This is especially true if the procedure is to be used on a production line where individual items may miss inspection. In a plant where LPG tanks were being constructed of a high-tensile-strength material, for example, the manufacturer had difficulty in performing satisfactory bend tests on an offset joint for the head and girth seam After various changes in welding procedure, a test finally satisfied Code requirements. Because of the history the plant had had with this material and type of welded joint, however, the Inspector realized that one successful test might not be sufficient.   Knowing that any repairs would be costly and failure could be disastrous, additional bend tests before approving the procedure were requested.

A qualified procedure is the most important factor in sound welding.   If one has been properly established, welders need only demonstrate their ability to make a weld in the position to be welded that will pass root and face bend or side bend tests in order to gain qualification.   It should be remembered that welding excellence depends not only on good welding electrodes but also on the way in which they are used.   Moreover, an Inspector should be satisfied not only that a manufacturer's procedure is acceptable and the operators qualified but also that the welding procedure is followed during actual fabrication.

RECORD OF QUALIFICATION

Once a welding procedure has been approved, it should be filed for future review by Authorized Inspectors or customers' representatives.   Some manufacturers are not as careful as they should be in keeping these records.   Not long ago an Inspector asked to see the written procedure of a certain shop and also the test results for material similar to that used on the job that was to be inspected.   The written procedure could not be located for several days.   After the Inspector explained the importance of proper filing and the unnecessary expense of having to requalify procedures, the manufacturer, eventually, had all records brought up to date and sets filed with both the shop superintendent and the engineering office.   When the representative of a large refining company made a survey of the plant a few months later, the general manager felt very pleased at being able to show the procedures promptly upon request and felt that this favorably helped the company's business position.

Another company with many approved welding procedures lost the file containing them and was forced into an expenditure of several thousand dollars to replace them.   Needless to say, this firm now keeps duplicate copies of all its welding qualifications and test results.

WELDING PROCEDURES AND QUALIFICATION SIMPLIFIED

Welding procedure specification (WPS), and *procedure qualification record* (PQR) are the most essential factors in contributing to sound welding.   A welding procedure specification records the ranges of variables that can be used with the WPS and gives direction to the welder using the procedure.

The procedure qualification record states what was used in qualifying the WPS, shows test results, and furnishes proof of the weldablity of the variables described in the WPS.   If these have been properly established, welders need only demonstrate their ability to make a weld that will pass root and face or side bend tests in the position to be welded in order to gain qualification.   In building welded ASME Code vessels, one requirement is to have written welding procedure specifications of the materials to be welded (see Code Par.

QW-201), properly completed records of qualification of this procedure (Code Par. QW-201), and records of qualification of the welders working on the vessel (Code Par. QW-301.2).

These procedures should be similar to the ones to be used in production by the fabricator. The wording of the procedure should be such that the meaning is clear and easily understood. The fabricator need only fill in the procedure in writing on the recommended WPS Code Form QW-482 and make test plates recorded on the PQR Code Form QW-483 in order to prove that welding is sound. Once a welding procedure has been approved, it should be filed for future review by Authorized Inspectors or customers' representatives. These records are one of the first things an Inspector of ASME vessels will examine. Code Par. QW-201 requires each manufacturer or contractor to have a qualified welding procedure specification. Code Par. QW-301 requires each welder or welding operator doing Code welding to be qualified in accordance with the welding procedure specification and procedure qualification record. Proven sound welding procedures are a prerequisite to the construction of ASME Code vessels.

Section IX of the *ASME Boiler and Pressure Vessel Code* presently consists of two parts. Part QW covers requirements for welding, Part QB contains requirements for brazing; the previous distinction between ferrous and nonferrous metals has been eliminated. Where a specific metal has a special requirement, such requirements are listed in the appropriate variables.

By far the greatest percentage of fabrication involves welding. For this reason, most examples given here apply to Part QW, "Welding." For brazing procedures, see Section IX, Part QB, "Brazing."

There are some differences in ferrous and nonferrous test plate thicknesses. For example, the minimum thickness of ferrous material qualified by a test plate $1/16$ to $3/8$ in thick inclusive is $1/16$ in, whereas for nonferrous material, the minimum thickness is one-half the thickness of the plate. Also, face, root, and some side bends for nonferrous grouping of base metals listed in Code Table QW-422.25 and QW-422.35 may be $1/8$ in thick instead of $3/8$ in as required by most material. [See Code Section IX, Table and Code Fig. QW-462.2 and QW 462.3. See also Fig. 6.4.] It is therefore clear that each Code welding procedure must be carefully read and separately evaluated. Also note, when welding clad or lined material, the procedure shall be qualified in accordance with the Section IX of the Code, with modified provisions as required in the referencing section, Part UCL for clad vessels, in Section VIII, Division 1, of the *ASME Boiler and Pressure Vessel Code.*

As stated previously, before a manufacturer can build or repair ASME Code vessels, proof must first be given that welding procedures and welders or welding operators qualify under the provisions of welding qualifications in Section IX (Code Par. QW-201). WPS Code Form QW-482 for recording welding procedure specifications and PQR Code Form QW-483 for procedure qualification record make it easy to fill in the required information. The following suggestions should be observed in preparing a WPS Code Form QW-482.

WELDING PROCEDURE
SPECIFICATION FORMS

The procedure should be titled and dated and should have an identifying WPS number and supporting PQR number.

WELDING PROCESSES

The procedure should state the welding processes and the type of equipment —automatic, semiautomatic, or manual. If inert gas metal-arc welding is used, it should be specified if a consumable or nonconsumable electrode is used and the material of which the electrode is made.

JOINTS (CODE PAR. QW-402)

Preparation of base material should include group design, i.e., double butt, single butt with or without a backing strip, or any type of backing material used (e.g., steel, copper, argon, flux, etc.), showing a sketch of groove design on the PQR. The groove design should include the number of passes for each thickness of material and whether the welds are strings or weaving beads. Sketches should be comprehensive and cover the full range of base metal thickness to be welded, the details of welding groove, spacing position of pieces to be welded giving maximum fit-up gap, the welding wire or electrode size, the number of weld passes, and the electrical characteristics; if automatic welding is used, the rate of travel and the rate of wire feed should be recorded.

BASE METALS (CODE PAR. QW-403)

The type of material to be welded, giving the ASME specification number, the P number and group number in which it is listed, and the thickness range should be specified. On the PQR, show the ASME materials specification, type, and grade number; P numbers to be welded together; and the thickness and diameter of pipes or tubes. If this procedure is one for an unlisted material, either the ASTM or other Code designations should be specified, or the chemical analysis and physical properties. This should be done only when you intend to work with a special material permitted by a Code interpretation case.

FILLER METAL (CODE PAR. QW-404)

State SFA specification number and the AWS classification when the type of ASME filler metal is given with the filler metal group F number (Code Table QW-432) and the weld metal analysis A number. Any manufacturer's product bearing these classifications can be used; if no A or F, state manufacturer's trade name. In submerged arc welding, the nominal composition of the flux must be stated. For inert arc welding, the shielding gas must be described. The size of the electrode or wire, amperage, voltage number of passes, rate of travel for automatic welding, and rate of shield gas flow for inert gas arc welding must be listed. State if a consumable insert is used on single butt welds.

POSITION (CODE PAR. QW-405)

State the position in which the welding will be done, and if it is done in the vertical position, state whether uphill or downhill in the progression specified for any pass of a vertical weld. The inclusion of more than one welding position is permitted as long as the intent for each position is clear and shown by attaching sketches to the procedure. However, it is best to confine a specification to one position.

PREHEAT (CODE PAR. QW-406)

Preheating and temperature control should be stated only if they are to be used, and any preheating and control of interpass temperature during welding that will be done should be described. For example, materials in the P-4 and P-5 groups must be preheated. Carbon steels for pressure vessels in the P-1 group, if preheated to 200°F, can be welded to 1½-in thickness before postweld heat treatement is required, whereas if not preheated, the limit is 1¼-in thickness.

POSTWELD HEAT TREATMENT (CODE PAR. QW-407; SEE ALSC SECTION VIII, DIVISION I, CODE TABLE UCS-56)

Describe any postweld heat treatment given, the temperature, time range, and any other treatment.

GAS (CODE PAR. QW-408)

State shielding gas type (e.g., argon, CO_2, etc.) and the composition of gas mixture used (e.g., 75 percent CO_2, 25 percent argon), as well as the shielding gas flow-rate range. State whether gas backing or trailing shielding gas is used and the gas backing or trailing shielding gas flow rate.

ELECTRICAL CHARACTERISTICS (CODE PAR. QW-409)

When describing electrical characteristics, it should be stated whether the current is ac or dc, and if dc current is used, it should be specified whether the base material will be on the negative or positive side of the line. Give the mean voltage and current for each pass. Mean should be the average of the specified upper or lower limits within 15 percent of mean; also give the travel speed.

TECHNIQUE (CODE PAR. QW-410)

The joint welding procedure may be shown on the same sketches with the preparation of base material or on a separate sketch. The sketch should show the number of passes for each thickness of material, whether the welds are strings or weaving beads, single or multiple electrodes.

Give orifice or gas cup size range, list oscillation parameter ranges covered by WPS (i.e., width, frequency, etc.), and give contact tube to work distance range.

Describe initial and interpass cleaning, including the method and extent of defect removal. Explain the method of back gouging. It is important to state clearly the method of defect removal. For example, flame gouging is detrimental to some materials. If this is so, make sure another method of defect removal is used.

Peening, if mentioned at all, should be used only when the desired result cannot be obtained by postweld heat treatment. If peening is used, clearly state all details of the peening operation such as not peening first and last pass. (See Section VIII, Division I, Code Par. UW-39.)

Treatment of the underside of the welding groove should be clearly indicated. For example, "Chipped to clean metal and welding applied as shown in sketch number. . . ."

VARIABLES (CODE PAR. QW-400, QW-250, QW-350, AND TABLES QW-415 AND QW-416)

Any variable which would affect the quality of characteristics of the finished weld may require a new welding procedure. If any such variables exist, they should be clearly indicated. Essential variables will require requalification of the procedure if any are changed from those qualified.

Additional requirements in the Code are supplemental, essential variables. These variables must be included only when notch toughness is required and are then in addition to the essential variables.

Changes in details can be made without setting up new procedures if the changes are recorded. The nonessential variables can be changed without requalification, but the changes must be indicated.

When writing a welding procedure specification, it is important to remember that qualification in any position for procedure qualification will qualify for all positions except where notch toughness or stud welding is a requirement. (See Code Par. QW-203 and QW-250.) Where notch toughness is a factor, qualification in the vertical, uphill position will qualify all positions (Code Par. QW-405). If the weld must also meet notch toughness requirements as required by another section of the Code, the supplementary essential variable must be included. [See Code Tables QW-253.1(a) to QW-259.1(a) and Code Par. QW-403.]

Test plates are required for all procedure qualifications. Welders' performance test plates must be welded in the positions required in the welding to be done. The six positions are shown in Fig. 6.1. The order of removal and the number and types of the test specimen from the welded test plates is shown in Fig. 6.2, transverse, and Fig. 6.3, longitudinal test specimens. The preparation of the transverse test specimens are shown in Figs. 6.4 and 6.5.

If the filled-in WPS Code Form QW-482 and the PQR Form QW-483 are clear and the test plates are approved and recorded, the manufacturer will be qualified to weld pressure vessels of the same type of material used in making the test plate and also any other material in the P-number grouping. Starting with the 1974 Code, P-number listing for base metals has been divided into groups. The purpose of the groupings is to cover qualifications where notch toughness is a requirement. When notch toughness is required, each group must be separately qualified. Many shops use a simple jig with a hydraulic jack similar to the one shown in Fig. 6.6 to make their own root, face, and side bends on the assumption that if the bend tests satisfy the requirements of Section IX, Code Par. QW-163 with no defect exceeding ⅛ in (or 1/16 in for corrosion resistance weld overlay cladding) then the tensile test should present no difficulty. The advantage of conducting your own bend tests is that it keeps a shop from having to go to the expense of sending tensile bars to be machined and pulled if the root, face, or side bends should fail. Moreover, this jig can always be used to qualify other welders.

As an alternative to bend tests, radiographic examinations may be employed to prove the performance qualification of welders to make a sound weld, using welding groove welds with the manual shielded metal-arc welded process with covered electrodes and the gas tungsten arc welding (Code Par. QW-304 and QW-305).

A record of the welders' and welding operators' test results should be kept on forms similar to recommended Code Form QW-484, Section IX. The record of the tests shall be certified by the manufacturer and accessible to the Authorized Inspector.

The qualification of a procedure is affected by the type of electrode used. If a manufacturer, for example, has been welding Code vessels of SA-285 grade C materials using an E-6020 electrode classed as type F1 (Section IX, Code Table QW-432, F numbers) but wants to change to an E-6015 or E-6016 electrode classed as type F4, the manufacturer must have this procedure qualified separately. (See Section IX, Code Par. QW-404.)

The welder who is qualified to use type F4 electrodes (E-6015 and E-6016), however, is also qualified to use type F1 (E-6020 and E-6030), type F2 (E-6012, E-6013, and E-6014), and type F3 (E-6010 and E-6011). (See Section IX, Table QW-432 and Code Par. QW-404.) The F-number grouping of electrodes and welding rods is based on their usability qualities, which essentially determine the ability of welders to make dependable welds with a given filler metal. For specific welding processes, shielded metal arc, submerged arc, gas-shielded metal arc, gas-shielded tungsten arc, and plasma arc where the melting technique is used, a greater latitude is permitted when qualifying for the welding of carbon and chrome-moly steel of P numbers 1, 3, 4, or 5, with a maximum nominal chromium content of 3 percent. Qualifications in the higher P-number metal to itself shall also qualify for the welding of that metal to a lower P-number metal but not vice versa. The only change permitted is the change in base metal to a lower P number. If the notch toughness is a requirement, the requirements applied to P numbers and group numbers are in addition to the preceding.

A welder is automatically qualified for a particular welding position if the test

WELDING QUALIFICATION POSITIONS

Qualification test Basic position	Also qualifies	
	Groove welds in	Fillet welds in
Flat position – 1G Plate and pipe horizontal / Pipe rolled while welding		Flat (Throat of weld vertical, 45°)
Horizontal position – 2G Plate and pipe vertical / Welds horizontal	Flat / Rolled	Flat / Weld horizontal / Horizontal
Vertical position – 3G Plate vertical / Weld vertical	Flat / Rolled	Flat / Weld vertical / Vertical

158

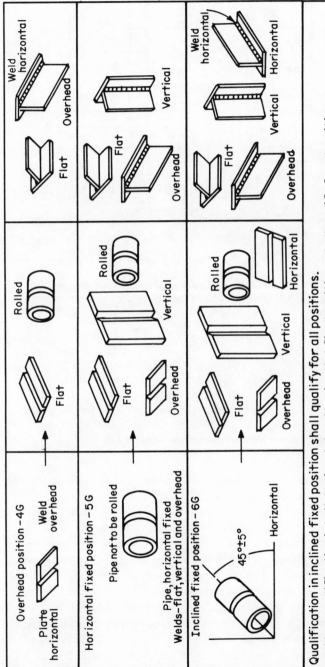

Overhead position – 4G

Plate horizontal → Weld overhead
Flat
Rolled

Horizontal fixed position – 5G

Pipe not to be rolled

Pipe, horizontal fixed Welds – flat, vertical and overhead
Flat → Overhead
Flat
Vertical
Rolled

Inclined fixed position – 6G

45°±5° Horizontal
Flat → Overhead
Overhead
Vertical
Horizontal
Rolled

Weld horizontal Overhead
Flat
Flat
Overhead
Vertical

Weld horizontal Horizontal
Flat
Overhead
Vertical

Qualification in inclined fixed position shall qualify for all positions.
Welders qualification in both horizontal and horizontal fixed positions shall qualify for all positions.
Procedure qualification of both groove and fillet welds shall be made on groove welds.
Welders passing the groove weld test need not make fillet weld test.
Welders passing the tests for fillet welds only, are qualified to make fillet welds only.

Fig. 6.1 Welding qualification positions.

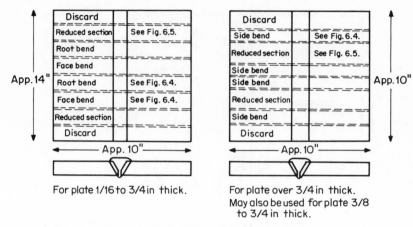

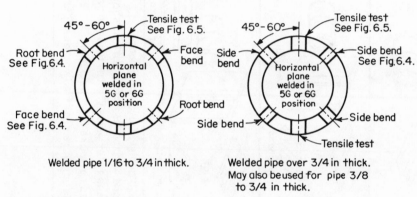

Fig. 6.2 Order of removal of welded transverse test specimens.

plates prepared for the welding procedure qualification pass the requirements. (See Section IX, Code Par. QW-301.2 and QW-303.)

A differentiation must be made here between a welder and a welding operator (Section IX, Code Par. QW-492 contains the definitions). A *welder* is one who has the ability to perform a manual or semiautomatic welding job. A *welding operator* is one who operates machine (automatic) welding equipment.

Welders passing the groove weld test need not make the fillet weld test. Welders passing the test for fillet welds only are qualified to make only fillet welds [Section IX, Code Par. QW-303.1, Code Fig. QW-462.4(b), and Code Par. QW-180]. Procedure qualification for both groove welds and fillet welds is made on groove welds (Code Par. QW-202.2).

The welding operator may be qualified with a test plate or pipe from which bend tests are removed. Although most shop welding is accomplished in the flat position with the weld metal deposited from above, this procedure is often not possible. Therefore, welds must be made in the position used in actual

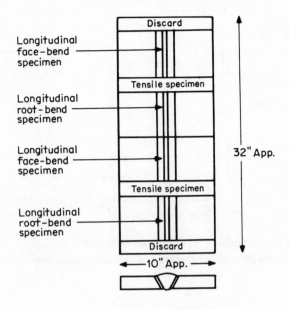

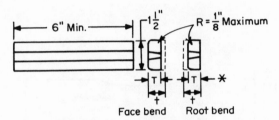

*For T see Code Fig. QW-462.3(b).

Fig. 6.3 Longitudinal test specimens—plates.

fabrication in order to qualify the welder (Section IX, Code Par. QW-303 and Code Fig. QW-461).

Some fabricators set up their procedures (Section IX, Code Par. QW-303 and Code Par. QW-461) for position 6G, pipe with the axis inclined to 45° to horizontal. Qualification in the inclined fixed position (6G) will also qualify for all positions. For performance qualification only, all position qualification for pipe may also be done by welding for the horizontal position (2G), the horizontal fixed position (5G), or the 6G position. (See Fig. 6.1.)

Qualification in these positions will also qualify the flat, vertical, and overhead positions (Section IX, Code Par. QW-303). It is to the fabricator's advantage to have one welder qualify in position 2G. Should this welder pass the test, this position also brings qualification for welding in the flat position (1G). Another welder then qualifies in the horizontal fixed position (5G). Should the tests be passed, this position also qualifies the welder in the flat, vertical, and overhead positions.

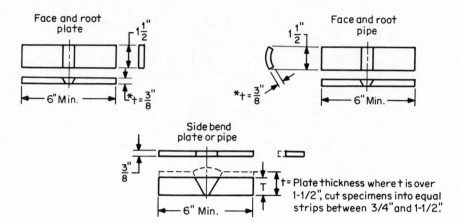

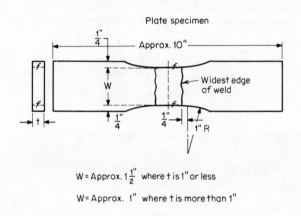

*For material less than 3/8", see Code Fig. QW-462.3a.

Fig. 6.4   Preparation of face, root, and side bend specimens.

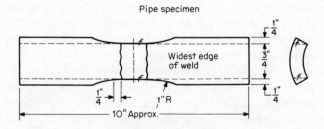

W = Approx. $1\frac{1}{2}$" where t is 1" or less

W = Approx. 1" where t is more than 1"

(Note that the pipe specimen and the plate specimen differ in size.)

Fig. 6.5   Preparation of reduced section tension test specimens.

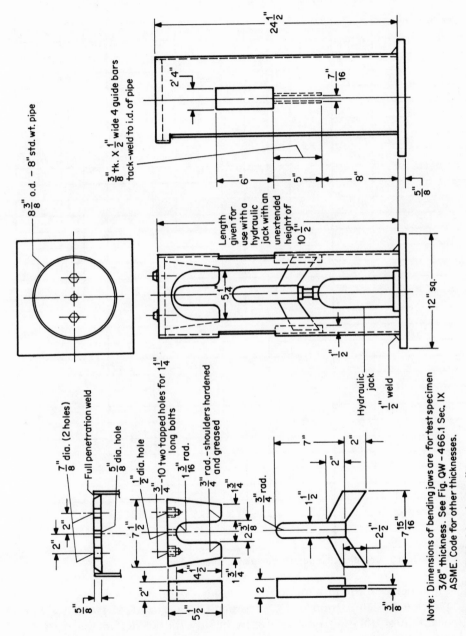

$8\frac{3}{8}''$ o.d. — 8'' std. wt. pipe

$\frac{3}{8}$ tk. X $1\frac{1}{2}''$ wide 4 guide bars
tack-weld to i.d. of pipe

$24\frac{1}{2}''$

2' 4''

$\frac{7}{16}''$

6''

5''

8''

$\frac{5}{8}''$

Length given for use with a hydraulic jack with an unextended height of $10\frac{1}{2}''$

$5\frac{1}{4}''$

12" sq.

$\frac{1}{2}''$

Hydraulic jack

$\frac{1}{2}''$ weld

$\frac{7}{8}''$ dia. (2 holes)

Full penetration weld

$\frac{5}{8}''$ dia. hole

$\frac{1}{2}''$ dia. hole

$\frac{3}{4}''$–10 two tapped holes for $1\frac{1}{4}''$ long bolts

$1\frac{3}{16}''$ rad.

$\frac{3}{4}''$ rad. – shoulders hardened and greased

$1\frac{3}{4}''$

2''

2''

$\frac{5}{8}''$

$7\frac{1}{2}''$

$2\frac{3}{8}''$

2''

$4\frac{1}{2}''$

$1\frac{3}{4}''$

$5\frac{1}{2}''$

$\frac{3}{4}''$ rad.

7''

2''

2''

$1\frac{1}{2}''$

$2\frac{1}{2}''$

$7\frac{15}{16}''$

2

$\frac{3}{8}''$

Note: Dimensions of bending jaws are for test specimen 3/8" thickness. See Fig. QW – 466.1 Sec. IX ASME. Code for other thicknesses.

Fig. 6.6 Guided-bend test jig.

163

WELDERS QUALIFICATIONS & PERFORMANCE RECORD LOG

Welder's name	Welder's symbol	Welding process qualified	Jan.	Feb.	Mar.	Apr.	May	June	July	Aug.	Sept.	Oct.	Nov.	Dec.

Fig. 6.7 Welders' Qualification and Performance Record Log.

It is also better to qualify in a position other than flat because the inclined fixed position (6G), horizontal (2G), vertical (3G), overhead (4G), and horizontal fixed position (5G) also automatically qualify the welder in the flat (1G) position. In the long run it is more economical to do this. Figure 6.1 indicates welding qualification positions at a glance.

Continuity of welders' qualification records should be maintained and a log kept of all welders' activities to indicate their continuity in specific welding procedures. If there is a gap of time in excess of 6 months in a specific procedure, the welder shall be requalified. Figure 6.7 gives an example of a "Welders' Qualifications and Performance Record Log."

If a manufacturer must qualify a welder on hard-to-get or costly material, such as stainless steel or alloy steel, qualification can be obtained by using type F5 austenitic electrodes on carbon steel plate (Section IX, Code Par. QW-310.5). Carbon steel may also be used in place of alloy steel. However, the total alloy content of the material for which carbon steel is substituted shall not exceed 6 percent. Where the total alloy content exceeds 6 percent, welders' qualification must be made on material of the same grade as required in the procedure specification. The Code allows single-welded butt joints without backing strips for some circumferential joints in positions 1G and 2G. If a manufacturer intends to use such joints on shells or pipes, the welders (Section IX, Code Par. QW-303) must be qualified. For this reason, some manufacturers have their welders qualified in metal-arc welding without a backing strip (Section IX, Code Fig. QW-469.2) so that they will also be qualified to metal-arc weld with a backing strip.

Many vessels used for cryogenic service require an impact test. Pressure vessels built of austenitic chrome-nickel steels have shown a reliable service record plus consistent good quality as demonstrated by the record of the impact tests required by the Code for vessels in service below $-20°F$. Therefore, Section VIII, Division 1, Code Par. UG-84 does not require impact tests of some vessels fabricated of these steels if impact tests are conducted as part of the welding procedure qualification tests. (See Fig. 6.8 on zone locations for removal of impact test specimens.)

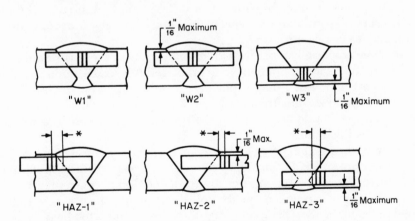

* Locate after etching so that maximum amount of heat – affected zone material is at the notch.

For welding procedure the heat – affected zone need not be impact tested when base material is exempted from impact tests. (CODE PAR. UG 84)

Fig. 6.8 Weld metal and heat-affected zone locations for impact test specimens.

To summarize, we have listed several suggestions to be followed in the preparation of welding procedures.

1. A WPS Code Form QW-482, should be written by a person with knowledge of production-shop needs and of the engineering-performance requirements.

2. The PQR Code Form QW-483 should be filed with the WPS Code Form QW-482.

3. Sketches and tables should be shown.

4. The existing Code should be reliably interpreted and the Code or specification translated to shop terminology.

5. Written procedures should be checked by reliable shop personnel who are to use these procedures.

Writing a welding procedure specification requires contributions of knowledge and experience from the shop personnel as well as the engineering department. If this is considered, a successful procedure can be used by the designing engineer to assist in designing a better weld at a lower cost.

WELDING DETAILS AND SYMBOLS

Welding details and symbols are an essential part of every shop drawing used in the design of welded pressure vessels; they guide the welder in the type of weld required, the proper size of weld metal to be deposited, and the size of bevel. Despite the rapid advances in welding proficiency in the past 25 years, many fabricating shops surprisingly still do not use either the details or the symbols.

There is a tendency in some pressure vessel shops toward overwelding, which not only increases production costs but also can result in distortion of the part welded, especially on stainless steel and aluminum vessels around nozzles, flanges, and accessory fittings.

Other shops, especially those with an incentive rate or piecework rate, have a tendency to underweld. When an Authorized Inspector discovers underwelding on a Code vessel, the proper size of weld will be insisted upon. The extra handling and welding this entails will add to the cost of the vessel. Underwelding that goes undetected, on the other hand, will result in a weakened vessel and possible failure.

Without realizing the possible consequences, many vessel manufacturers leave the amount of weld to be deposited to the discretion of the shop supervisor. As a result, weld sizes do not always meet Code requirements. Some welders or supervisors given the responsibility to produce a safe weld are inclined to make the weld "heavy enough." Usually, this is too heavy. Others, especially those on an incentive rate, take advantage of the fact and underweld. In the long run, underwelding may be the more costly because of the possibility of failure and the loss of future business.

Responsibility for correct design instructions to the shop must be taken by the engineering department. The designer knows the exact thicknesses required for the shell, heads, nozzles, and flanges and the stress loads to be carried by the welds.   The type and size of the welds required must be stated and clearly shown on the shop drawings either as a separate detail or by use of the appropriate symbol.   Both underwelding and overwelding may often be attributed to inadequate welding information on shop drawings.

The Code gives fillet weld sizes for nozzles and flanges in terms of throat dimensions (to facilitate load calculations).   The American Welding Society's "Standard Welding Symbols" gives fillet weld sizes in terms of leg dimensions, and it is this measurement that should be given to the shop rather than the throat dimensions.   Calculations for leg sizes are illustrated and explained in Fig. 6.9.   For example, if the Code-required throat dimension of a nozzle is 0.75 in, the leg size would be 0.75 divided by 0.707, or 1.06 in, and it is this last figure that should appear on the drawing.   If this is not done when the designer calls for a throat measurement of 0.75 in, the welder may assume a leg size of 0.75 in.

The person in the shop certainly cannot take the time to calculate the leg sizes of welds and should not be asked to do so.   It can easily be seen why all dimensions given to the shop should be prepared in advance.   Some Code

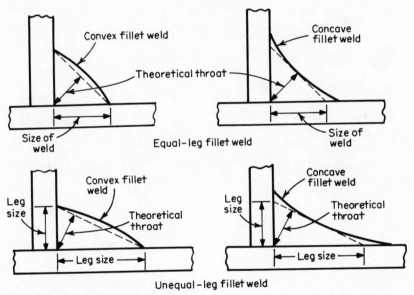

Fig. 6.9   Fillet weld size.

For equal-leg fillet welds, weld size is measured by the leg length of the largest inscribed right isosceles (two sides of equal length) triangle. Therefore,

$$Size = \frac{theoretical\ throat}{0.707}$$

For unequal-leg fillet welds, weld size is measured by both legs lengths of the largest right triangle that can be inscribed within the weld cross section. For calculating the allowable load that can be carried by a fillet weld, use the throat dimension. For shop drawings, use the leg dimensions.

AMERICAN WELDING SOCIETY

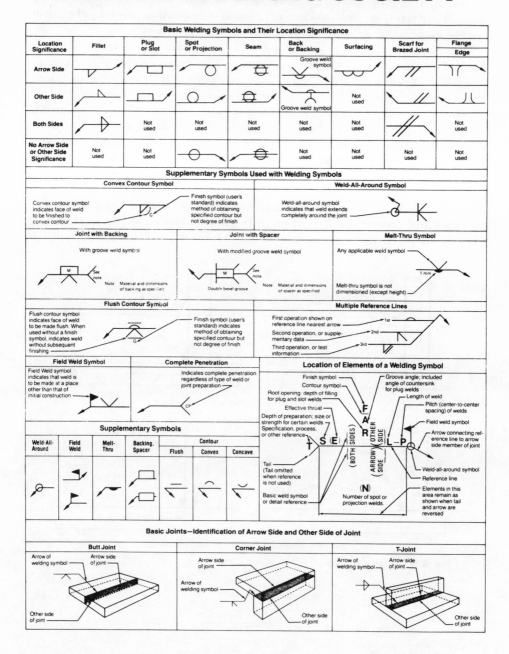

STANDARD WELDING SYMBOLS

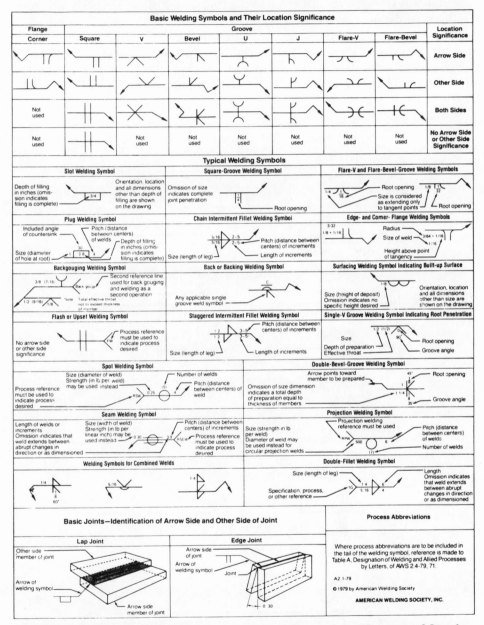

Fig. 6.10 **American Welding Society's standard welding symbols.** *(Courtesy of American Welding Society)*

vessel shops use throat dimensions as if they were the leg sizes, and the resulting welds naturally do not meet Code requirements. Correct weld design will be assured only if weld sizes are given to the shop as leg dimensions (shown either in the weld detail or weld symbol).

The American Welding Society's "Standard Welding Symbols" are standard for almost all codes and companies, although some companies make slight variations to meet their particular needs. A chart illustrating their use (Fig. 6.10) can be obtained from the American Welding Society (P. O. Box 351040,

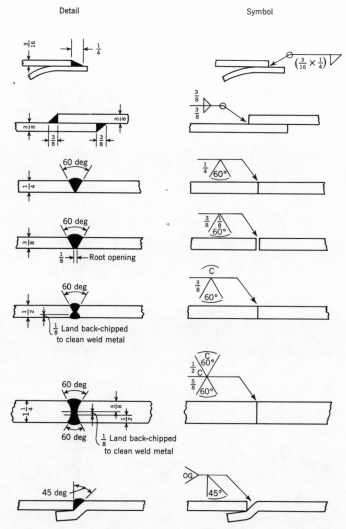

Fig. 6.11 Welding details and symbols for shell plates and heads.

Miami FL 33125). Many of these symbols, which are intended to cover all types of welding, are not used in the fabrication of pressure vessels. In fact, most pressure vessel manufacturers require only a few of them.

Figures 6.11 to 6.13 — which make no attempt to explain the function of all the AWS symbols — include details of the welded joints most often used in pressure vessel construction and the proper symbols to indicate those joints.

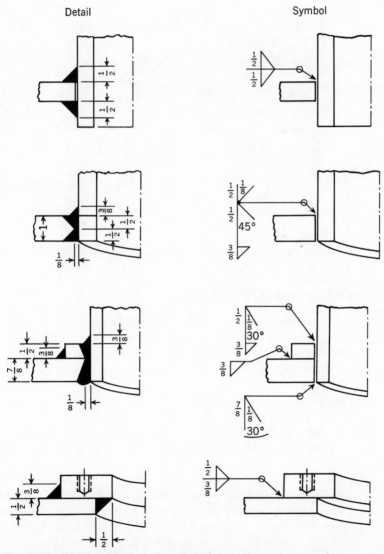

Fig. 6.12 Welding details and symbols for nozzles and connections.

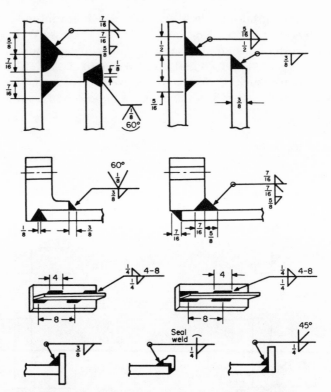

Fig. 6.13 **Miscellaneous welding details and symbols.**

JOINT PREPARATION AND FIT-UP

A sound weld is almost impossible without good fit-up. The importance of proper joint preparation of welding edges on longitudinal seams, girth seams, heads, nozzles, and pipe cannot be overemphasized. A good weld also cannot be obtained if the plate adjacent to the welding groove is covered with paint, rust, scale, grease, or oil.

The Code states that if flame cutting is used to cut plates, the edges must be uniformly smooth and free of all loose scale and slag accumulations. Before welding is undertaken, all joint surfaces for a distance of at least ½ in from the edge of the welding groove on ferrous material, and at least 2 in on nonferrous material, should be free of all paint, scale, rust, grease, or oil.

The Code permits a maximum offset, after welding, of one-quarter the plate thickness at the joint, with a maximum allowable offset of ⅛ in in the longitudinal joints. For circumferential seams, greater tolerances are permitted on plates more than ½ in thick. In pressure vessels of larger diameter in which a weld may be chipped to clean metal from the opposite side and welded the full thickness of the plate, this permitted offset is satisfactory. If single-welded butt joints are used, however, special care must be taken in aligning and separating the edges to allow sufficient clearance for complete penetration and fusion at the bottom of the joint.

On smaller vessels and pipe that cannot be welded from the inside, special care must be taken when fitting-up the abutting sections. This is especially true in pipe of uneven thickness, in which the proper fit-up of inside diameters is almost impossible to obtain. If there is sufficient thickness in the pipe wall, however, machining can provide the proper fit-up. If the inside diameters are not equal, it is difficult to achieve full penetration at the root of a weld. As such penetration is needed to prevent root pass defects and cracks, alignment of these diameters is very important.

Good joint preparation is especially essential in vessels intended for high-pressure and critical service. Probes of welds that appear sound often reveal cracks that failed to show up in the radiographs made when the joint was first welded. As time went on, the small, unseen crack opened up and became larger — the probe showing the crack to have started at the edge of the plate at the backing ring because of a poor fit or a tack weld that was not properly removed. This sort of thing has happened so often that large companies regularly examine pipe that is in service at critical pressures and temperatures. They have also set up more rigid standards of inspection for their new construction, requiring a fit-up tolerance of 0.005 in and the use of heliarc welding on the first pass to eliminate any incipient cracking that might develop into a major defect. Not all companies have inspection departments to make such field inspections, however, and often the unit cannot be shut down long enough to permit such examinations.

When investigation reveals poor fit-up or a cracked tack weld before welding as the cause of an accident, the manufacturer of the vessel or pipe is often unaware that such practices are permitted in the shop. Although care may be taken to prevent like accidents in the future, preventive measures should always be in practice beforehand.

EFFECTS OF WELDING HEAT

Welding metallurgy is the study of the properties of welded metals and the thermal and mechanical effects occurring during the welding process, including the movements of the atoms in the liquid state and the formation of the grain structure of the metal in changing from the liquid to the solid state. Some understanding of the behavior of metal during welding is useful and necessary to the designer and fabricator of pressure vessels in order to help prevent dangerous residual stresses or defects in the welded joints.

Many metallurgists think of metals in terms of a "life cycle," that is, a metal is "born" in the melting operation, "dies" when it fails in service or is no longer used, and is even subject in the interim to "diseases." It is known that heat treatment can change grain structure, which in turn may affect the physical properties. A metal can be hardened or softened, toughened or made more ductile.

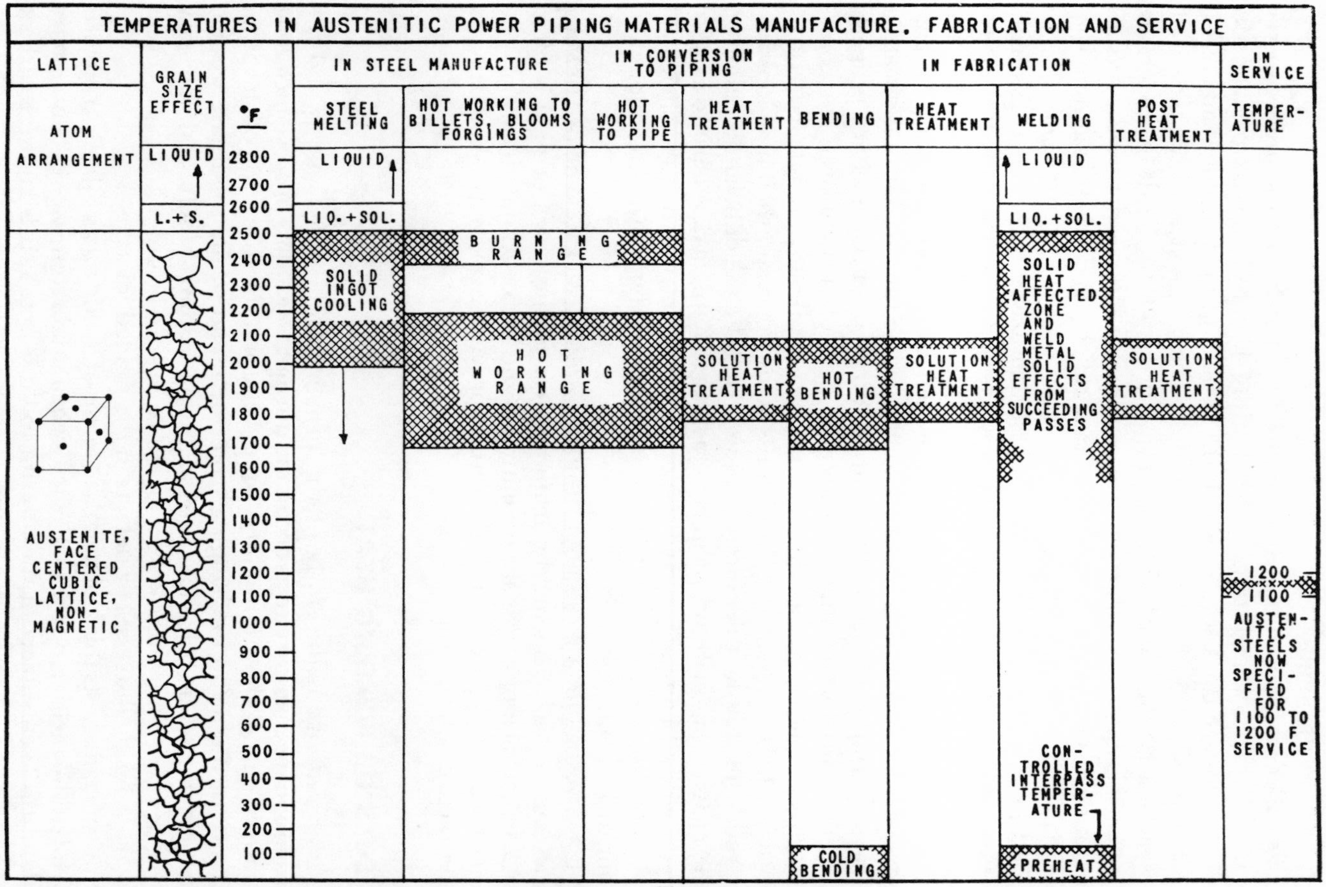

Fig. 6.14 Metallurgical changes in austenitic materials at different fabrication and service temperatures.

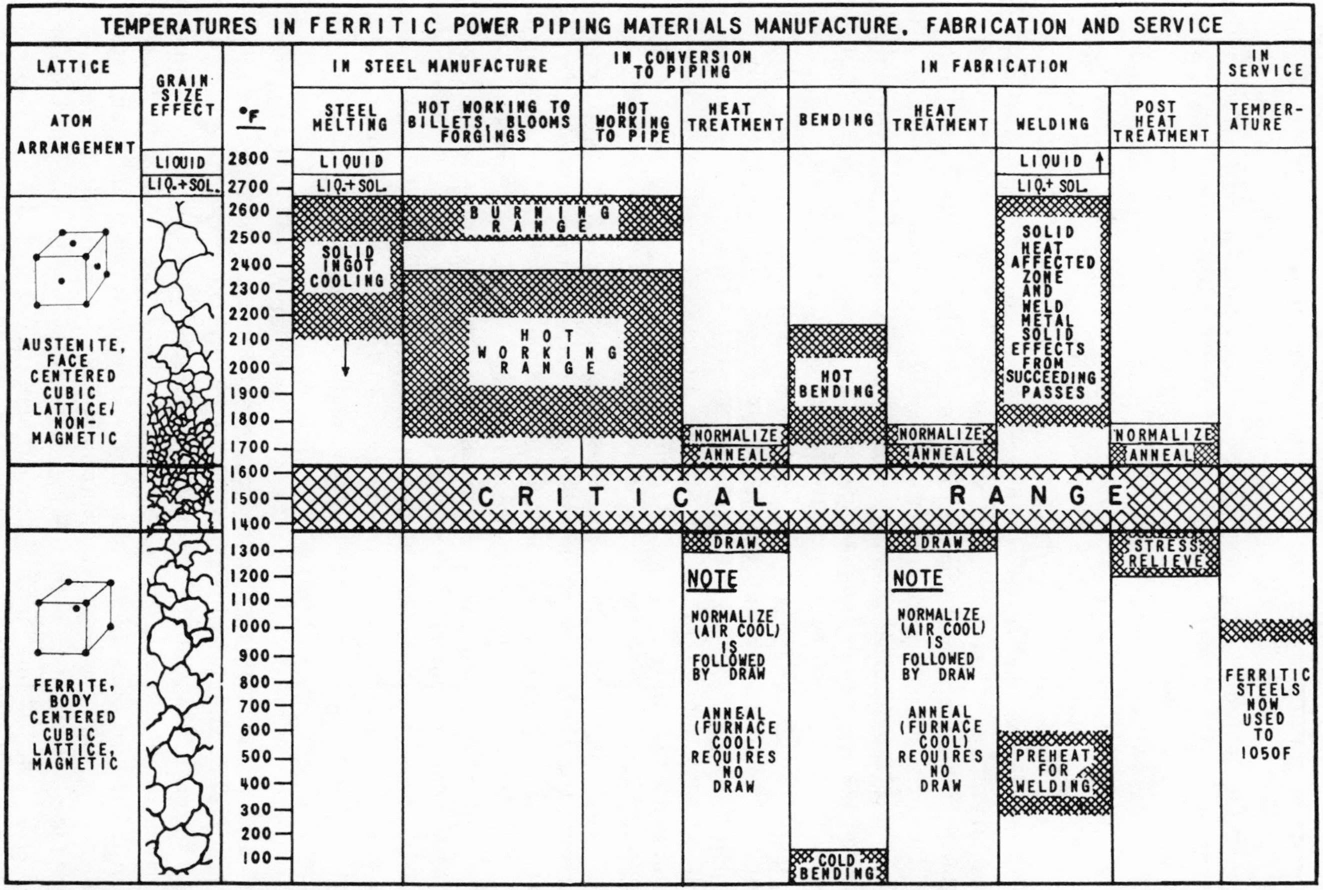

TEMPERATURES IN FERRITIC POWER PIPING MATERIALS MANUFACTURE, FABRICATION AND SERVICE

LATTICE	GRAIN SIZE EFFECT	°F	IN STEEL MANUFACTURE		IN CONVERSION TO PIPING		IN FABRICATION				IN SERVICE
ATOM ARRANGEMENT			STEEL MELTING	HOT WORKING TO BILLETS BLOOMS FORGINGS	HOT WORKING TO PIPE	HEAT TREATMENT	BENDING	HEAT TREATMENT	WELDING	POST HEAT TREATMENT	TEMPERATURE

AUSTENITE, FACE CENTERED CUBIC LATTICE/ NON MAGNETIC

FERRITE, BODY CENTERED CUBIC LATTICE, MAGNETIC

°F scale: 2800, 2700, 2600, 2500, 2400, 2300, 2200, 2100, 2000, 1900, 1800, 1700, 1600, 1500, 1400, 1300, 1200, 1100, 1000, 900, 800, 700, 600, 500, 400, 300, 200, 100

LIQUID
LIQ.+SOL.

SOLID INGOT COOLING

BURNING RANGE

HOT WORKING RANGE

SOLID HEAT AFFECTED ZONE AND WELD METAL SOLID EFFECTS FROM SUCCEEDING PASSES

LIQUID
LIQ.+SOL.

NORMALIZE ANNEAL

NORMALIZE ANNEAL

NORMALIZE ANNEAL

HOT BENDING

CRITICAL RANGE

DRAW

DRAW

STRESS RELIEVE

COLD BENDING

PREHEAT FOR WELDING

FERRITIC STEELS NOW USED TO 1050F

NOTE

NORMALIZE (AIR COOL) IS FOLLOWED BY DRAW

ANNEAL (FURNACE COOL) REQUIRES NO DRAW

Fig. 6.15 Metallurgical changes in ferritic materials at different fabrication and service temperatures.

Welding is an operation in which separate pieces of metal are fused together to produce an integral section. Some metals are more difficult to weld than others. However simple the welding operation may appear, it is noteworthy that certain phenomena that occur in the making of steel also take place in the electric arc welding of steel products (plate, pipe, and the like), but more rapidly and on a smaller scale. Substantially the same processes and metallurgical changes are involved. The heat of the arc melts a quantity of plate metal and the electrode. Chemical ingredients may be added by the coating on the electrode or by fluxes. The heat of the arc also melts or vaporizes the flux or coating, causing a gaseous shield to surround the arc stream and pool of molten metal and do the important job of keeping out all the elements of the air (including nitrogen, hydrogen, and oxygen).

During the cooling and solidification process, the behavior of the molten metal in the weld joint is similar to that of molten metal poured into a mold to form a steel casting. The mold in the welded joint is the solid parent metal; the melt consists of the molten metal of the electrode plus the liquified part of the parent metal, the two combining to form a casting within the solid parent metal. When the metal is in the liquid state, its atoms can move about at random. When it cools to the temperature at which it starts to solidify, the atoms arrange themselves into a regular pattern (for iron and steel the pattern is a cube). This atomic arrangement and the corresponding effect on grain size are influenced by the rate of cooling from the liquid to the solid state and also by the composition.

Mr. H. S. Blumberg, a New York consulting metallurgist, has prepared two charts on the changes that take place at different temperatures in ferritic (carbon and low-alloy steels) and austenitic (chromium-nickel stainless steels that are nonmagnetic) materials (Figs. 6.14 and 6.15). These charts show some of the simple fundamentals of atom arrangement and the grain size effects that take place when the metals change from the liquid to the solid state. They also show the important grain size changes in the "critical range" (the temperature range, with transition points, in which the structural changes take place) of ferritic carbon and low-alloy steels, including the temperatures used to normalize, anneal, draw, and stress-relieve these steels.

Grain size is influenced by the maximum temperature reached by the metal, the length of time the metal is held at a specific temperature, and the rate of cooling. For example, annealing, which involves heating to temperatures above the critical range followed by slow cooling, is used to refine the grain size to privide ductility and softness. Stresses can be relieved by heating beneath the critical range, but unless temperatures exceed that range, full grain refinement does not take place. Normalizing is the process by which the steel is heated to approximately 100°F above the critical range (as in annealing) and then cooled in still air. This heat treatment also refines the grain size and improves certain mechanical properties. Refinement of ferritic steel can be easily achieved by heat treatment because of the movement of the atoms in the solid metal. Note that the austenitic steels do not have a critical range and that consequently their grain size cannot be changed by heat treatment alone.

The production of a fusion weld naturally results in heating of some of the solid metal adjacent to the weld. The temperature reached by the metal in this zone varies from just below the melting point at the line of fusion of just above room temperature at a distance of a few inches from the weld. The effect of the welding heat on the parent metal depends on the temperature reached, the time this temperature is held, and the rate of cooling after welding. It is therefore necessary to understand the changes that take place in the zone where the metal is heated to very high temperatures without ever reaching the molten condition.

Usually, the weld metal zone is stronger than the parent metal but has less ductility. The heat-affected zone is the region where most cracking defects are likely to occur because the grain structure becomes coarse just beyond the fusion line and the ductility is lowest at this point. This reduction of ductility must be taken into consideration and measures taken to minimize the effects of total welding heat, for example: (1) the correct choice of parent material and welding electrode, (2) preheating before welding, (3) use of the proper welding procedure, and (4) postweld heat treatment after welding.

PREHEATING

Preheating prior to welding will eliminate the danger of formation of cracks, reduce hardness, reduce distortion, and reduce or prevent shrinkage stresses. Basically, preheating is a means of raising the temperature of a metal to a desired level above the surrounding ambient temperature before welding. It is usually applied between 125 and 500°F but may involve temperatures up to 1200°F, depending upon the material to be welded.

Preheating reduces the thermal strains set up during welding that may crack a weld. Preheating is especially necessary in thicker carbon steel plates in order to guard against the high thermal conductivity of steel when the heat of the welding arc is absorbed rapidly by the metal adjacent to the weld. The rate of cooling is slowed by preheating and the hardness is decreased. This slowing of the cooling rate allows more time for metal transformation to a favorable structure minimizing hard zones and cracks; thus preheating retards the formation of undesirable metallurgical structures in metals having high hardness and susceptibility to cracking. Preheating dries up moisture, reducing the incidence of hydrogen and oxygen accumulation in a weld zone; consequently, it minimizes porosity, embrittlement, and cracking in the weld and heat-affected zone. Even though most medium- and low-carbon steels are welded without preheat, the metal adjacent to the start of a weld should be preheated thoroughly.

Preheating is an essential requirement when welding low-alloy steels in the P-3, P-4, and P-5 material groups. Most alloy steels permit higher stress values at high temperatures and have greater strength at low temperatures and allow higher notch toughness. In these steels, the required preheat temperature is stated and may be mandatory in order to comply with Code requirements. The higher yield strength of these metals allows the use of higher stress values.

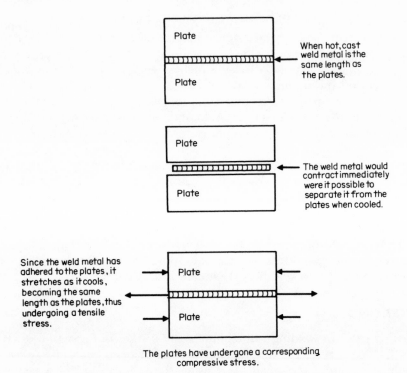

Fig. 6.16   Good design provides for expansion and contraction.

These groups require careful planning of the welding procedure specifications in order to prevent cracking in the heat-affected zone, where the microstructure of the steel has been modified by the heat of welding. This heat-affected zone is comparatively hard and is formed during the welding operation; in some steels it may be necessary to hold the preheat temperature until the weld can be postweld heat-treated to avoid cracking.

Preheating the metal before welding has proved a dependable method of securing a sound weld. Preheating helps equalize and distribute the stresses evenly by distributing the expansion and contraction forces over an enlarged area to which the stress differential is applied due to the heat of welding. (See Fig. 6.16.)

Appendix R, "Preheating," in Section VIII, Division 1, refers the reader to other Code paragraphs and to Section IX of the Code, "P Numbers and Groups," where preheating is required.

POSTWELD HEAT TREATMENT

The process of postweld heat treatment consists of uniform heating of a vessel or part of a vessel to a suitable temperature for the material below the critical range of the base metal, followed by uniform cooling. This process is used to release the locked-up stresses in a structure or weld in order to "stress-relieve" it. Due to the great ductility of steel at high temperatures, usually above 1100°F, heating the material to such a temperature permits the stresses caused by deformation or straining of the metal to be released. Thus postweld heat treatment provides more ductility in the weld metal and a lowering of hardness in the heat-affected zone (HAZ). It also improves the resistance to corrosion and caustic embrittlement. The proper heat treatment before and after fabrication is one that renews the material as near as possible to its original state.

Heat treatment procedures require careful planning and will depend on a number of factors: temperatures required and time control for material; thickness of material and weldment sizes; and contour, or shape, and heating facilities. To be effective, heat treatment must be carefully watched and recorded. Code Par. UW-49 requires that the Inspectors themselves be assured that postweld heat treatment is properly performed.

Furnace charts should be reviewed to determine if the rate of heating, holding time, and cooling time are correct. If a vessel is to be fully postweld heat-treated, the furnace should be checked to make sure that the heat will be applied uniformly and without flame impingement on the vessel.

Vessels with thin walls or large diameters should be protected against deformation by internal bracing. The vessel should be evenly blocked in the furnace to prevent deformation at the blocks and sagging during the postweld heat treatment operation.

During this operation, both the increase and decrease in temperature must be gradual in order to allow uniform temperatures throughout. Where the difference in thickness of materials is greater, the rate of temperature change should be slower. The applicable requirements of Code Par. UW-10, UW-40, UW-49, UCS-56, UCS-85, Table UCS-56, Par. UHT-56 or Table UHT-56, Par. UHT-79, UW-80, Table UHA-32, Par. UHA-32, UHA-105, UCL-34, UNF-56, ULW-26, ULT-56, UF-32, and UF-52 must be followed in the postweld heat treatment of a Code vessel.

CONDITIONS REQUIRING POSTWELD HEAT TREATMENT

The Code requires postweld heat treatment of pressure vessels for certain design and service conditions (see Code Par. UW-40 for prescribed procedures). Postweld heat treatment requirements (according to materials, thicknesses, service, etc.) include the following:

1. All carbon steel weld joints must be postweld heat-treated if their thickness exceeds 1.25 in or exceeds 1.5 in if the material has been preheated to a minimum temperature of 200°F during welding (Code Par. UCS-56 and Table UCS-56).

2. Vessels containing lethal substances must be postweld heat-treated (Code Par. UW-2).

3. Carbon and low-alloy steel vessels for service at temperatures below

TABLE 6.1 Postweld heat treatment requirements

Carbon and Low-Alloy Steels

P number and materials	Postweld heat treatment required* when thickness exceeds
P-1 Carbon steels (40,000 to 75,000 psi tensile strength)	1.25 in or 1.50 in if preheated 200°F
P-3 Alloy steel (chromium content ¾ percent maximum) Alloy steels (total alloy content 2 percent maximum)	0.625 in Production weld thickness shall not exceed the procedure qualification test plate thickness
P-4 Alloy steels (chromium content between ¾ and 2 percent) Alloy steels (total alloy content 2¾ percent maximum)	0.0 in Not required for circumferential butt welds of tubes if chromium content does not exceed 3 percent. Nominal outside diameter of 4 in maximum, thickness ½ in. Minimum, preheat 250°F
P-5 Alloy steels (total alloy content 10 percent maximum)	0.0 in Same as in P-4 except that minimum preheat is 300°F
P-9 Nickel alloy steels	0.0 in

* These thicknesses may vary under certain conditions; check notes in Code Table UCS-56.

−20°F must be postweld heat-treated unless exempted from impact tests (Code Par. UCS-67).

4. Ferritic steel flanges must be given a normalizing or fully annealing heat treatment if the thickness of the flange section exceeds 3 in (Code Par. UA-46).

5. Unfired steam boilers with a design pressure exceeding 50 psi must be postweld heat-treated (Code Par. UW-2).

6. Castings repaired by welding to obtain a 90 or 100 percent casting factor must be postweld heat-treated (Code Par. UG-24).

7. Carbon and low-alloy steel plates formed by blows at a forging temperature must be postweld heat-treated (Code Par UCS-79).

8. Some vessels of integrally clad or applied corrosion-resistance lining material must be postweld heat-treated when the base plate is required to be postweld heat-treated (Code Par. UCL-34).

9. Some high-alloy chromium steel vessels must be postweld heat-treated (Code Par. UHA-32, UHA-105, and Table UHA-32).

10. Some welded repairs on forgings require postweld heat treatment (Code Par. UF-37). (See Table 6.1.)

SEVEN

Nondestructive Examinations

Nondestructive examinations allow trained personnel to evaluate the integrity of pressure vessel parts without damaging them. Various sections of the *ASME Boiler and Pressure Vessel Code* now refer to the ASME Code Section V, "Nondestructive Examinations," for nondestructive examination methods to detect internal and surface discontinuities in welds, materials, components, and fabricated parts.

Subsection A of Section V describes the methods of nondestructive examination to be used if referenced by other Code sections. They include the following:

General requirements
Radiographic examination
Ultrasonic examination
Liquid penetrant examination
Magnetic particle examination
Eddy current examination of tubular products
Visual examination
Leak testing

Subsection B lists American Society of Testing Material Standards covering nondestructive examination methods which have been accepted as Code standards. These are included for use or for reference, or as sources of technique details which may be used in preparation of manufacturer's procedures.

The acceptance standards for these methods and procedures are stated in the referencing Code sections. Subsection B includes radiographic, ultrasonic, liquid penetrant, magnetic particle, eddy current, leak testing, and visual examination standards.

Subsections A and B of Section V, "Nondestructive Examination," provide rules and instruction in the selection of nondestructive test methods, development of inspection procedures and evaulation of test indications. With the growing emphasis upon quality, nondestructive evaulation is becoming increasingly important in the construction of pressure vessels.

These examinations should be made according to a procedure that satisfies the requirements of the referencing Code and should be acceptable to the Authorized Inspector.

For the pressure vessel fabricator, nondestructive examinations provide a means of spotting faulty welds, plates, castings, and forgings before, during, and after fabrication. For the user, it offers a maintenance technique that more or less assures the safety of a vessel until the next scheduled inspection period. A serious flaw in a welded seam or a dangerously diminished plate thickness can cause a costly shut down, or worse, a serious accident. Such defects must be located as soon as possible. Visual inspection is not enough. A weld may look perfect on the outside and yet reveal cracks, lack of penetration, heavy porosity, or undercutting at the backing strip when the weld is given a nondestructive examination by radiography or ultrasonic tests. The internal condition of a material or weld can be accurately appraised only by use of the proper testing techniques.

Accident prevention requires many skills. Nondestructive testing is one that should be used more often. Since no weld or material is perfect, nondestructive testing is an excellent way to examine the internal conditions of a material. Nondestructive testing is often used on materials supplied to the fabricator of pressure vessels. Additional tests are made during fabrication and are also used during a vessel's operating life to assure safe operation and to gage plate thicknesses, thus enabling accurate prediction of the life of the vessel. Flaws are not uncommon in plate material and welds during any stage of fabricating a vessel. The point is to find them as soon as possible in order to prevent difficulties later. For example, a serious flaw in material or a welded seam may cause an accident. Many such accidents can be prevented with nondestructive tests.

In this chapter, some nondestructive test procedures are explained with their regard toward Code pressure vessel fabrication. The understanding of their use in addition to visual inspection is essential to safe pressure vessel fabrication and operation.*

Table 7.1 and Figs. 7.1–7.5 analyze and illustrate some nondestructive testing procedures commonly used in Code pressure vessel fabrication. An authorized Code vessel manufacturer's quality control manual must describe the system that is employed for assuring that all nondestructive examination personnel are qualified in accordance with the applicable Code.

There should be written procedures to assure that nondestructive examination requirements are maintained in accordance with the Code. If the nonde-

---

* For more complete instructions on the methods of nondestructive examinations, refer to Section V of the *ASME Boiler and Pressure Vessel Code.*

TABLE 7.1   Analysis of nondestructive testing techniques

Applications	Advantages	Remarks
*Radiographic inspection (x-ray)**		
For examination of internal soundness of welds, castings, forgings, and plate material (see Fig. 7.1)	1. Gives sharper definition and greater contrast for thickness up to 3 in 2. Gives a graphic and permanent record indicating size and nature of defects 3. Offers established standards of interpretation for guidance 4. Radiation source can be turned off when not in use	1. Protective precautions necessary to protect personnel in surrounding area 2. Trained technicians required to take and interpret film 3. See pages 141 to 145 for interpretation of radiography
Radiographic inspection (gamma ray)		
For examination of internal soundness of welds, castings, forgings, and plate material (see Fig. 7.2)	1. More suitable for heavy thickness 2.. Portable equipment 3. Source holder permits use through small openings 4. Lower initial cost 5. Requires no cooling mechanism 6. Gives a graphic and permanent record 7. No tube replacement	1. Government license is required for possession and use of isotopes 2. Protective precautions necessary 3. Sensitivity is not as sharp as x-ray, and persons not properly trained in interpreting film may underestimate seriousness of a defect 4. Energy cannot be adjusted, so isotope must be chosen to meet sensitivity requirements and thickness of material

* See Code Par. UW-51 and UW-52.
† See Code Appendix 6.

TABLE 7.1   Analysis of nondestructive testing techniques (cont.)

Applications	Advantages	Remarks
Magnetic particle inspection†		
For locating surface and subsurface defects that are not too deep (see Figs. 7.3 and 7.4) Technique employs either electric coils wound around the part or prods to create a magnetic field. A magnetic powder applied to the surface will show up defects as local magnetic fields. The nature of the defects will be revealed by the way the powder is attracted.	1. Useful for the inspection of nozzle and manhole welds for which radiography would be difficult at best and most often impossible 2. Reveals small surface defects that cannot be detected radiographically 3. Useful in weld repair to assure removal of defect before rewelding 4. Can be used to detect laminations at plate edges	1. Useful only for magnetic material 2. Not suitable for defects parallel to magnetic field 3. Training needed to evaluate visual indications of defects 4. Prod methods of testing may mar the surface of a machined or smooth part 5. Magnetic particle examination methods (see Code Appendix 6)
Penetrant inspection (dye)		
For locating surface defects A liquid dye penetrant is applied to a dry clean surface and allowed to soak long enough to penetrate any surface defects. After a time interval up to an hour, the excess penetrant is wiped off and a thin coating of developer applied. The penetrant entrapped in a defect will be drawn to the surface by the developer, and the defect will be indicated by the contrast between the color of the penetrant and that of the developer.	1. Especially applicable for nonmagnetic material 2. Useful for the inspection of nozzle and manhole welds when radiography is impossible 3. Easy to apply, accurate, fast, and low cost	1. Detects only defects that are open to the surface 2. Not practical on rough surfaces 3. Liquid penetrant examination methods (see Code Appendix 8) 4. Take care in applying dye penetrant to ensure that excessive amounts of vapors are not inhaled

Applications	Advantages	Remarks
Penetrant inspection (fluorescent)		
For locating defects that run through to surface Penetrant is applied by spraying, dipping, or brushing. Excess penetrant is washed from surface with a water spray and the surface allowed to dry. Dry powder or water-suspension developer is applied to part to draw penetrant to surface. Penetrant glows under black (ultraviolet) light. For leak test, penetrant is applied to one side and the other side is examined under black light for indications of glow.	1. Easy to perform 2. Defects show clearly under black light 3. Especially applicable for nonmagnetic material 4. Can be used on rough surfaces 5. Detects porosity and defects that would otherwise be hard to find	1. Detects only defects that are open to the surface 2. Black light requires a source of electricity 3. Not very effective as leak test for plates more than ¼ in thick

TABLE 7.1 Analysis of nondestructive testing techniques (cont.)

Applications	Advantages	Remarks
	Ultrasonic inspection	
For detecting defects in welds and plate material and for gaging thickness of plate (see Fig. 7.5) High frequency sound impulses are transmitted thrugh a search unit, usually a quartz crystal. This search unit is held in intimate contact with the part being tested, using an intermediary such as oil or glycerine to exclude air. The sound waves pass through the part being tested and are reflected from the opposite side or from a defect. The time of travel shows up the defect on a cathode-ray tube or scope. This scope can also measure depth of crack or defect. Although the impulse instrument used to detect defects will also give thickness measurements, the resonance instrument is actually an electronic micrometer designed to measure thickness from one side of material within tolerances of ±0.002 in.	1. Portable equipment 2. Access from one side of part being tested is possible 3. Reveals small root cracks and defects not indicated by radiographic film, especially in thick-walled vessels 4. Thickness measurements are rapidly made 5. Good for detection of laminated plates 6. Can be used in nuclear vessels, where induced radiation eliminates the use of radiography 7. Remote inspection can be made in hostile environment 8. Can be used to determine whether use or corrosion has affected the thickness of vessel walls and piping	1. Training required for interpretation of visual indications of defects 2. Rough surfaces must be made smooth if crystal is to make contact with part 3. Photographs must be taken to provide permanent records 4. Ultrasonic test methods when required (see Code Appendix 12) 5. Not very effective on welds with backing rings

Applications	Advantages	Remarks
	Eddy current testing	
Eddy current testing is based upon the principles involving circulating currents into an electrically conductive article and observing the interaction between the article and the currents	1. Detects small discontinuities 2. High-speed testing 3. Accurate measurement of conductivity 4. Checks variation in wall thickness 5. Noncontacting	1. Limited to use with conductive materials or conductive base materials 2. Depth of penetration restricts testing to depths of less than ¼ in in most cases 3. The presence of strong magnetic fields will cause erroneous readings 4. Testing of ferromagnetic metals is sometimes difficult 5. Training required for interpretations of visual defects and indications

structive examination is subcontracted, the person should be named who is responsible for checking the subcontractor's site, equipment, and qualification of procedures to assure that they are in conformance with the applicable Codes.

In a nondestructive examination certification program, the manufacturer has the responsibility for qualification and certification of NDE personnel. To obtain these goals, the manufacturer may train and certify personnel or may hire an independent laboratory that has qualified NDE procedures and personnel in accordance with the reference Code section and Section V to perform these duties. If others do the work, it must be stated in the quality control

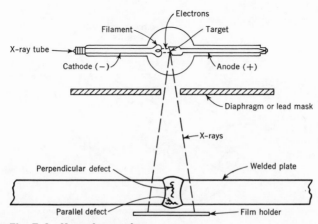

Fig. 7.1 X-ray inspection.

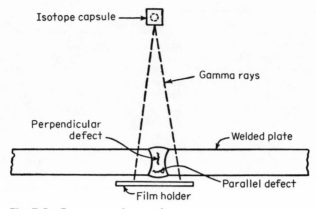

Fig. 7.2   **Gamma-ray inspection.**

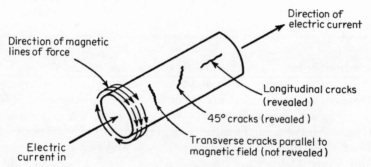

Fig. 7.3   **Magnetic particle inspection (circumferential magnetic field).**

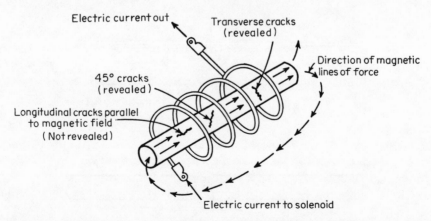

Fig. 7.4   **Magnetic particle inspection (longitudinal magnetic field).**

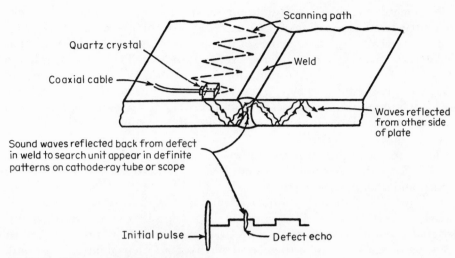

Fig. 7.5 Ultrasonic weld inspection.

manual what department and person (by name and title) are responsible for ensuring that the necessary quality control requirements are met, because the manufacturer's responsibility is the same as if the parent company had done the work.

When required by the referencing section of the pressure vessel Code, personnel performing nondestructive examinations must be qualified in accordance with SNT-TC-1A of the American Society for Nondestructive Testing. SNT-TC-1A, "Recommended Practice for Nondestructive Examining Personnel Qualification and Certification," is published by the American Society for Nondestructive Testing, Inc., P.O. Box 21142, Columbus, OH 43221.

It is the responsibility of the manufacturer to conduct training and qualification tests of nondestructive examination personnel in accordance with the requirements of SNT-TC-1A, to arrange for eye examinations for the personnel interpreting nondestructive examinations, and to have audits made of the system in order to assure that qualified procedures and personnel are being used. The manufacturer must also prepare and maintain test examinations as well as test procedures which comply with Section V of the Code and SNT-TC-1A. Only personnel qualified to perform examinations should be assigned this task.

Adequate nondestructive examining procedures are an essential part of any well-managed operation. Procedure format and content must be appropriate to the specific requirements of the shop. Without a written procedure, there is no guarantee as to the effectiveness or repeatability of the test. Nondestructive testing should be performed in accordance with a written procedure. Each manufacturer should follow the procedure in accordance with the requirements of the referencing section of the Code and ASME Code Section V, "Nondestructive Examinations."

The manufacturer is responsible for seeing that all personnel performing nondestructive examinations for Code vessels are competent and knowledgeable in applicable nondestructive examining requirements. They must be qualified in accordance with the requirements of SNT-TC-1A. The intent of SNT-TC-1A is to serve as a guide for the qualification and certification of nondestructive test personnel. This document establishes the general framework for a qualification and certification program. It explains the various levels of qualification. It gives the requirements and shows the type of examination required. Certification may take any one of several forms. The manufacturer may employ persons previously trained and certify these people. If this is done, the responsibility would be the same as if the manufacturer had trained and qualified these persons.

The manufacturer may conduct a training program to meet the requirements of SNT-TC-1A. This responsibility would be carried out through a designated employee who meets Level III requirements. This employee will train the persons, examine them, and certify as to their ability to do nondestructive examinations. Through the examinations, the employee will be able to judge the qualifications of these persons for Level I and Level II certification.

The American Society of Nondestructive Testing has programmed training courses to help a manufacturer write a training program to conduct the courses, which are usually given by a qualified Level III individual who uses questions derived from the ASNT course.

In addition to the required general examination, a specific written examination is necessary. The specific examination should reflect the procedures and equipment used in the plant.

In the practical examination, test objects should be similar to those fabricated by the manufacturer applying the written procedures and the same equipment used in production. The examinee must be able to use the written procedures and equipment.

In summarizing, a manufacturer must have qualified personnel. In addition, the manufacturer should have written procedures if the shop does nondestructive examinations. The Code requires that nondestructive examining personnel be trained and qualified, also that their qualifications be documented as required in the referencing Code and SNT-TC-1A with its supplements and appendixes. It cannot be too strongly emphasized that nondestructive examination of ASME Code pressure vessels must be performed by qualified personnel.

LEAK TESTING

Several leak-detecting methods are used in testing pressure vessels. One is the *halogen diode detector testing method* which is generally performed by pressurizing the vessel being tested with an inert gas, usually freon or a trace of freon, and examining the areas to be tested with a device that is sensitive to freon. The presence of a leak will give an audible alarm, be indicated on a meter, or may light up an indicator light.

Another method consists of pressurizing the vessel to be tested with air containing a freon gas. A halide torch, whose flame changes color when a leak appears, is employed to search out leaks.

Often a *helium mass spectrometer* leak detector is used. It measures a tracer medium, usually helium gas. If a leak is present, the leak detector detects the helium gas and gives an electrical signal. In the helium mass spectrometer test, the tracer medium helium gas passes through a leak from the pressurized side to the nonpressurized side. The test can be made under a vacuum, pressure test, or a combination of both.

A common method in pressure vessel testing is the *gas and bubble formation test*. Its purpose is to detect leaks in a vessel by applying a solution which will form bubbles at the leak area when gas passes through it.

ACOUSTIC EMISSION TESTING

Acoustic emission testing is a recent NDE method that has been successfully used in industry with hydrostatic tests and in service examination of pressure vessels and piping. The acoustical unit continuously cautions for possible flaws in pressure retaining parts.

The phenomena involve the emission of minute pulses of elastic energy in materials under stress. These emissions, when properly detected and analyzed, provide information concerning the discontinuities contained in the structure under stress.

Once detected, follow-up analysis of these signals verifies the location and meaning of a particular discontinuity.

Acoustic emissions system use piezo electric transducers to detect the stress waves generated by discontinuities in materials under stress. These transducers are attached directly to the structure being evaluated.

The number of transducers needed to test a structure is determined by its complexity and not its size.

Twelve to fifteen may be required to test a storage vessel. Sixty to ninety may be required to test the pressure containment system of a nuclear plant which includes reactor vessel, pumps, steam generator, pressurizer, main piping, and valves. Pipeline testing needs at the least 1 transducer every 1000 ft.

Acoustic emission technology is most beneficial where accessibility to a pressure containment system is limited, and because of its ability to conduct a complete structural integrity analysis with a minimum of shutdown time.

ASME Code Section VIII, Division 2, Alternative Rules

The *ASME Boiler and Pressure Vessel Code,* Section VIII, "Pressure Vessels," is published in two parts, Division 1 and Division 2, Alternative Rules. In comparison to Division 1, Division 2 vessels require more complex design procedures as well as more restrictive requirements on design, materials and nondestructive examinations; however, higher design stress values are permitted in Division 2, and it is possible to design pressure vessels with a factor of safety of three. This differs from rules which are designated as Division 1, requiring a theoretical factor of safety of four, if certain fabrication procedures are followed, such as radiography or spot radiography of the welded seams. If these are not followed, a factor of safety of five is required. *It should be noted that in this chapter, all Code paragraph references are to specific parts of* Division 2. The alternative rules apply only to vessels installed in a fixed (stationary) location. (See Code Par. AG-100.) An exception to fixed location may be taken in the diving industry where pressure vessels are used for human occupancy.

In Division 2 vessels, the requirements of Section VIII have been upgraded. A very detailed stress analysis of the vessel will be required. There are many restrictions in the choice of materials, more exacting design, and stress analysis. Closer inspection of the required fabrication details, material inspection, welding procedures and welding, and more nondestructive examinations will be necessary since design stresses are larger than those permitted in Division 1. There will, in reality, be a theoretical design factor of safety of three, based on the ultimate strength of materials. With the larger stresses allowable, a vessel can be built for the same pressure and diameter using thinner shells and heads, consequently less weld metal and lighter vessels. However, the cost of engineering and quality control inspection will be much greater and may exceed the savings made in material cost and fabrication.

Also, vessels which will be fabricated under Division 1 may not meet the requirements of Division 2. For example, only one quality vessel is provided for in Division 2, with all longitudinal and circumferential seams butt-welded and 100 percent radiographed. In another example, the allowable stress values are higher than in Division 1, being the lowest of (1) one-third of minimum tensile strength at room temperature, (2) two-thirds of minimum yield strength at room temperature, (3) two-thirds of minimum yield strength at temperature, (4) one-third of tensile strength at temperature.

A big feature of Division 2 is that the user of the vessel, or a representative agent, must provide the manufacturer with design specifications giving detailed information about the intended operating conditions. (See Code Par. AG-301 and AG-100.) It is the user's responsibility to give sufficient design information to evaluate whether or not a fatigue analysis of the vessel must be made for cyclic service. (See Code Par. AD-160.) If the vessel has to be evaluated for cyclic service, it should be designed as required in the Appendix. It is also the user's responsibility to state whether or not a corrosion and/or erosion allowance must be provided and the amount required. The user's design specifications must be certified by a registered professional engineer experienced in pressure vessel design. (See Code Par. AG-301.2.)

It is the responsibility of the vessel manufacturer to prepare a manufacturer's design report establishing the conformance with the rules of Division 2 as will be required to meet the user's design specifications. The manufacturer's design report must be certified by a registered professional engineer experienced in pressure vessel design. (See Code Par. AG-302.3.)

Division 2 of the Code does not hold the Authorized Inspector responsible for approving design calculations. The Inspector verifies the existence of the user's design specifications and the manufacturer's design reports and that these reports have been certified by a registered professional engineer. (See Code Par. AG-303.)

It is also the Authorized Inspector's duty to review the design, with respect to materials used, to see that the manufacturer has qualified welding procedures and welders to weld the material and to review the vessel geometry, weld details, nozzle attachment details, and nondestructive testing requirements. The manufacturer should obtain the Inspector's review of these details before ordering materials and starting fabrication so that the Inspector can point out the parts that may not meet the intent of the Code and avoid later rejection of the material or vessel.

Material controls will be tighter for Division 2 vessels. For example, where Division 1, under Par. UG-10, allow tests of materials not identified, Division 2 requires identification of all materials and is more restrictive than Division 1 in the choice of materials that can be used. Because of this it permits higher allowable stress values; therefore more precise design procedures are required. Special inspection and testing requirements of the materials have added to the reliability of the material. For example, plates and forgings over 4 in thick require ultrasonic inspection. (See Code Par. AM-203.) Cut edges of base materials over 1½ in thick also must be inspected for defects by liquid penetrant or magnetic particle methods. In addition, welding requires specific examination. (See Code Table AF-241.1.) On some materials, the location of test coupons have been made more specific when compared to the standard material specifications.

Steel-casting requirements and inspection in Division 2 are more demanding than those in Division 1. Division 2 requires inspection and tests for welded joints that are not covered in several welded pipe specifications. Therefore, welded pipe is not covered in the new division. (See Code Par. AG-120.)

In Division 2, because of the higher stress allowed for materials and for protection against brittle fracture of some steels like SA-285, thicknesses of over 1 in will be limited to 65°F or more unless they are impact tested. (See Code Par. AM 211 and Fig. AM-218.1.) In Division 1, with less stress allowed, these commonly used plate steels, such as SA-285, allow temperatures as low as −20°F.

The tighter restrictions can be seen in the plate steels, such as SA-516, which will permit a plate thickness of 1 in at −25°F when normalized or a temperature of 50°F in the as-rolled condition. In Division 1, because of the lower allowable stresses, the above-stated materials can be used to −20°F in the as-rolled condition.

Heat treatment must be tightly controlled. (See Code Table AF-402.1; Code Par. AF-635 and AF-640 for quenched and tempered materials; Code Par. AF-730 and AF-776 for forged vessels; and Code Par. AL-820 for layered vessels.)

Division 2 requires more nondestructive examinations. See Code Par. AF-220 and AF-240 and Code Table AF-241.1, which summarize the required examination methods for the type of weld joints. The table indicates the applicable Code paragraphs and Code figures, gives helpful remarks, and shows when NDE examination methods may be substituted. Magnetic particle and dye penetrant examiners need only be certified by the manufacturer (Code Par. AI-311 and Appendixes 9-1 and 9-2). Radiographers and ultrasonic examiners must be qualified (Code Par. AI-501 and Appendix 9-3), their qualifications documented as specified when reference is made by Division 2 to the ASME Code, Section V and SNT-TC-1A, which is the "Recommended Practice for Nondestructive Examining Personnel Qualification and Certification" published by the American Society for Nondestructive Examination. Code Par. AI-511 gives unacceptable defects and repair requirements.

Hydrostatic testing requires a minimum of 1.25 times the design pressure, corrected for temperature. Division 2 differs from Division 1 in that it establishes an upper limit on test pressure. This upper limit must be specified by the design engineer (Code Par. AT-302 and AD-151). When pneumatic tests are used, the minimum test pressure is 1.15 times the design pressure.

The Manufacturer's Data Report, required by Division 2, has a space provided for recording the names of the professional engineer who certified the user's design specifications and the professional engineer certifying the manufacturer's design report.

The manufacturer must also sign the ASME data report before submitting it to the Authorized Inspector for the Inspector's review and signature. This data report must be retained for the expected life of the vessel or 10 years (see Code Par. AS-301), whichever is longer.

The manufacturer is also required to have a system for the maintenance of records for 5 years of specifications, drawings, materials used, fabrication details, repairs, nondestructive examinations and inspections, heat treatments, and testing.

It may be seen, with the above examples, that a Division 1 vessel, though thicker and heavier, may offset the cost of special design engineering and quality control that are required in Division 2, especially in smaller vessels.

Division 2 will allow more flexibility in design, which will enable the use of thinner materials. This will be helpful to the user in the manufacture of large-diameter, high-pressure vessels where the additional cost of design and inspection is secondary to the weight of the vessel and where additional cost of engineering and quality control can be offset by the use of these thinner materials, thereby saving in weight of the vessel and in handling.

It must also be remembered that the quality control system required by Code Par. AG-304 and mandatory Appendix 18 will be more rigorous than the quality control system for Division 1 vessels. The quality control system for Division 2 vessels should be established more on the methods used in that part of Section III on nuclear vessels' quality assurance systems (see Chap. 4 and Chap. 10 for a more detailed explanation of a quality control system for pressure vessels and a quality assurance system for nuclear vessels), because vessels designed to ASME Code Section VIII, Division 2, have special requirements. These requirements, such as how changes are made in either specification or design and their reconciliation via the registered professional engineer who certifies the manufacturer's design report, must be stated in the quality control program.

The quality control manual should also state that it is the duty of the Authorized Inspector to see that the certified documents are on file (code Par. AG-303).

In addition, with respect to design, drawings, and specifications control, the quality control manual should state that, before engineering and fabrication starts, the user's design specifications must be certified by a registered professional engineer experienced in pressure vessel design who must assume the responsibility for its use.

The user's design report must interface with the manufacturer's design report.

The quality control manual with regard to "user's design specifications" should cover, as a minimum, those conditions set forth in Code Par. AG-301.

The manufacturer's design report, including calculations and drawings certified by a registered professional engineer experienced in pressure vessel design, must be furnished by the vessel manufacturer, stating that the user's design specifications have been satisfied (Code Par. AG-302).

A copy of the manufacturer's report must be given to the user and a copy, along with a copy of the user's design specifications, shall be kept on file for 5 years.

Power Boilers, Guide for the ASME Code Section I Power Boilers

The ASME Boiler and Pressure Vessel Code Section I, "Power Boilers," includes rules and general requirements for all methods of construction of power, electric, and miniature boilers and high-temperature water boilers used in stationary service. It also includes power boilers used in locomotive, portable, and traction service.

The rules of Section I are applicable to boilers in which steam or other vapor is generated at a pressure more than 15 psig and high-temperature water boilers intended for operation at pressures exceeding 160 psig and/or temperatures exceeding 250°F. Super heaters, economizers, and other pressure parts connected directly to the boiler without intervening valves are considered as part of the scope of Section I. (See Code Section I, Power Boilers Preamble.)

References in this chapter are for "Power Boiler Code Section I".

The 1914 ASME Boiler Code was the original pressure vessel code which now includes eleven sections. Section VIII, Division I, Pressure Vessel Code, is used to build more vessels than all the others combined. This book was written mainly for users of Section VIII, Division I. However, many engineers, fabricators, and inspectors must work with all sections of the Code.

Those who use the ASME Code, Section I, "Power Boilers," regularly in construction and inspection may not have difficulty in finding the information needed. It is the engineer, fabricator, or inspector who consults it infrequently who may have difficulty.

A "Quick Reference Guide" (Fig. 9.1) illustrates some of the types of boiler construction which are provided for under Section I of the ASME Code and furnishes direct reference to the applicable rule in the Code.

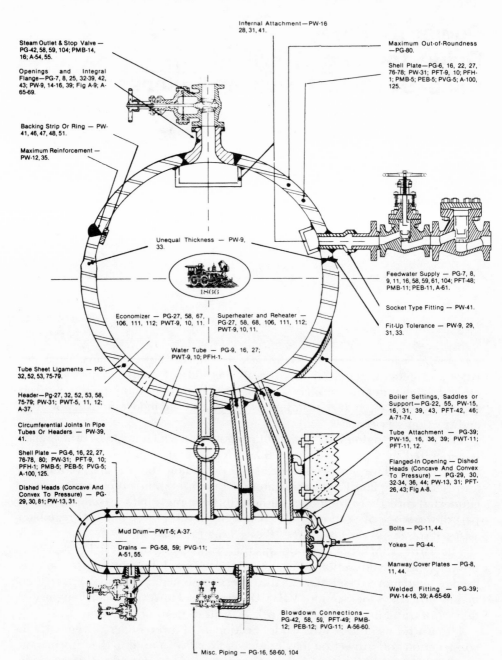

Internal Attachment—PW-16
28, 31, 41.

Maximum Out-of-Roundness
—PG-80.

Steam Outlet & Stop Valve —
PG-42, 58, 59, 104; PMB-14,
16; A-54, 55.

Shell Plate—PG-6, 16, 22, 27,
76-78; PW-31; PFT-9, 10; PFH-
1; PMB-5; PEB-5; PVG-5; A-100,
125.

Openings and Integral
Flange—PG-7, 8, 25, 32-39, 42,
43; PW-9, 14-16, 39; Fig A-9; A-
65-69.

Backing Strip Or Ring — PW-
41, 46, 47, 48, 51.

Maximum Reinforcement —
PW-12, 35.

Unequal Thickness — PW-9,
33.

Feedwater Supply — PG-7, 8,
9, 11, 16, 58, 59, 61, 104; PFT-48;
PMB-11; PEB-11, A-61.

Socket Type Fitting — PW-41.

Economizer — PG-27, 58, 67,
106, 111, 112; PWT-9, 10, 11.

Superheater and Reheater —
PG-27, 58, 68, 106, 111, 112;
PWT-9, 10, 11.

Fit-Up Tolerance — PW-9, 29,
31, 33.

Water Tube — PG-9, 16, 27;
PWT-9, 10; PFH-1.

Tube Sheet Ligaments — PG-
32, 52, 53, 75-79.

Boiler Settings, Saddles or
Support—PG-22, 55, PW-15,
16, 31, 39, 43; PFT-42, 46;
A-71-74.

Header—Pg-27, 32, 52, 53, 58,
75-79; PW-31; PWT-5, 11, 12;
A-37.

Tube Attachment — PG-39;
PW-15, 16, 36, 39; PWT-11;
PFT-11, 12.

Circumferential Joints In Pipe
Tubes Or Headers — PW-39,
41.

Flanged-In Opening — Dished
Heads (Concave And Convex
To Pressure) — PG-29, 30,
32-34, 36, 44; PW-13, 31; PFT-
26, 43; Fig A-8.

Shell Plate — PG-6, 16, 22, 27,
76-78, 80; PW-31; PFT-9, 10;
PFH-1; PMB-5; PEB-5; PVG-5;
A-100, 125.

Dished Heads (Concave And
Convex To Pressure) — PG-
29, 30, 81; PW-13, 31.

Mud Drum—PWT-5; A-37.

Bolts — PG-11, 44.

Yokes — PG-44.

Drains — PG-58, 59; PVG-11;
A-51, 55.

Manway Cover Plates — PG-8,
11, 44.

Welded Fitting — PG-39;
PW-14-16, 39; A-65-69.

Blowdown Connections—
PG-42, 58, 59, PFT-49; PMB-
12; PEB-12; PVG-11; A-56-60.

Misc. Piping — PG-16, 58-60, 104

Fig. 9.1 Quick Reference Guide: *ASME Boiler and Pressure Vessel Code,* **Section I,** **"Power Boilers."** *(Courtesy of The Hartford Steam Boiler Inspection Insurance Company)*

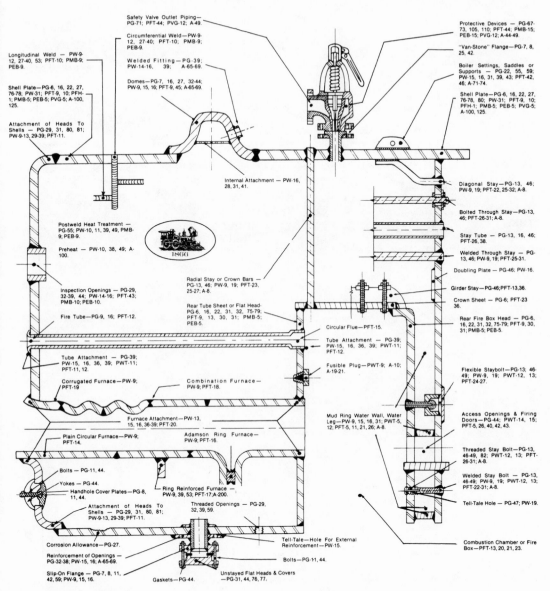

Safety Valve Outlet Piping—PG-71; PFT-44; PVG-12; A-49.

Circumferential Weld—PW-9-12, 27-40; PFT-10; PMB-9; PEB-9.

Welded Fitting—PG-39; PW-14-16, 39; A-65-69.

Domes—PG-7, 16, 27, 32-44; PW-9, 15, 16; PFT-9, 45; A-65-69.

Longitudinal Weld — PW-9-12, 27-40, 53; PFT-10; PMB-9; PEB-9.

Shell Plate—PG-6, 16, 22, 27, 76-78; PW-31; PFT-9, 10; PFH-1; PMB-5; PEB-5; PVG-5; A-100, 125.

Attachment of Heads To Shells — PG-29, 31, 80, 81; PW-9-13, 29-39; PFT-11.

Protective Devices — PG-67-73, 105, 110; PFT-44; PMB-15; PEB-15; PVG-12; A-44-49.

"Van-Stone" Flange—PG-7, 8, 25, 42.

Boiler Settings, Saddles or Supports — PG-22, 55, 59; PW-15, 16, 31, 39, 43; PFT-42, 46; A-71-74.

Shell Plate—PG-6, 16, 22, 27, 76-78, 80; PW-31; PFT-9, 10; PFH-1; PMB-5; PEB-5; PVG-5; A-100, 125.

Internal Attachment — PW-16, 28, 31, 41.

Diagonal Stay—PG-13, 46; PW-9, 19; PFT-22, 25-32; A-8.

Bolted Through Stay—PG-13, 46; PFT-26-31; A-8.

Postweld Heat Treatment — PG-55; PW-10, 11, 39, 49; PMB-9; PEB-9.

Preheat — PW-10, 38, 49; A-100.

Stay Tube — PG-13, 16, 46; PFT-26, 38.

Welded Through Stay — PG-13, 46; PW-9, 19; PFT-25-31.

Doubling Plate — PG-46; PW-16.

Radial Stay or Crown Bars — PG-13, 46; PW-9, 19; PFT-23, 25-27; A-8.

Girder Stay—PG-46; PFT-13, 36.

Crown Sheet — PG-6; PFT-23 36.

Inspection Openings — PG-29, 32-39, 44; PW-14-16; PFT-43; PMB-10; PEB-10.

Fire Tube—PG-9, 16; PFT-12.

Rear Tube Sheet or Flat Head-PG-6, 16, 22, 31, 32, 75-79; PFT-9, 13, 30, 31; PMB-5; PEB-5.

Rear Fire Box Head — PG-6, 16, 22, 31, 32, 75-79; PFT-9, 30, 31; PMB-5; PEB-5.

Circular Flue—PFT-15.

Tube Attachment — PG-39; PW-15, 16, 36, 39; PFT-11; PFT-12.

Tube Attachment — PG-39; PW-15, 16, 36, 39; PWT-11; PFT-11, 12.

Fusible Plug—PWT-9; A-10; A-19-21.

Flexible Staybolt—PG-13; 46-49; PW-9, 19; PWT-12, 13; PFT-24-27.

Corrugated Furnace—PW-9; PFT-19

Combination Furnace—PW-9; PFT-18.

Furnace Attachment—PW-13, 15, 16, 36-39; PFT-20.

Mud Ring Water Wall, Water Leg—PW-9, 15, 16, 31; PWT-5, 12; PFT-5, 11, 21, 26; A-8.

Access Openings & Firing Doors—PG-44; PWT-14, 15; PFT-5, 26, 40, 42, 43.

Plain Circular Furnace—PW-9; PFT-14.

Adamson Ring Furnace—PW-9; PFT-16.

Threaded Stay Bolt—PG-13, 46-49, 82; PWT-12, 13; PFT-26-31; A-8.

Bolts — PG-11, 44.

Yokes — PG-44.

Handhole Cover Plates—PG-8, 11, 44.

Ring Reinforced Furnace — PW-9, 39, 53; PFT-17; A-200.

Welded Stay Bolt — PG-13, 46-49; PW-9, 19; PWT-12, 13; PFT-22-31; A-8.

Attachment of Heads To Shells — PG-29, 31, 80, 81; PW-9-13, 29-39; PFT-11.

Threaded Openings — PG-29, 32, 39, 59.

Tell-Tale Hole — PG-47; PW-19.

Corrosion Allowance—PG-27.

Reinforcement of Openings — PG-32-38; PW-15, 16; A-65-69.

Slip-On Flange — PG-7, 8, 11, 42, 59; PW-9, 15, 16.

Gaskets—PG-44.

Unstayed Flat Heads & Covers — PG-31, 44, 76, 77.

Tell-Tale—Hole For External Reinforcement—PW-15.

Bolts — PG-11, 44.

Combustion Chamber or Fire Box—PFT-13, 20, 21, 23.

Fig. 9.1 (Cont.)

Sling Stay—PG-13, 46; PFT-13, 36.

Wrapper Sheet—PG-6, 27; . PFT-9, 11, 23, 26, 36, 41.

Radial Stay or Crown Bars— PG-13, 46; PW-9, 19; PFT-23, 25-27; A-8.

Furnace Sheet & Throat Sheet —PG-6; PFT-9, 11, 13, 20.

Firebox Welding—PW-9, 19-40; PFT-21, 26.

Furnace Staying — PG-17, 46- 49; PFT-23-28; A-8.

Mud Ring, Water Wall, Water Leg—PW-9, 15, 16, 31; PWT-5, 12; PFT-5, 11, 21, 26; A-8.

Base Plate—PG-6.

Combustion Chamber or Fire Box—PFT-13, 20, 21, 23.

Crown Bar — PG-6; PFT-13, 36.

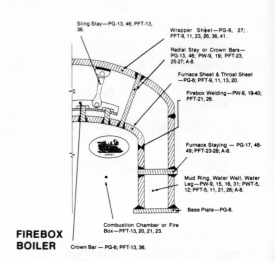

Truncated Cone Shaped Com- bustion Chamber—PW-9, 29- 39; PFT-14, 23, 37.

Radial Stay—PG-13, 46; PW-9, 19; PFT-23, 25-27; A-8.

Shell Plate—PG-6, 16, 22, 27, 76-78, 80; PW-31; PFT-9, 10; PFH-1; PMB-5; PEB-5; PVG-5; A-100, 125.

Upper Tube Sheet and Lower Tube Sheet—PG-6, 16, 22, 31, 32, 75-79; PW-9; PFT-9, 11, 31, 31; PMB-5; PEB-5.

Fire Tube—PG-9, 16; PFT-12.

Mud Ring, Water Wall, Water Leg—PW-9, 15, 16, 31; PWT- 5, 12, PFT-5, 11, 21, 26; A-8.

Furnace Sheet & Throat Sheet —PG-6;PFT-9, 11, 13, 20.

Furnace Staying—PG-17, 46- 49; PFT-23-28; A-8.

Combustion Chamber or Fire Box—PFT-13, 20, 21, 23.

Ogee Attachment — PG-46; PFT-5, 20, 26.

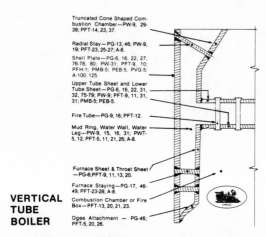

VERTICAL TUBE BOILER

FIREBOX BOILER

Fig. 9.1   (Cont.)

General Notes

Quality Control System	—A-300
Material General:	—PG-5
a. Plate	—PG-6, 77
b. Forgings	—PG-7
c. Castings	—PG-8, 25
d. Pipe, Tubes &	
Pressure Containing Parts	—PG-9
e. Misc. Pressure Parts	—PG-11
f. Stays	—PG-13
Loadings	—PG-22; PW-43; A-71 - 74
Stress—Max. Allowable	—Tables PG-23.1 and 23.2; A-150
Manufacturer's Responsibility	—PG-104; PW-1; PEB-18
Inspector's Responsibility	—PG-90, 91
Pressure Tests	—PG-99, 100; PW-54; PMB-21; PEB-17; A-22-23, 63
Safety Devices	—PG-67 - 73; PMB-15; PEB-15; PVG-12
Trim:	
a. Gage Glasses	—PG-60; PFT-47; PMB-13; PEB-13; A-50 - 52
b. Pressure Gages	—PG-60; A-23, 53
c. Gage Cocks	—PG-60; A-51
d. Water Columns	—PG-8, 60; PW-42; A-52
Service Restrictions	—PG-2; PMB-2; PEB-2
Welded Construction	—Part PW
Nameplates, Stamping, & Reports	—PG-104 - 113; PFH-1; PEB-19
NDE Requirements:	
a. MT & PT	—PG-25
b. RT	—PG-25; PW-51
c. UT	—PW-52
Porosity Charts	—A-250
Code Jurisdiction	—PG-58
Minimum Thickness	—PG-16;PFT-9
Feedwater Heater (Optional Requirements)	—PFH
Miniature Boilers	—PMB
Electric Boilers	—PEB
Organic Fluid Vaporizer Generators	—PVG

Special Reference to Particular Boiler Types

Traction, Stationary, or Portable Return Tube Firetube (Typically the Locomotive Type)	PG-60.1.5, 111.3; PFT-5.2, 23.2, 24, 42, 43.1, 43.3, 45, 45.3, 45.7
Horizontal Tube & Horizontal Return Tube	PG-59.3.7, 60.1.4, 111.1, 111.2; PFT-11.4.7, 26.5, 31.4, 42, 43.1, 43.2, 43.3, 44, 45.3, 46.3, 46.4, 46.5, 47, 48.1
Vertical Tube	PG-111.4 PFT-5.2, 11.4, 20.5, 21.2, 21.3, 23.3, 23.5, 26.6, 42. 43.1, 43.5, 48.2
Fire Box	PFT-21, 26, 36.2, 42, 43.1
Scotch Marine	PG-111.6 PFT-20.2, 26.5, 43.1, 43.4
Waste Heat	PG-106.4.1 PFT-43.1, 44
Forced Flow Steam Generator	PG-16.2, 21.1, 21.2, 58.3.3, 58.3.5, 58.3.6, 59.3.3, 60.1.2, 60.6.2, 61.5, 67.4, 106.3, 111.5.2, 112.1

Riveted Boilers
ASME Section I, Part PR, 1971 Edition

Fig. 9.1   (Cont.)

In the event of a discrepancy, the rules in the current edition of the Code shall govern. This chart should be used only as a quick reference. The current edition of the Code should always be referenced.

There are also many conditions in the boiler Code where all boiler manufacturers, assemblers, and Authorized Inspectors have to take special precautions, especially when large boilers are fabricated partly in the shop and partly in the field by one manufacturer and the piping and components are manufactured by others.

This is particularly important if all the parts are being assembled in the field by the boiler manufacturer or an assembler who is certified by the ASME to assemble boilers.

When construction of a boiler is to be completed in the field, either the manufacturer of the boiler or the assembler has the responsibility of designing the boiler unit, fabricating or obtaining all the parts that make up the entire boiler unit including all piping and piping components within the scope of the boiler code, and for inclusion of these components in the "master data report" to be completed for the boiler unit. (See Code Par. PG-107.)

Possible arrangements include the following:

1. The boiler manufacturer assumes full responsibility by supplying the personnel to complete the job.

2. The boiler manufacturer assumes full responsibility by completing the job with personally supervised local labor.

3. The boiler manufacturer assumes full responsibility but engages an outside erection company to assemble the boiler to Code procedures specified by the manufacturer. With this arrangement, the manufacturer must make inspections to be sure these procedures are being followed.

4. The boiler may be assembled by an assembler with the required ASME certification and Code symbol stamp and quality control system. This indicates that the organization is familiar with the boiler Code and that the welding procedures and operators are Code-qualified. Such an assembler, if maintaining an approved quality control system, may assume full responsibility for the completed boiler assembly.

Other arrangements are also possible. If more than one organization is involved, the Authorized Inspector should find out on the first visit which one will be responsible for the completed boiler. If a company other than the shop fabricator or certified assembler is called to the job, the extent of association must be determined. This is especially important with a power boiler when an outside contractor is hired to install the external piping. This piping, often entirely overlooked, lies within the jurisdiction of the Code and must be fabricated in accordance with the applicable rules of the ANSI B31-1 Code for pressure piping by a certified manufacturer or contractor.

It must be inspected by the Authorized Inspector who conducts the final tests before the master data report is signed.

Once the responsible company has been identified, the Inspector must contact directly the person who is responsible for the erection to agree upon a schedule that will not interfere with the erection but will still permit a proper examination of the important fit-ups, chip-outs, and the like.

The first inspection will include a review of the company's quality control manual, welding procedures, and test results to make sure they cover all positions and materials used on the job. Performance qualifications records of all welders, working on parts of the boiler and piping under the jurisdiction of the Code, must also be examined.

It is the manufacturer's, contractor's, or assembler's responsibility to submit drawings, calculations, and all pertinent design and test data so that the Inspector may make comments on the design or inspection procedures before the field work is started and thereby expedite the inspection schedule.

New material received in the field for field fabrication must be acceptable Code material and carry proper identification markings for checking against the original manufacturer's certified mill test reports. Material may be examined at this time for thicknesses, tolerances, and possible defects.

The manufacturer, who completes the boiler, has the responsibility for the structural integrity of the whole or the part fabricated. Included in the responsibilities of the manufacturer are documentation of the quality control program to the extent specified in the Code.

Nuclear Vessels and Required Quality Assurance Systems

What is the difference between a fossil fuel power plant and a nuclear power plant? In both types of plants, heat is used to produce steam in order to drive a turbine that turns a generator that, in turn, creates electricity. The basic difference between fossil fuel power plants and nuclear power plants is the source of heat.

In a fossil fuel power plant, coal, oil, or gas is burned in a furnace to heat water in a boiler in order to create steam. In nuclear plants no burning or combustion takes place. Nuclear fission is used, the fission reaction generates heat, and this heat is transferred, sometimes indirectly, to the water that provides the steam.

In a nuclear energy plant, the furnace is replaced by a reactor which contains a core of nuclear fuel consisting primarily of uranium. Energy is produced in the reactor by fission. A tiny particle, called a neutron, strikes the center or nucleus of a uranium atom, which splits into fragments releasing more neutrons. These fragments fly apart at great speed, generating heat as they collide with other atoms. The free neutrons collide with other uranium atoms causing a fission, which creates more neutrons that continue to split more atoms in a controlled chain reaction.

The chain fission reaction heats the water which moderates the reaction in the reactor creating steam, whose force turns the turbine-generator. From this point on, nuclear energy plants operate essentially the same as fossil fuel steam-electric plants.

POWER PLANT CYCLES

The purpose of this chapter is not to discuss nuclear technology but to let the reader know more about atomic power plant heat cycles so as to give a better understanding of the classifications and rules for construction and inspection of nuclear plant components. Code Classes 1, 2, 3, and MC allow a choice of rules that provide different levels of structural integrity and quality and explain why these components are subjected to a rigorous quality assurance inspection, bearing upon the relative importance assigned to the individual components of the nuclear power system to ensure their lasting performance.

There are several types of reactor coolant systems that are being used. The two plant cycles most often found in utility service at present are the *pressurized water reactor* and the *boiling water reactor.*

PRESSURIZED WATER REACTOR PLANTS (PWR)

The pressurized water reactor, which is inside the reactor pressure vessel, is subjected to a high coolant water pressure. Here the pressurized water is heated. A pump circulates the heated water through a heat exchanger where steam for the turbine is generated. (See Fig. 10.1.)

The part of the pressurized water nuclear power plant which contains the reactor coolant is called the *primary circuit.* Included in the primary circuit is an important vessel called the *pressurizer.* The coolant volume varies when load changes require coolant temperature changes. When this occurs, the pressurizer serves as the expansion tank in the primary system allowing the water to undergo thermal expansion and contraction holding the primary circuit pressure nearly constant. If pressures were allowed to fluctuate too far, steam

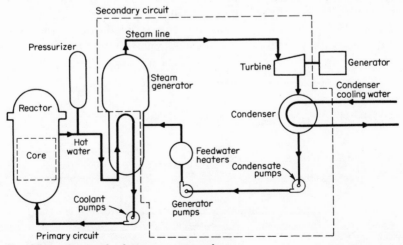

Fig. 10.1 Pressurized water reactor plant.

bubbles might form at the reactor heating surfaces; these bubbles, or voids, if formed inside the reactor core, greatly alter reactor power output. The pressurizer has electric heating elements located low inside to provide the vapor needed to cushion the flowing liquid coolant. All of the above-mentioned items are located in the primary circuit.

The rest of the plant, called the *secondary circuit,* is essentially the same as a fossil fuel steam-electric plant. The heat exchangers make steam that passes through the turbine, condenser, condensate pumps, feed pump, feedwater heaters, and back to the heat exchanger.

BOILING WATER REACTOR PLANTS (BWR)

In the boiling water reactor, the cooling water is inside the reactor core. The reactor makes steam that passes through the turbine, condenser, condensate pumps, feedwater heaters, and back to the reactor vessel.

In a boiling water reactor plant, steam generators and the pressurizer are not needed, as the plant has no secondary circuit, only a primary circuit. No primary to secondary heat exchange loss takes place. (See Fig. 10.2.)

The above explanation of the composition and operation of the boiling water plant (BWR) and the pressurized water plant (PWR) is meant to give the reader a basic understanding of nuclear power plant systems.

In addition to the two described, there are other types of plants, such as gas-cooled and liquid-cooled reactors, which have fewer plants in operation.

GAS-COOLED REACTOR PLANTS (HTGR)

In this case, gas, such as helium, is the reactor coolant. Gas passes through the reactor core then around steam-generator heating surfaces. The steam the generator makes passes through the conventional power plant components and returns as feedwater to the steam generator.

Even though the secondary circuit in the gas-cooled reactor plants includes little that is new, the primary circuit merits special comment. First, the entire

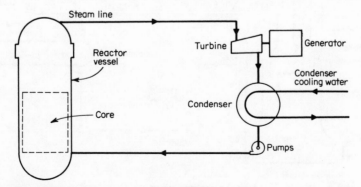

Fig. 10.2   Boiling water reactor plant.

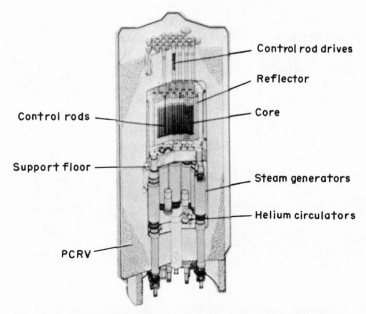

Control rod drives

Reflector

Core

Control rods

Support floor

Steam generators

Helium circulators

PCRV

Fig. 10.3   Reactor arrangement within prestressed concrete reactor vessel (PCRV). *(Courtesy of General Atomic Company)*

gas system, including the reactor, the gas circulators, and steam generators, are inside one vessel.   Second, the vessel is made of concrete.   Third, the concrete is prestressed with steel tendons.   Each tendon is made up of about 175 one-quarter-inch wires, and each vessel requires about 500 tendons.

This *prestressed concrete reactor vessel* (calld PCRV) is more than a reactor vessel; it is a nuclear steam supply system.   Nuclear fuel is placed inside producing superheated steam to be used in the secondary circuit.   Figures 10.3 and 10.4 give an example of the high-temperature gas-cooled reactor (HTGR) system built by General Atomic Company for the Public Service Company of Colorado at their Fort St. Vrain plant.

The plant consists of a reactor building, which houses the nuclear portion of the station — the reactor, the prestressed concrete reactor vessel (PCRV), the steam generators, the helium circulators, and the nuclear auxiliary systems — and a turbine building, which contains equipment such as the turbine generator, feedwater system, and condenser.

As shown in the accompanying flow schematic, the helium coolant, at a pressure of about 700 psia, flows through the reactor core, where it is heated to 1430°F.   From the reactor, the helium flows to the steam generators.   After passing through the steam generators, the coolant is returned to the reactor by helium circulators.   Two identical loops are used, each including a six-module steam generator and two helium circulators.   Each loop contributes half the total output of the nuclear steam system, which produces steam at 2400 psig and 1000°F with single reheat to 1000°F.

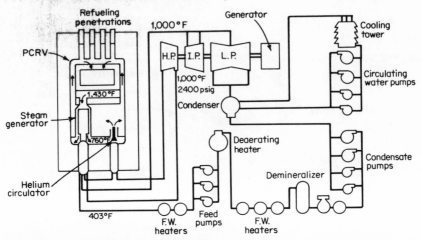

Fig. 10.4 High-temperature gas-cooled reactor (HTGR), power plant flow schematic. *(Courtesy of General Atomic Company)*

REACTOR DESIGN DETAILS

The 841 Mw(t) HTGR is a thermal reactor employing U^{235} as fissile material; Th^{232} as fertile material; graphite as moderator, cladding, structure, and reflector; and helium as coolant. The active core is a cylinder 19.6 ft in diameter and 15.6 ft high containing 1482 fuel elements, 74 control-rod elements, and 37 control-rod drives. These elements are arranged vertically in a closely packed hexagonal array, and a graphite reflector completely surrounds the core. Coolant gas passes downward through vertical passages in each of the fuel elements into the plenum area under the core. Coolant gas then passes through the core support floor into the steam generators, where it gives up its heat to water and steam. The gas then passes to the suction side of the helium circulators, where the pressure is raised approximately one atmosphere. Discharge from the circulators is then routed to the top of the reactor core via an annular space provided between the core barrel and the liner insulation.

PRESTRESSED CONCRETE REACTOR VESSEL (PCRV)

The prestressed concrete reactor vessel (PCRV) is 31 ft in internal diameter and 75 ft in internal height. The upper and lower heads are nominally 15 ft thick, and the walls have a nominal thickness of 9 ft. Thus the PCRV provides the dual function of containing the coolant operating pressure and also providing radiological shielding. The exterior, vertical surface of the vessel is an approximate hexagonal prism, 61 ft across flats and 106 ft high.

The top head has refueling penetrations which also house the control-rod

drives. It also incorporates penetrations and wells to house the helium purifica-
tion system and neutron chambers. The bottom head has penetrations for
each of the steam-generator modules and helium circulators plus a large central
opening for access to the main cavity.

All the penetrations through the PCRV are provided with two independent
closures. The PCRV inner cavity and the primary closures serve as primary
containment for the reactor; the massive PCRV and the secondary closures acts
as the secondary containment.

The concrete walls and heads of the PCRV are constructed around a carbon
steel liner which is nominally ¾ in thick. The liner is anchored to the concrete
and provides a helium-tight membrane. A system of water-cooled tubes
welded to the concrete side of the liner provides a heat-removal system to
control concrete temperature. In conjunction with the coolant tubes, a thermal barrier is provided on the inside surface of the liner to impede the flow of
heat to the liner walls.

Tendons through steel tubes placed in the concrete serve to place the
concrete structure compression prior to service.

The heat cycles that have been described are only some of the ideas that have
been tested.

Nuclear energy will only survive if practical plants that make more nuclear
fuel than they consume are built. These plants are called *breeders*. If the
breeder reactor is soon perfected and used, the energy bank of the United

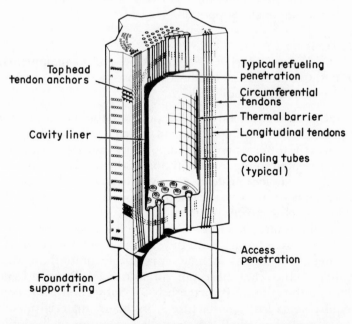

Fig. 10.5   Cutaway view of prestressed concrete reactor vessel.
(Courtesy of General Atomic Company)

States' uranium can be extended until a system is developed where energy is produced by fusion similar to the heat production method of the sun and other stars. If not, our uranium may be used up within a century.

Nuclear fusion is the joining together or fusion of atoms to form an atomic nucleus of higher atomic number and weight. Nuclear fission is the splitting apart of atoms. Both release enormous amounts of energy. All the nuclear plants today operate by nuclear fission. Extensive research is being done to make electricity by nuclear fusion, which represents our best alternative for an abundant, accessible, and long-term energy source.

All nuclear vessels must be manufactured to the appropriate part of the ASME Code Section III, "Nuclear Vessels," Division 1, "Nuclear Power Plant Components" or Division 2, "Concrete Reactor Vessels and Containments." Both divisions are broken down into subsections which are designated by capital letters preceded by the letter N for Division 1 and by the letter C for Division 2.

The ASME Code Section III, Division 1, "Nuclear Power Plant Components," is divided into seven subsections covering all aspects of safety-related design and construction:

SUBSECTION NCA: General Requirements for Division 1 and Division 2
SUBSECTION NB:Class 1 Components
SUBSECTION NC: Class 2 Components
SUBSECTION ND: Class 3 Components
Subsection NE: Class MC Components (Metal Containment)
Subsection NF: Component Supports
Subsection NG: Core Support Structures
Section III Division I Appendices
Division 2. Code for Concrete Reactor Vessel and Containments, includes Subsection CB, Concrete Reactor Vessels, Subsection CC, Concrete Containments, and Division 2 Appendices.

These Code classes are intended to be applied to the classification of items of a nuclear power system and containment system. Within these systems the Code recognizes the different levels of importance associated with the functions of each item as related to the safe operation of the nuclear power plant. The Code classes allow a choice of rules that provide assurance of structural integrity and quality commensurate with the relative importance assigned to the individual items of the nuclear power plant.

Each subsection is intended to be a separate document with no cross-referencing of subsections necessary, with the exception of Subsection NCA, "General Requirements," which is a mandatory requirement for those who intend to use one or more of the other subsections of Section III, Division 1 and Division 2. It includes Subsections NB through NG. Subsection NCA comprises all general requirements for manufacturers, installers, material manufacturers, suppliers, and owners of nuclear power plant components including the requirements for quality assurance, inspectors' duties, and stamping.

Section III, Division 2, "Concrete Reactor Vessels and Containments" defines the requirements for the design and construction of concrete reactor vessels and concrete containments and Division 2 Appendices. It covers requirements for materials, parts, and appurtenances of components that are designed to provide a pressure-retaining or containing barrier.

The following explanation covers only Section III, Division 1, and does not include *Division 2:*

To avoid confusion it should be noted that all Code references in this chapter, unlike those in the rest of this book, refer to the ASME Code Section III, Division 1, "Nuclear Vessels," rather than Section VIII, Division 1, "Pressure Vessels."

These seven subsections are intended to be applied to the classification of pressure-retaining or pressure-containing components of a nuclear power system and containment system, component supports, and core support structures.

The *ASME Boiler and Pressure Vessel Code* Section III, Division 1, "Nuclear Power Plant Components," differs from the other sections of the Code in that it requires a very detailed stress analysis of vessels classified as Class 1, Class 2, and Class MC. Class 1 includes reactor and primary circuit vessels, containing coolant for the core that cannot be separated from the core and its radioactivity.

Class 2 components are those which are not part of the reactor-coolant boundary but are used to remove the heat from the reactor-coolant pressure boundary.

Class 3 components support Class 2 without being part of them, while class MC includes containment vessels.

All nuclear vessels and components require constant vigilance during fabrication and operation. The manufacturer is required to prepare, or have prepared, design specifications, drawings, and a complete stress analysis of pressure parts. The stress report that will include all this material has to be certified by a registered professional engineer. Copies of the certified stress report are required to be made available to the Authorized Inspector at the manufacturing site as well as with the enforcement authority at the point of installation.

It is the duty of the Authorized Inspector to verify that the equipment is fabricated according to the design specifications and drawings as submitted. Another duty is to make all inspections as specified by the Code rules.

Quality controls adequate in the past are unacceptable today. Pressure vessel fabricators view this period as the age of quality assurance. Today, the industry faces a heavy demand for test reports and other supporting data. This is especially so when constructing nuclear vessels and parts. Fundamentally, there should be no difference in the level of quality of a high-pressure, high-temperature fossil fuel boiler and a nuclear vessel. The basic differences are the amount of testing and documentation required in nuclear fabrication, whereas the power boiler code requires a quality control system ("Power Boiler Code," Appendix A-300).

The nuclear code requires that manufacturers of nuclear equipment have a more rigorous quality assurance program.*   Excellence of each item produced must be the aim of each manufacturer.

The responsibility for the program lies with the manufacturer.   A manufacturer who wants to fabricate nuclear vessels or parts, covered by Section III, obligates the company with respect to quality assurance and documentation. The manufacturer must have an understanding of the many Code details and a written quality assurance manual describing the system of design, material, procurement, fabrication, testing, and documentation.   The manufacturer must also have a clear understanding of nuclear quality assurance standards being used in every aspect of nuclear power plant construction from the design and construction phases to full operation.

The nuclear industry's safety record is due to the rigorous standards of the quality assurance programs applied to all phases of information, creation, planning, and fabrication as well as to the design, construction, testing, inspection, and corrective action taken when nonconformities are discovered during construction and later when implementing the quality assurance programs in the operation and maintenance of a nuclear power plant.

QUALITY ASSURANCE

The quality assurance control program must be evaluated and approved for compliance by the ASME before and upon each renewal of the Certificate of Authorization.   This is done at the manufacturer's request and expense.   The evaluation of the manufacturer's quality assurance capability will be made by a survey team authorized by the ASME.   The team will be made up of one or more consultants to the ASME as designated by the secretary of the Boiler and Pressure Vessel Committee, which includes the ASME team leader, a team member, a representative of the National Board of Boiler and Pressure Vessel Inspectors, a representative of the jurisdictional authority in which the manufacturing facility is located, an inspection specialist from the manufacturer's authorized inspection agency, the Authorized Inspector doing the Code inspection at the plant being surveyed, and a representative of a utility using nuclear energy.

The survey team's purpose will be to compile a report of the manufacturer's quality assurance capabilities at the plant or site.   This report will be submitted to the secretary of the Boiler and Pressure Vessel Committee, who will carry out the necessary procedures leading to the issuance of a Certificate of Authorization to the manufacturer for the use of a Code N Symbol for the types of fabrication to be manufactured.

The intent of the Code survey team will be to verify that the manufacturer has an inspection contract or agreement with an authorized inspection agency; to review the manufacturer's quality assurance program and its compliance with the intent of Section II, Section III, Section V, and Section IX, where the

* Note that portions of nuclear Code Article NCA-4000 of Section III, Divisions 1 and 2, General Requirements, 1980 edition, were deleted and replaced by reference to ANSI/ASME NQA-1, "Quality Assurance Requirements for Nuclear Power Plants."

requirements for materials, nondestructive examination methods, and the qualifications of nondestructive examination procedures and for welding qualification procedures are found; to examine the means and methods for the documentation of the manufacturer's quality assurance program, to review the personnel involved in the manufacturer's quality assurance activities, and to examine their status with respect to production and management functions; to examine the manufacturer's relationship with the Authorized Inspector with respect to the organization and the means of keeping the Inspector informed and up-to-date on quality control matters; to review all procedures as required by the Code with respect to qualifications of processes and individuals; and to examine equipment and facilities employed for testing and carrying out quality assurance work.

It is the manufacturer's responsibility to aid the Code survey team so that it can do its work. It would be advisable for the manufacturer to provide qualified personnel who are able to reply to the necessary inquiries generally made by the individuals of the survey team. The manufacturer should also provide a suitable workroom for the survey team. Facilities must be made accessible to the survey team so that an understanding of the manufacturer's operational capabilities is possible.

A management organization chart should be available as well as the organization chart of the quality assurance department. Also available to the survey team should be the manufacturer's qualification records of welding procedures and welders, nondestructive examination procedures and testing personnel, traveler or process sheet check-off lists, and other exhibits showing the methods used with respect to quality control of production. All nondestructive examining equipment and measuring and testing facilities must be available for examination. A written procedure of the training and indoctrination program for personnel performing work affecting quality control with names of persons involved and documentation records of examinations should be included.

Most important, the quality assurance manual, where the company explains how they control fabrication to meet Code requirements, will form the basis for understanding the manufacturer's system. It will be reviewed by the survey team. The manual will serve as the basis for continuous auditing of the system by the Authorized Inspectors and by later survey reviews when the manufacturer's Certificate of Authorization is to be reviewed. All manuals should be classed as either controlled or noncontrolled, as follows:

1. A controlled manual is one that is given a number, assigned to one individual, and kept up-to-date in accordance with the approved system by the party responsible for manuals in the plant.

2. A noncontrolled manual is a copy of the manual, which is distributed to customers and other interested parties, which will only be considered up-to-date when distributed and should be marked as such by rubber stamping the first as well as several other pages throughout the book indicating it will not be kept up-to-date. All pages of the manual should have the name of the manufacturer and a revision block including a space for dates of revisions.

An outline of the procedures to be used by the manufacturer in describing a quality assurance system should include the following items:

1. MANAGEMENT'S STATEMENT OF AUTHORITY AND RESPONSIBILITY
   a. This statement should specify management's participation in the program — that the quality assurance manager shall have the authority and responsibility to establish and maintain a quality assurance program, and organizational freedom to recognize quality control problems in order to provide solutions to these problems. This statement of authority should be signed by the president of the company or someone else in top management.

2. ORGANIZATION
   a. Manufacturer's organization chart giving titles showing the relationship between management, quality assurance, purchasing, engineering, fabrication, testing, and inspection.
   b. Quality assurance manual should give a program for indoctrination and training, which includes program description, listing of personnel involved, and records of their performance. All personnel involved with quality control must be given training, with excellence of performance in mind.

3. DESIGN SPECIFICATIONS AND STRESS REPORT
   a. Verification by the manufacturer that design specifications certified by a registered professional engineer are the owner's or the agent's certified design specifications, a copy of which is to be made available to the Authorized Inspector before construction begins and to the jurisdictional authority at the location of their installation before components are placed in service.
   b. Responsibility for structural integrity and stress reports for Class 1, Class 2 vessels, Class MC, and Class CS should be certified by a registered professional engineer who makes the stress report and the owner or agent.
   c. Load capacity data sheet for Class 1 and Class MC component supports must be certified by a registered professional engineer stating that the load capacity is rated in accordance with Code Subsection NF; the data sheet specifying the organization responsible for retaining data substantiating stated load capacity.
   d. When class of vessel or part does not require a stress report, the design calculations must be made available to the Authorized Inspector.
   e. The complete certified stress report by the manufacturer and owner shall be made available to the Authorized Inspector.

4. DESIGN CONTROL
   a. Manufacturer should have an order entry system which provides for competent review and remarks by applicable persons to assure correct translation of design specifications requirements into specifications, drawings, procedures, and instructions.

b. Design reviews and checking shall be made by others than those who performed the original design.

c. Document control of revisions to assure that documents and changes are reviewed for adequacy, approved, documented, and released by authorized personnel and distributed and used where required.

5. DOCUMENT CONTROL

a. Establish procedures identifying responsible personnel to assure that the required documents are reviewed and approved before release.

b. The revision system must provide for engineering review by persons, including the registered professional engineer responsible for stress report or design calculations.

c. All revisions and replacements must be controlled and documented.

d. Documents must be made available to the Authorized Inspector.

6. MATERIAL PROCUREMENT CONTROL

a. The manufacturer is responsible for surveying and qualifying vendor quality system programs, or the material manufacturer must hold a quality system certificate issued by the ASME.

b. The system of material control used in purchasing to assure identification of all material used, and to check compliance with specifications, should be explained.

c. Details of vendor's evaluation should include a sample of the survey questions asked by the manufacturer when visiting a vendor and how often the vendor is audited.

d. How is a vendor qualified and listed? Is the accepted vendor's list properly revised and controlled? How is vendor removed from accepted vendor's list?

e. Control of material at source — who issues and approves purchase documents.

f. Material manufacturer's and subcontractor's corrective action — how verified and controlled.

g. Receiving inspection control — identification of material with receiving check list.

h. Nonconformity of purchased items — how controlled; state corrective action system.

i. Manufacturer's certified materials test report, including welding and brazing materials. State how examined and signed off to assure correlation with material used for traceability.

7. EXAMINATION AND INSPECTION PROGRAM

a. Identification and control of material heat number transfer.

b. Use of a "traveler" process control check list indicating fabricating processes. (See Chap. 4, Fig. 4.1.)

c. Description of the preparation and use of traveler or process control check list with sign-off spaces for use by manufacturer.

d. Mandatory hold points (Code Par. NCA-5241).

e. Nonconformities, material, parts or components.   How controlled?

f. Methods used for corrective action and nonconformities.

8. WELDING

a. Describe system for assuring that all welding procedures and welders or welding operators are properly qualified (Code Par. NCA-5250).

b. Welding qualification records and identifying stamps, including the method of recording the continuance of valid performance qualification by production welding.

c. Describe the method of welding electrode control, i.e., electrode ordering, inspection when received, storage, how they are issued, and the in-process protection of the electrode.

d. Method used on drawings to tell type and size of weld and electrode to use, i.e., whether the drawings show weld detail in full cross section or use welding symbols.

e. Explain fit-up and weld inspection.

9. NONDESTRUCTIVE EXAMINATIONS

a. Describe system for assuring that all nondestructive examination personnel are qualified.

b. Show written procedures to assure that nondestructive examination requirements are maintained.

c. Describe required written training program.

d. Show examination methods in required areas.
  (1) General
  (2) Specific
  (3) Practical
  (4) Physical

e. Show personnel qualification record and Level III support documents for examinations, experience, education, and physical.

f. Describe equipment qualification and calibration.

g. Where NDE operations are subcontracted, show similar controls as above and the contract terms with the NDE organization.

10. HEAT TREATMENT

a. Explain procedures for preheating postweld heat treatment, hot forming, etc.

b. Describe how furnace is loaded.   Describe thermocouple placement, how and where they are placed, also recording system for time and temperature, and how recorded for each thermocouple.

c. Describe the methods of heating and cooling, and how temperature is controlled.

d. Describe time and temperature recordings; are they reviewed by Authorized Inspector.

11. CONTROL OF MEASUREMENT AND TEST EQUIPMENT

    a. Describe established and documented procedures used for assuring that gages and measuring and testing devices are calibrated and controlled.
    b. Calibration should be to national standards where such standards exist. Designate other standards when used.
    c. Identification and calibration control. Establish procedures to assure that measuring and test equipment are calibrated and adjusted at specified periods.
    e. Nonconforming equipment control. When discrepancies in examination or testing equipment are found, state corrective action to be taken.

12. QUALITY ASSURANCE RECORDS

    a. Records retention. Establish written procedure for handling of records. Permanent and nonpermanent for time specifid in Code. Show how records will be kept. Determine early and develop list to serve as index to the document file.
    b. Permanent records. The owner has the responsibility for continued maintenance of all records and the traceability of records for the life of the plant at the power plant site, the manufacturer's plant, or other locations determined by mutual agreement.
    c. Nonpermanent records. State responsibility for, where they will be kept, and how long they will be kept.
    d. The quality assurance system must verify records of materials, manufacturing, examination, and tests before and during construction. State responsibility for and location of all records.
    e. State how protected from damage and how records are stored.

13 NONCONFORMING MATERIAL CONTROL

    a. Either correct nonconforming materials, components, and parts that do not meet the required standards or do not use them.
    b. Describe how part is identified and records kept.
    c. Describe methods of corrective action, return of material for use, and how records of traceability are kept.
    d. Give names of appropriate levels of management involved with responsibilities and authority for control of nonconformities.
    e. When disposition of nonconformity involves design, stress report, or dimensional changes, provide for review by responsible engineering person or group. (See "Correction of Nonconformities," Chap. 4, Item #7.)

14. AUTHORIZED INSPECTOR

    a. The manufacturer should be carefully acquainted with the provisions in the Code that establish the duties of the Authorized Inspector. This third party in the manufacturer's plant, by virtue of authorization by the state to do Code inspections, is the legal representative. This party's presence permits the manufacturer to fabricate under state laws. Under the ASME Code rules, the Authorized Inspector has certain duties as are specified in the Code.

It is the obligation of the manufacturer to see that the Inspector has the opportunity to perform as required by law. The details of authorized inspection at the plant or site are to be worked out between the Inspector and the manufacturer. The manufacturer's quality assurance manual should state provisions for assisting the Authorized Inspector in performing duties.

b. The manual should state that the Authorized Inspector and inspection agency have free access to areas and documents as necessary to carry out the duties specified in the Code, such as monitoring the manufacturer's quality assurance system, reviewing qualification records, verifying materials, inspecting and verifying in-process fabrication, witnessing final tests, and certifying data reports.

c. The Inspector should be given the opportunity to indicate "hold points" and to discuss with the quality assurance manager arrangements for inspections.

d. The Authorized Inspector must be provided with the current quality assurance manual and latest revisions which have been accepted by the Authorized Inspection Specialists. The final hydrostatic or pneumatic tests required shall be witnessed by the Authorized Inspector, and any examinations that are necessary before and after these tests shall be made by the Inspector.

e. The manufacturer's data reports shall be carefully reviewed by a responsible representative of the manufacturer or assembler. If the reports are properly completed, the representative shall sign them for the company. The Authorized Inspector, after examining all documents and feeling assured that the requirements of the Code have been met and the manufacturer's data sheets properly completed, may sign these data sheets.

15. HANDLING, STORAGE, SHIPPING, AND PRESERVATION

a. Describe procedures used to control methods of handling storage, cleaning and preservation of materials, packaging, and shipping to prevent deterioration, damage, and loss.

b. Procedures for examination and inspection should be included on traveler check-off list.

16. AUDITS

a. Explain system of planned, periodic audits to determine the effectiveness of the quality assurance program.

b. Perform audits in accordance with written procedure using check lists developed for each audit in order to assure that a consistent approach is used by audit personnel.

    c. Give report of audit results to top management. Also indicate what corrective action will be taken if necessary.

    d. Follow up corrective action taken by the quality assurance manager to resolve deficiencies, if any, revealed by the audits.

    e. Make results of audit available to the Authorized Inspector.

    f. Give name and title of person responsible for the audit system.

When constructing nuclear pressure vessels according to the ASME Code, complete quality assurance is a must. In this chapter, an attempt has been made to acquaint the reader with the basic elements of a total quality assurance manual and system — from design specifications and control to auditing of the system, which includes document control, purchase control, also vendor surveys and control, receiving inspection, identification and control of materials, control of manufacturing and/or installations, control of examination and tests, control of measuring and test equipment, handling, storage, shipping and preservation, examination of process status, nonconformity control of materials and items, and corrective action to take, as well as quality assurance record retention and authorized inspection.

If a manufacturer uses this chapter as a guide, it will help in preparing for an ASME survey, or if one is contemplated, it will help in understanding the rigorous standards an organization and manufacturing system must meet to enable nuclear power plant components to be fabricated.

Pursuit of excellence in the manufacture of nuclear vessel components must be the aim of each manufacturer seeking ASME certification to fabricate nuclear vessel components.

Cylinder Volume Tables and Diagrams

The most economical method of constructing cylindrical pressure vessels is to keep the length-to-diameter ratio between 2 and 5 while using a diameter that will provide the lowest number of courses.

Occasionally a customer will order a vessel of a given capacity but allow the fabricator some leeway in its diameter and length. The fabricator may then select the size that will be easiest and least expensive to construct. If a 1000-gal tank has been ordered, for example, a course 8 ft long with two hemispherical heads 48 in in diameter might be used. The cylinder would then have a capacity of 752 gal and the two heads a capacity of 125 gal each, for a total of 1002 gal.

By fabricating the vessel in one course rather than two, moreover, the manufacturer will save time in laying out the shell and in handling as well as the cost of joint preparation and welding on the girth seam. It is important to choose a length and diameter that will give the required capacity with as few seams as possible.

The volume tables for cylinders and heads of Appendix A are based on the diameters and lengths most used in the pressure vessel industry. To use the cylinder tables, locate the inside diameter of the vessel in the first column on the left side and move across horizontally to the column headed by the length of the cylinder. Read the capacity in gallons at this point. If the vessel has dished heads, the length of the cylinder is measured between the tangent lines of each head.

The accompanying volume diagram allows for larger diameters and longer lengths and also for rapid estimation of tank lengths and diameters for a known capacity. To use the diagram, simply rotate a straight edge around the required capacity in the middle column and read the desired length in the right-hand column and the corresponding diameter for this length and capacity in the left-hand column. To illustrate: A tank is wanted that will hold 8000 gal. If the shell is made 20 ft long from tangent line to tangent line of two torispherical

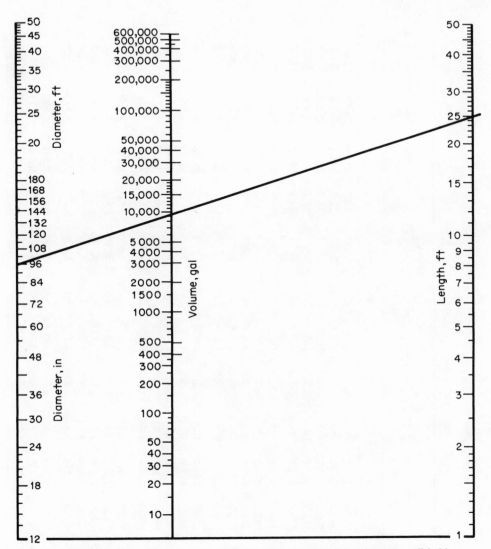

Fig. A.1 Cylinder volume diagram.   A cylinder 25 ft long and 96 in in diameter will hold 9400 gal.

heads and 96 in in inside diameter, the capacity will be 8016 gal (7520 gal for the cylinder and 298 gal for each head).

The volume diagram, Fig. A.1, can also be used to determine the length of a tank of which the capacity and the diameter are both fixed.   In fact, as long as any two factors are known, the third can easily be found.   For this reason, many prefer to use the diagram rather than the tables.

Table A.1 Cylinder volume

Volume $= D^2 \times L \times 0.0034$, in which D = internal diameter, in inches, and L = length, in inches.

Inside diameter, in.	Cubic feet per foot of cylinder	Gallons per inch of cylinder	Gallons per foot of cylinder	Length of shell, ft — Volume, gal										
				3	4	5	6	7	8	9	10	11	12	13
12	0.79	0.49	5.88	17.64	23.52	29.40	35.28	41.16	47.04	52.92	58.80	64.68	70.56	76.44
18	1.77	1.10	13.22	39.66	52.88	66.10	79.32	92.54	105.7	118.9	132.2	145.4	158.6	171.8
24	3.14	1.96	23.58	70.74	94.32	117.9	141.4	165.0	188.6	212.2	235.8	259.3	282.9	306.5
30	4.9	3.06	38.72	116.1	154.8	193.6	232.3	271.0	309.7	348.4	387.2	425.9	464.6	503.3
36	7.1	4.41	52.88	158.6	211.5	264.4	317.2	370.1	423.0	475.9	528.8	581.6	634.5	687.4
42	9.6	6.00	71.97	215.9	287.8	359.8	431.8	503.7	575.7	647.3	719.7	791.6	863.6	935.6
48	12.6	7.83	94.00	282.0	376.0	470.0	564.0	658.0	752.0	846.0	940.0	1,034	1,128	1,222
54	15.9	9.91	118.9	356.9	475.8	594.8	713.8	832.7	951.7	1,070	1,189	1,308	1,427	1,546
60	19.6	12.24	146.8	440.6	587.5	734.4	881.2	1,028	1,175	1,321	1,468	1,615	1,762	1,909
66	23.8	14.81	177.7	533.1	710.8	888.6	1,066	1,244	1,421	1,599	1,777	1,954	2,132	2,310
72	28.3	17.63	211.5	634.5	846.0	1,057	1,269	1,480	1,692	1,903	2,115	2,326	2,538	2,749
78	33.2	20.69	248.2	744.6	992.8	1,241	1,489	1,737	1,985	2,233	2,482	2,730	2,978	3,226
84	38.5	23.99	287.8	863.6	1,151	1,439	1,727	2,015	2,303	2,591	2,878	3,166	3,454	3,742
90	44.2	27.54	330.4	991.4	1,321	1,652	1,982	2,313	2,643	2,974	3,304	3,635	3,965	4,296
96	50.3	31.33	376.0	1,128	1,504	1,880	2,256	2,632	3,008	3,384	3,760	4,136	4,512	4,888
102	56.8	35.37	424.4	1,273	1,697	2,122	2,546	2,971	3,395	3,820	4,244	4,669	5,093	5,518
108	63.6	39.66	475.8	1,427	1,903	2,379	2,855	3,331	3,807	4,283	4,758	5,234	5,710	6,186
114	70.9	44.19	530.2	1,590	2,120	2,651	3,181	3,711	4,241	4,772	5,302	5,832	6,362	6,893
120	78.5	48.96	587.5	1,762	2,350	2,937	3,525	4,112	4,700	5,287	5,875	6,462	7,050	7,637
126	86.3	53.98	647.7	1,943	2,590	3,238	3,886	4,534	5,181	5,829	6,477	7,125	7,772	8,420
132	95.	59.24	710.9	2,132	2,843	3,554	4,265	4,976	5,687	6,398	7,109	7,819	8,530	9,241
138	103.6	64.75	777.0	2,331	3,108	3,885	4,662	5,439	6,216	6,993	7,770	8,547	9,324	10,101
144	113.1	70.50	846.0	2,538	3,384	4,230	5,076	5,922	6,768	7,614	8,460	9,306	10,152	10,998

Table A.1 Cylinder volume (cont.)

Inside diameter, in.	Cubic feet per foot of cylinder	Gallons per inch of cylinder	Gallons per foot of cylinder	Length of shell, ft										
				14	15	16	17	18	19	20	25	30	35	40
				Volume, gal										
12	0.79	0.49	5.88	82.32	88.20	94.08	99.96	105.8	111.7	117.6	147.0	176.4	205.8	235.2
18	1.77	1.10	13.22	185.0	198.3	211.5	224.7	237.9	251.1	264.4	330.5	396.6	462.7	528.8
24	3.14	1.96	23.58	330.1	353.7	377.2	400.8	424.4	448.0	471.6	589.5	707.4	825.3	943.2
30	4.9	3.06	38.72	542.0	580.8	619.5	658.2	696.9	735.6	774.4	968.0	1,161	1,355	1,548
36	7.1	4.41	52.88	740.3	793.2	846.0	898.9	951.8	1,004	1,057	1,322	1,586	1,850	2,115
42	9.6	6.00	71.97	1,007	1,079	1,151	1,223	1,295	1,367	1,439	1,799	2,159	2,518	2,878
48	12.6	7.83	94.00	1,316	1,410	1,504	1,598	1,692	1,786	1,880	2,350	2,820	3,290	3,760
54	15.9	9.91	118.9	1,665	1,784	1,903	2,022	2,141	2,260	2,379	2,974	3,569	4,163	4,758
60	19.6	12.24	146.8	2,056	2,203	2,350	2,496	2,643	2,790	2,937	3,672	4,406	5,140	5,875
66	23.8	14.81	177.7	2,488	2,665	2,843	3,021	3,198	3,376	3,554	4,443	5,331	6,220	7,108
72	28.3	17.63	211.5	2,961	3,172	3,384	3,595	3,807	4,018	4,230	5,287	6,345	7,402	8,460
78	33.2	20.69	248.2	3,475	3,723	3,971	4,219	4,467	4,716	4,964	6,205	7,446	8,687	9,928
84	38.5	23.99	287.8	4,030	4,318	4,606	4,894	5,182	5,469	5,757	7,197	8,636	10,076	11,515
90	44.2	27.54	330.4	4,626	4,957	5,287	5,618	5,948	6,279	6,609	8,262	9,914	11,566	13,219
96	50.3	31.33	376.0	5,264	5,640	6,016	6,392	6,768	7,144	7,520	9,400	11,280	13,160	15,040
102	56.8	35.37	424.4	5,942	6,367	6,791	7,216	7,640	8,489	8,489	10,612	12,734	14,856	16,979
108	63.6	39.66	475.8	6,662	7,138	7,614	8,090	8,566	9,041	9,517	11,897	14,276	16,656	19,035
114	70.9	44.19	530.2	7,432	7,953	8,483	9,014	9,544	10,074	10,604	13,256	15,907	18,558	21,209
120	78.5	48.96	587.5	8,225	8,812	9,400	9,987	10,575	11,162	11,750	14,688	17,625	20,563	23,500
126	86.3	53.98	647.7	9,068	9,716	10,363	11,011	11,659	12,307	12,954	16,193	19,432	22,670	25,909
132	95.	59.24	710.9	9,952	10,663	11,374	12,085	12,796	13,507	14,218	17,772	21,327	24,881	28,436
138	103.6	64.75	777.0	10,878	11,655	12,432	13,209	13,986	14,763	15,540	19,425	23,310	27,195	31,080
144	113.1	70.50	846.0	11,844	12,690	13,536	14,382	15,228	16,074	16,920	21,150	25,380	29,611	33,841

Table A.2 Head volume

Volumes listed are approximate volumes of dished portion of head; straight flanges are not included.

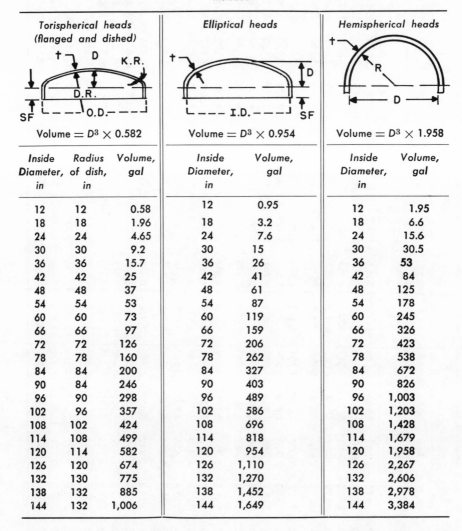

Torispherical heads (flanged and dished) Volume = $D^3 \times 0.582$			Elliptical heads Volume = $D^3 \times 0.954$		Hemispherical heads Volume = $D^3 \times 1.958$	
Inside Diameter, in	Radius of dish, in	Volume, gal	Inside Diameter, in	Volume, gal	Inside Diameter, in	Volume, gal
12	12	0.58	12	0.95	12	1.95
18	18	1.96	18	3.2	18	6.6
24	24	4.65	24	7.6	24	15.6
30	30	9.2	30	15	30	30.5
36	36	15.7	36	26	36	53
42	42	25	42	41	42	84
48	48	37	48	61	48	125
54	54	53	54	87	54	178
60	60	73	60	119	60	245
66	66	97	66	159	66	326
72	72	126	72	206	72	423
78	78	160	78	262	78	538
84	84	200	84	327	84	672
90	84	246	90	403	90	826
96	90	298	96	489	96	1,003
102	96	357	102	586	102	1,203
108	102	424	108	696	108	1,428
114	108	499	114	818	114	1,679
120	114	582	120	954	120	1,958
126	120	674	126	1,110	126	2,267
132	130	775	132	1,270	132	2,606
138	132	885	138	1,452	138	2,978
144	132	1,006	144	1,649	144	3,384

Circumferences of Cylinders

The length of a shell plate to be cut for a cylindrical vessel is measured by the mean or bend-line circumference of the shell so that the plate will equal a required inside diameter after being rolled. The circumferences given in Table B.1 of Appendix B are for the inside diameter.

When plates are to be cut to provide a given inside diameter, an amount equal to the plate thickness multiplied by 3.1416 must be added to the circumference shown in the table for that diameter. For example, suppose a vessel is to be 60 in in inside diameter and have a shell thickness of 1 in. The 60-in inside diameter generates a circumference of 188.496 in. To this figure must be added the plate thickness multiplied by 3.1416, or 3.1416 in. The length of plate to be cut to make a cylinder 60 in in inside diameter is therefore 188.496 plus 3.1416, or 191.6376 in.

A simpler method is to add the thickness of plate to the inside diameter and obtain the circumference directly from the table. For example, for a vessel requiring a 60-in inside diameter and a 1-in shell plate, add 1 in to 60 in for a total of 61 in. The circumference given in the table for a 61-in diameter equals 191.637 in.

Plate for cylinders with dished heads should never be cut until the heads have been taped if a good fit is to be secured at the head joint. To illustrate: If the inside circumference of a head has been taped at 188.5 in and the thickness of the shell and head is 1 in, an amount equal to the plate thickness multiplied by 3.1416 will have to be added to the taped measurement for the shell to meet the inside diameter of the head when the plate is cut. Therefore, the plate will have to be cut to 191.64 in.

If the head has been taped on the outside circumference, subtract an amount equal to the plate thickness multiplied by 3.1416 from the total measurement before cutting the shell plate.

Some may find it easier to use Table B.1 to arrive at a measurement. For example, if the outside circumference of a head has been taped at 194.78 in, the diameter shown for this measurement is 62 in. If the shell plate is 1 in thick, subtract 1 in from 62 in, for a diameter of 61 in. The circumference given in the table for this diameter equals 191.637 in. This figure represents the mean or bend-line circumference and therefore the length of plate required to make a cylinder 60 in in inside diameter.

Table B.1 Circumferences and areas of circles

Areas are approximate but sufficiently accurate for estimating.

Dia.	Circum.	Area	Dia.	Circum.	Area	Dia.	Circum.	Area	Dia.	Circum.	Area
3.	9.4248	7.0686	11.	34.558	95.033	19.	59.690	283.53	27.	84.823	572.56
1/8	9.8175	7.6699	1/8	34.950	97.205	1/8	60.083	287.27	1/8	85.216	577.87
1/4	10.210	8.2958	1/4	35.343	99.402	1/4	60.476	291.04	1/4	85.608	583.21
3/8	10.603	8.9462	3/8	35.736	101.62	3/8	60.868	294.83	3/8	86.001	588.57
1/2	10.996	9.6211	1/2	36.128	103.87	1/2	61.261	298.65	1/2	86.394	593.96
5/8	11.388	10.321	5/8	36.521	106.14	5/8	61.654	302.49	5/8	86.786	599.37
3/4	11.781	11.045	3/4	36.914	108.43	3/4	62.046	306.35	3/4	87.179	604.81
7/8	12.174	11.793	7/8	37.306	110.75	7/8	62.439	310.24	7/8	87.572	610.27
4.	12.566	12.566	12.	37.699	113.10	20.	62.832	314.16	28.	87.965	615.75
1/8	12.959	13.364	1/8	38.092	115.47	1/8	63.225	318.10	1/8	88.357	621.26
1/4	13.352	14.186	1/4	38.485	117.86	1/4	63.617	322.06	1/4	88.750	626.80
3/8	13.744	15.033	3/8	38.877	120.28	3/8	64.010	326.05	3/8	89.143	632.36
1/2	14.137	15.904	1/2	39.270	122.72	1/2	64.403	330.06	1/2	89.535	637.94
5/8	14.530	16.800	5/8	39.663	125.19	5/8	64.795	334.10	5/8	89.928	643.55
3/4	14.923	17.728	3/4	40.055	127.68	3/4	65.188	338.16	3/4	90.321	649.18
7/8	15.315	18.665	7/8	40.448	130.19	7/8	65.581	342.25	7/8	90.713	654.84
5.	15.708	19.635	13.	40.841	132.73	21.	65.973	346.36	29.	91.106	660.52
1/8	16.101	20.629	1/8	41.233	135.30	1/8	66.366	350.50	1/8	91.499	666.23
1/4	16.493	21.648	1/4	41.626	137.89	1/4	66.759	354.66	1/4	91.892	671.96
3/8	16.886	22.691	3/8	42.019	140.50	3/8	67.152	358.84	3/8	92.284	677.71
1/2	17.279	23.758	1/2	42.412	143.14	1/2	67.544	363.05	1/2	92.677	683.49
5/8	17.671	24.850	5/8	42.804	145.80	5/8	67.937	367.28	5/8	93.070	689.30
3/4	18.064	25.967	3/4	43.197	148.49	3/4	68.330	371.54	3/4	93.462	695.13
7/8	18.457	27.109	7/8	43.590	151.20	7/8	68.722	375.83	7/8	93.855	700.98
6.	18.850	28.274	14.	43.982	153.94	22.	69.115	380.13	30.	94.248	706.86
1/8	19.242	29.465	1/8	44.375	156.70	1/8	69.508	384.46	1/8	94.640	712.76
1/4	19.635	30.680	1/4	44.768	159.48	1/4	69.900	388.82	1/4	95.033	718.69
3/8	20.028	31.919	3/8	45.160	162.30	3/8	70.293	393.20	3/8	95.426	724.64
1/2	20.420	33.183	1/2	45.553	165.13	1/2	70.686	397.61	1/2	95.819	730.62
5/8	20.813	34.472	5/8	45.946	167.99	5/8	71.079	402.04	5/8	96.211	736.62
3/4	21.206	35.785	3/4	46.338	170.87	3/4	71.471	406.49	3/4	96.604	742.64
7/8	21.598	37.122	7/8	46.731	173.78	7/8	71.864	410.97	7/8	96.997	748.69

Table B.1 Circumference and areas of circles (cont.)

Dia.	Circ.	Area		Dia.	Circ.	Area		Dia.	Circ.	Area		Dia.	Circ.	Area
7.	21.991	38.485		15.	47.124	176.71		23.	72.257	415.48		31.	97.389	754.77
1/8	22.384	39.871		1/8	47.517	179.67		1/8	72.649	420.00		1/8	97.782	760.87
1/4	22.776	41.282		1/4	47.909	182.65		1/4	73.042	424.56		1/4	98.175	766.99
3/8	23.169	42.718		3/8	48.302	185.66		3/8	73.435	429.13		3/8	98.567	773.14
1/2	23.562	44.179		1/2	48.695	188.69		1/2	73.827	433.74		1/2	98.960	779.31
5/8	23.955	45.664		5/8	49.087	191.75		5/8	74.220	438.36		5/8	99.353	785.51
3/4	24.347	47.173		3/4	49.480	194.83		3/4	74.613	443.01		3/4	99.746	791.73
7/8	24.740	48.707		7/8	49.873	197.93		7/8	75.006	447.69		7/8	100.138	797.98
8.	25.133	50.265		16.	50.265	201.06		24.	75.398	452.39		32.	100.531	804.25
1/8	25.525	51.849		1/8	50.658	204.22		1/8	75.791	457.11		1/8	100.924	810.54
1/4	25.918	53.456		1/4	51.051	207.39		1/4	76.184	461.86		1/4	101.316	816.86
3/8	26.311	55.088		3/8	51.444	210.60		3/8	76.576	466.64		3/8	101.709	823.21
1/2	26.704	56.745		1/2	51.836	213.82		1/2	76.969	471.44		1/2	102.102	829.58
5/8	27.096	58.426		5/8	52.229	217.08		5/8	77.362	476.26		5/8	102.494	835.97
3/4	27.489	60.132		3/4	52.622	220.35		3/4	77.754	481.11		3/4	102.887	842.39
7/8	27.882	61.862		7/8	53.014	223.65		7/8	78.147	485.98		7/8	103.280	848.83
9.	28.274	63.617		17.	53.407	226.98		25.	78.540	490.87		33.	103.673	855.30
1/8	28.667	65.397		1/8	53.800	230.33		1/8	78.933	495.79		1/8	104.065	861.79
1/4	29.060	67.201		1/4	54.192	233.71		1/4	79.325	500.74		1/4	104.458	868.31
3/8	29.452	69.029		3/8	54.585	237.10		3/8	79.718	505.71		3/8	104.851	874.85
1/2	29.845	70.882		1/2	54.978	240.53		1/2	80.111	510.71		1/2	105.243	881.41
5/8	30.238	72.760		5/8	55.371	243.98		5/8	80.503	515.72		5/8	105.636	888.00
3/4	30.631	74.662		3/4	55.763	247.45		3/4	80.896	520.77		3/4	106.029	894.62
7/8	31.023	76.589		7/8	56.156	250.95		7/8	81.289	525.84		7/8	106.421	901.26
10.	31.416	78.540		18.	56.549	254.47		26.	81.681	530.93		34.	106.814	907.92
1/8	31.809	80.516		1/8	56.941	258.02		1/8	82.074	536.05		1/8	107.207	914.61
1/4	32.201	82.516		1/4	57.334	261.59		1/4	82.467	541.19		1/4	107.600	921.32
3/8	32.594	84.541		3/8	57.727	265.18		3/8	82.860	546.35		3/8	107.992	928.06
1/2	32.987	86.590		1/2	58.119	268.80		1/2	83.252	551.55		1/2	108.385	934.82
5/8	33.379	88.664		5/8	58.512	272.45		5/8	83.645	556.76		5/8	108.778	941.61
3/4	33.772	90.763		3/4	58.905	276.12		3/4	84.038	562.00		3/4	109.170	948.42
7/8	34.165	92.886		7/8	59.298	279.81		7/8	84.430	567.27		7/8	109.563	955.25

Table B.1 Circumferences and areas of circles (cont.)

Dia.	Circum.	Area	Dia.	Circum.	Area	Dia.	Circum.	Area	Dia.	Circum.	Area
35.	109.956	962.11	43.	135.088	1452.2	51.	160.221	2042.8	59.	185.354	2734.0
1/8	110.348	969.00	1/8	135.481	1460.7	1/8	160.614	2052.8	1/8	185.747	2745.6
1/4	110.741	975.91	1/4	135.874	1469.1	1/4	161.007	2062.9	1/4	186.139	2757.2
3/8	111.134	982.84	3/8	136.267	1477.6	3/8	161.399	2073.0	3/8	186.532	2768.8
1/2	111.527	989.80	1/2	136.659	1486.2	1/2	161.792	2083.1	1/2	186.925	2780.5
5/8	111.919	996.78	5/8	137.052	1494.7	5/8	162.185	2093.2	5/8	187.317	2792.2
3/4	112.312	1003.8	3/4	137.445	1503.3	3/4	162.577	2103.3	3/4	187.710	2803.9
7/8	112.705	1010.8	7/8	137.837	1511.9	7/8	162.970	2113.5	7/8	188.103	2815.7
36.	113.097	1017.9	44.	138.230	1520.5	52.	163.363	2123.7	60.	188.496	2827.4
1/8	113.490	1025.0	1/8	138.623	1529.2	1/8	163.756	2133.9	1/8	188.888	2839.2
1/4	113.883	1032.1	1/4	139.015	1537.9	1/4	164.148	2144.2	1/4	189.281	2851.0
3/8	114.275	1039.2	3/8	139.408	1546.6	3/8	164.541	2154.5	3/8	189.674	2862.9
1/2	114.668	1046.3	1/2	139.801	1555.3	1/2	164.934	2164.8	1/2	190.066	2874.8
5/8	115.061	1053.5	5/8	140.194	1564.0	5/8	165.326	2175.1	5/8	190.459	2886.6
3/4	115.454	1060.7	3/4	140.586	1572.8	3/4	165.719	2185.4	3/4	190.852	2898.6
7/8	115.846	1068.0	7/8	140.979	1581.6	7/8	166.112	2195.8	7/8	191.244	2910.5
37.	116.239	1075.2	45.	141.372	1590.4	53.	166.504	2206.2	61.	191.637	2922.5
1/8	116.632	1082.5	1/8	141.764	1599.3	1/8	166.897	2216.6	1/8	192.030	2934.5
1/4	117.024	1089.8	1/4	142.157	1608.2	1/4	167.290	2227.0	1/4	192.423	2946.5
3/8	117.417	1097.1	3/8	142.550	1617.0	3/8	167.683	2237.5	3/8	192.815	2958.5
1/2	117.810	1104.5	1/2	142.942	1626.0	1/2	168.075	2248.0	1/2	193.208	2970.6
5/8	118.202	1111.8	5/8	143.335	1634.9	5/8	168.468	2258.5	5/8	193.601	2982.7
3/4	118.596	1119.2	3/4	143.728	1643.9	3/4	168.861	2269.1	3/4	193.993	2994.8
7/8	118.988	1126.7	7/8	144.121	1652.9	7/8	169.253	2279.6	7/8	194.386	3006.9
38.	119.381	1134.1	46.	144.513	1661.9	54.	169.646	2290.2	62.	194.779	3019.1
1/8	119.773	1141.6	1/8	144.906	1670.9	1/8	170.039	2300.8	1/8	195.171	3031.3
1/4	120.166	1149.1	1/4	145.299	1680.0	1/4	170.431	2311.5	1/4	195.564	3043.5
3/8	120.559	1156.6	3/8	145.691	1689.1	3/8	170.824	2322.1	3/8	195.957	3055.7
1/2	120.951	1164.2	1/2	146.084	1698.2	1/2	171.217	2332.8	1/2	196.350	3068.0
5/8	121.344	1171.7	5/8	146.477	1707.4	5/8	171.609	2343.5	5/8	196.742	3080.3
3/4	121.737	1179.3	3/4	146.869	1716.5	3/4	172.002	2354.3	3/4	197.135	3092.6
7/8	122.129	1186.9	7/8	147.262	1725.7	7/8	172.395	2365.0	7/8	197.528	3104.9

Table B.1 Circumferences and areas of circles (cont.)

Dia.	Circumf.	Area		Dia.	Circumf.	Area		Dia.	Circumf.	Area
39.	122.522	1194.6	47.	147.655	1734.9	63.	197.920	3117.2
1/8	122.915	1202.3	1/8	148.048	1744.2	1/8	198.313	3129.6
1/4	123.308	1210.6	1/4	148.440	1753.5	1/4	198.706	3142.0
3/8	123.700	1217.7	3/8	148.833	1762.7	3/8	199.098	3154.5
1/2	124.093	1225.4	1/2	149.226	1772.1	1/2	199.491	3166.9
5/8	124.486	1233.2	5/8	149.618	1781.4	5/8	199.884	3179.4
3/4	124.878	1241.0	3/4	150.011	1790.8	3/4	200.277	3191.9
7/8	125.271	1248.8	7/8	150.404	1800.1	7/8	200.669	3204.4
40.	125.664	1256.6	48.	150.796	1809.6	64.	201.062	3217.0
1/8	126.056	1264.5	1/8	151.189	1819.0	1/8	201.455	3229.6
1/4	126.449	1272.4	1/4	151.582	1828.5	1/4	201.847	3242.2
3/8	126.842	1280.3	3/8	151.975	1837.9	3/8	202.240	3254.8
1/2	127.235	1288.2	1/2	152.367	1847.5	1/2	202.633	3267.5
5/8	127.627	1296.2	5/8	152.760	1857.0	5/8	203.025	3280.1
3/4	128.020	1304.2	3/4	153.153	1866.5	3/4	203.418	3292.8
7/8	128.413	1312.2	7/8	153.545	1876.1	7/8	203.811	3305.6
41.	128.805	1320.3	49.	153.938	1885.7	65.	204.204	3318.3
1/8	129.198	1328.3	1/8	154.331	1895.4	1/8	204.596	3331.1
1/4	129.591	1336.4	1/4	154.723	1905.0	1/4	204.989	3343.9
3/8	129.983	1344.5	3/8	155.116	1914.7	3/8	205.382	3356.7
1/2	130.376	1352.7	1/2	155.509	1924.4	1/2	205.774	3369.6
5/8	130.769	1360.8	5/8	155.902	1934.2	5/8	206.167	3382.4
3/4	131.161	1369.0	3/4	156.294	1943.9	3/4	206.560	3395.3
7/8	131.554	1377.2	7/8	156.687	1953.7	7/8	206.952	3408.2
42.	131.947	1385.4	50.	157.080	1963.5	66.	207.345	3421.2
1/8	132.340	1393.7	1/8	157.472	1973.3	1/8	207.738	3434.2
1/4	132.732	1402.0	1/4	157.865	1983.2	1/4	208.131	3447.2
3/8	133.125	1410.3	3/8	158.258	1993.1	3/8	208.523	3460.2
1/2	133.518	1418.6	1/2	158.650	2003.0	1/2	208.916	3473.2
5/8	133.910	1427.0	5/8	159.043	2012.9	5/8	209.309	3486.3
3/4	134.303	1435.4	3/4	159.436	2022.8	3/4	209.701	3499.4
7/8	134.696	1443.8	7/8	159.829	2032.8	7/8	210.094	3512.5

Dia.	Circumf.	Area
55.	172.788	2375.8
1/8	173.180	2386.6
1/4	173.573	2397.5
3/8	173.966	2408.3
1/2	174.358	2419.2
5/8	174.751	2430.1
3/4	175.144	2441.1
7/8	175.536	2452.0
56.	175.929	2463.0
1/8	176.322	2474.0
1/4	176.715	2485.0
3/8	177.107	2496.1
1/2	177.500	2507.2
5/8	177.893	2518.3
3/4	178.285	2529.4
7/8	178.678	2540.6
57.	179.071	2551.8
1/8	179.463	2563.0
1/4	179.856	2574.2
3/8	180.249	2585.4
1/2	180.642	2596.7
5/8	181.034	2608.0
3/4	181.427	2619.4
7/8	181.820	2630.7
58.	182.212	2642.1
1/8	182.605	2653.5
1/4	182.998	2664.9
3/8	183.390	2676.4
1/2	183.783	2687.8
5/8	184.176	2699.3
3/4	184.569	2710.9
7/8	184.961	2722.4

Table B.1 Circumferences and areas of circles (cont.)

Dia.	Circum.	Area	Dia.	Circum.	Area	Dia.	Circum.	Area	Dia.	Circum.	Area
67.	210.487	3525.7	75.	235.619	4417.9	83.	260.752	5410.6	91.	285.885	6503.9
1/8	210.879	3538.8	1/8	236.012	4432.6	1/8	261.145	5426.9	1/8	286.278	6521.8
1/4	211.272	3552.0	1/4	236.405	4447.4	1/4	261.538	5443.3	1/4	286.670	6539.7
3/8	211.665	3565.2	3/8	236.798	4462.2	3/8	261.930	5459.6	3/8	287.063	6557.6
1/2	212.058	3578.5	1/2	237.190	4477.0	1/2	262.323	5476.0	1/2	287.456	6575.5
5/8	212.450	3591.7	5/8	237.583	4491.8	5/8	262.716	5492.4	5/8	287.848	6593.5
3/4	212.843	3605.0	3/4	237.976	4506.7	3/4	263.108	5508.8	3/4	288.241	6611.5
7/8	213.236	3618.3	7/8	238.368	4521.5	7/8	263.501	5525.3	7/8	288.634	6629.6
68.	213.628	3631.7	76.	238.761	4536.5	84.	263.894	5541.8	92.	289.027	6647.6
1/8	214.021	3645.0	1/8	239.154	4551.4	1/8	264.286	5558.3	1/8	289.419	6665.7
1/4	214.414	3658.4	1/4	239.546	4566.4	1/4	264.679	5574.8	1/4	289.812	6683.8
3/8	214.806	3671.8	3/8	239.939	4581.3	3/8	265.072	5591.4	3/8	290.205	6701.9
1/2	215.199	3685.3	1/2	240.332	4596.3	1/2	265.465	5607.9	1/2	290.597	6720.1
5/8	215.592	3698.7	5/8	240.725	4611.4	5/8	265.857	5624.5	5/8	290.990	6738.2
3/4	215.984	3712.2	3/4	241.117	4626.4	3/4	266.250	5641.2	3/4	291.383	6756.4
7/8	216.377	3725.7	7/8	241.510	4641.5	7/8	266.643	5657.8	7/8	291.775	6774.7
69.	216.770	3739.3	77.	241.903	4656.6	85.	267.035	5674.5	93.	292.168	6792.9
1/8	217.163	3752.8	1/8	242.295	4671.8	1/8	267.428	5691.2	1/8	292.561	6811.2
1/4	217.555	3766.4	1/4	242.688	4686.9	1/4	267.821	5707.9	1/4	292.954	6829.5
3/8	217.948	3780.0	3/8	243.081	4702.1	3/8	268.213	5724.7	3/8	293.346	6847.8
1/2	218.341	3793.7	1/2	243.473	4717.3	1/2	268.606	5741.5	1/2	293.739	6866.1
5/8	218.733	3807.3	5/8	243.866	4732.5	5/8	268.999	5758.3	5/8	294.132	6884.5
3/4	219.126	3821.0	3/4	244.259	4747.8	3/4	269.392	5775.1	3/4	294.524	6902.9
7/8	219.519	3834.7	7/8	244.652	4763.1	7/8	269.784	5791.9	7/8	294.917	6921.3
70.	219.911	3848.5	78.	245.044	4778.4	86.	270.177	5808.8	94.	295.310	6939.8
1/8	220.304	3862.2	1/8	245.437	4793.7	1/8	270.570	5825.7	1/8	295.702	6958.2
1/4	220.697	3876.0	1/4	245.830	4809.0	1/4	270.962	5842.6	1/4	296.095	6976.7
3/8	221.090	3889.8	3/8	246.222	4824.4	3/8	271.355	5859.6	3/8	296.488	6995.3
1/2	221.482	3903.6	1/2	246.615	4839.8	1/2	271.748	5876.5	1/2	296.881	7013.8
5/8	221.875	3917.5	5/8	247.008	4855.2	5/8	272.140	5893.5	5/8	297.273	7032.4
3/4	222.268	3931.4	3/4	247.400	4870.7	3/4	272.533	5910.6	3/4	297.666	7051.0
7/8	222.660	3945.3	7/8	247.793	4886.2	7/8	272.926	5927.6	7/8	298.059	7069.6

Table B.1 Circumferences and areas of circles (cont.)

	Circumference	Area
71. 1/8	223.053	3959.2
1/4	223.446	3973.1
3/8	223.838	3987.1
1/2	224.231	4001.1
5/8	224.624	4015.2
3/4	225.017	4029.2
7/8	225.409	4043.3
	225.802	4057.4
72. 1/8	226.195	4071.5
1/4	226.587	4085.7
3/8	226.980	4099.8
1/2	227.373	4114.0
5/8	227.765	4128.2
3/4	228.158	4142.5
7/8	228.551	4156.8
	228.944	4171.1
73. 1/8	229.336	4185.4
1/4	229.729	4199.7
3/8	230.122	4214.1
1/2	230.514	4228.5
5/8	230.907	4242.9
3/4	231.300	4257.4
7/8	231.692	4271.8
	232.085	4286.3
74. 1/8	232.478	4300.8
1/4	232.871	4315.4
3/8	233.263	4329.9
1/2	233.656	4344.5
5/8	234.049	4359.2
3/4	234.441	4373.8
7/8	234.834	4388.5
	235.227	4403.1

	Circumference	Area
79. 1/8	248.186	4901.7
1/4	248.579	4917.2
3/8	248.971	4932.7
1/2	249.364	4948.3
5/8	249.757	4963.9
3/4	250.149	4979.5
7/8	250.542	4995.2
	250.935	5010.9
80. 1/8	251.327	5026.5
1/4	251.720	5042.3
3/8	252.113	5058.0
1/2	252.506	5073.8
5/8	252.898	5089.6
3/4	253.291	5105.4
7/8	253.684	5121.2
	254.076	5137.1
81. 1/8	254.469	5153.0
1/4	254.862	5168.9
3/8	255.254	5184.9
1/2	255.647	5200.8
5/8	256.040	5216.8
3/4	256.433	5232.8
7/8	256.825	5248.9
	257.218	5264.9
82. 1/8	257.611	5281.0
1/4	258.003	5297.1
3/8	258.396	5313.3
1/2	258.789	5329.4
5/8	259.181	5345.6
3/4	259.574	5361.8
7/8	259.967	5378.1
	260.359	5394.3

	Circumference	Area
87. 1/8	273.319	5944.7
1/4	273.711	5961.8
3/8	274.104	5978.9
1/2	274.497	5996.0
5/8	274.889	6013.2
3/4	275.282	6030.4
7/8	275.675	6047.6
	276.067	6064.9
88. 1/8	276.460	6082.1
1/4	276.853	6099.4
3/8	277.246	6116.7
1/2	277.638	6134.1
5/8	278.031	6151.4
3/4	278.424	6168.8
7/8	278.816	6186.2
	279.209	6203.7
89. 1/8	279.602	6221.1
1/4	279.994	6238.6
3/8	280.387	6256.1
1/2	280.780	6273.7
5/8	281.173	6291.2
3/4	281.565	6308.8
7/8	281.958	6326.4
	282.351	6344.1
90. 1/8	282.743	6361.7
1/4	283.136	6379.4
3/8	283.529	6397.1
1/2	283.921	6414.9
5/8	284.314	6432.6
3/4	284.707	6450.4
7/8	285.100	6468.2
	285.492	6486.0

	Circumference	Area
95. 1/8	298.451	7088.2
1/4	298.844	7106.9
3/8	299.237	7125.6
1/2	299.629	7144.3
5/8	300.022	7163.0
3/4	300.415	7181.8
7/8	300.807	7200.6
	301.200	7219.4
96. 1/8	301.593	7238.2
1/4	301.986	7257.1
3/8	302.378	7276.0
1/2	302.771	7294.9
5/8	303.164	7313.8
3/4	303.556	7332.8
7/8	303.949	7351.8
	304.342	7370.8
97. 1/8	304.734	7389.8
1/4	305.127	7408.9
3/8	305.520	7428.0
1/2	305.913	7447.1
5/8	306.305	7466.2
3/4	306.698	7485.3
7/8	307.091	7504.5
	307.483	7523.7
98. 1/8	307.876	7543.0
1/4	308.269	7562.2
3/8	308.661	7581.5
1/2	309.054	7600.8
5/8	309.447	7620.1
3/4	309.840	7639.5
7/8	310.232	7658.9
	310.625	7678.3

Table B.1 Circumferences and areas of circles (cont.)

Dia.	Circum.	Area	Dia.	Circum.	Area	Dia.	Circum.	Area
99.	311.018	7697.7	107.	336.15	8992	115.	361.28	10387
1/8	311.410	7717.1	1/8	336.54	9014	1/8	361.68	10410
1/4	311.803	7736.6	1/4	336.94	9035	1/4	362.07	10432
3/8	312.196	7756.1	3/8	337.33	9056	3/8	362.46	10455
1/2	312.588	7775.6	1/2	337.72	9077	1/2	362.86	10477
5/8	312.981	7795.2	5/8	338.12	9098	5/8	363.25	10500
3/4	313.374	7814.8	3/4	338.51	9119	3/4	363.64	10522
7/8	313.767	7834.4	7/8	338.90	9140	7/8	364.03	10545
100.	314.16	7854	108.	339.29	9161	116.	364.43	10568
1/8	314.55	7873	1/8	339.69	9183	1/8	364.82	10590
1/4	314.95	7893	1/4	340.08	9204	1/4	365.21	10613
3/8	315.34	7913	3/8	340.47	9225	3/8	365.60	10636
1/2	315.73	7933	1/2	340.86	9246	1/2	366.00	10659
5/8	316.12	7952	5/8	341.26	9268	5/8	366.39	10682
3/4	316.52	7972	3/4	341.65	9289	3/4	366.78	10705
7/8	316.91	7992	7/8	342.04	9310	7/8	367.18	10728
101.	317.30	8012	109.	342.43	9331	117.	367.57	10751
1/8	317.69	8032	1/8	342.83	9353	1/8	367.96	10774
1/4	318.09	8052	1/4	343.22	9374	1/4	368.35	10798
3/8	318.48	8071	3/8	343.61	9396	3/8	368.75	10821
1/2	318.87	8091	1/2	344.01	9417	1/2	369.14	10844
5/8	319.27	8111	5/8	344.40	9439	5/8	369.53	10867
3/4	319.66	8131	3/4	344.79	9460	3/4	369.92	10890
7/8	320.05	8151	7/8	345.18	9481	7/8	370.32	10913
102.	320.44	8171	110.	345.58	9503	118.	370.71	10936
1/8	320.84	8191	1/8	345.97	9525	1/8	371.11	10960
1/4	321.23	8211	1/4	346.36	9546	1/4	371.49	10983
3/8	321.62	8231	3/8	346.75	9568	3/8	371.89	11007
1/2	322.01	8252	1/2	347.15	9589	1/2	372.28	11030
5/8	322.41	8272	5/8	347.54	9611	5/8	372.67	11053
3/4	322.80	8292	3/4	347.93	9633	3/4	373.07	11076
7/8	323.19	8312	7/8	348.33	9655	7/8	373.46	11099

Dia.	Circum.	Area
123.	386.42	11882
1/8	386.81	11907
1/4	387.20	11931
3/8	387.60	11956
1/2	387.99	11980
5/8	388.38	12004
3/4	388.77	12028
7/8	389.17	12052
124.	389.56	12076
1/8	389.95	12101
1/4	390.34	12125
3/8	390.74	12150
1/2	391.13	12174
5/8	391.52	12199
3/4	391.92	12223
7/8	392.31	12248
125.	392.70	12272
1/8	393.09	12297
1/4	393.49	12321
3/8	393.88	12346
1/2	394.27	12370
5/8	394.66	12395
3/4	395.06	12419
7/8	395.45	12444
126.	395.84	12469
1/8	396.23	12494
1/4	396.63	12518
3/8	397.02	12543
1/2	397.41	12568
5/8	397.81	12593
3/4	398.20	12618
7/8	398.59	12643

Table B.1 Circumferences and areas of circles (cont.)

Dia.	Circum.	Area	Dia.	Circum.	Area	Dia.	Circum.	Area	Dia.	Circum.	Area
103.	323.59	8332	111.	348.72	9677	119.	373.85	11122	127.	398.98	12668
1/8	323.98	8352	1/8	349.11	9698	1/8	374.24	11146	1/8	399.38	12693
1/4	324.37	8372	1/4	349.50	9720	1/4	374.64	11169	1/4	399.77	12718
3/8	324.76	8393	3/8	349.90	9742	3/8	375.03	11193	3/8	400.16	12743
1/2	325.16	8413	1/2	350.29	9764	1/2	375.42	11216	1/2	400.55	12768
5/8	325.55	8434	5/8	350.68	9786	5/8	375.81	11240	5/8	400.95	12793
3/4	325.94	8454	3/4	351.07	9808	3/4	376.21	11263	3/4	401.34	12818
7/8	326.33	8474	7/8	351.47	9830	7/8	376.60	11287	7/8	401.73	12843
104.	326.73	8495	112.	351.86	9852	120.	376.99	11310	128.	402.13	12868
1/8	327.12	8515	1/8	352.25	9874	1/8	377.39	11334	1/8	402.52	12893
1/4	327.51	8536	1/4	352.65	9897	1/4	377.78	11357	1/4	402.91	12919
3/8	327.91	8556	3/8	353.04	9919	3/8	378.17	11381	3/8	403.30	12944
1/2	328.30	8577	1/2	353.43	9941	1/2	378.56	11404	1/2	403.70	12970
5/8	328.69	8597	5/8	353.82	9963	5/8	378.96	11428	5/8	404.09	12995
3/4	329.08	8618	3/4	354.22	9985	3/4	379.35	11451	3/4	404.48	13020
7/8	329.48	8638	7/8	354.61	10007	7/8	379.74	11475	7/8	404.87	13045
105.	329.87	8659	113.	355.00	10029	121.	380.13	11499	129.	405.27	13070
1/8	330.26	8679	1/8	355.39	10052	1/8	380.53	11522	1/8	405.66	13096
1/4	330.65	8700	1/4	355.79	10074	1/4	380.92	11546	1/4	406.05	13121
3/8	331.05	8721	3/8	356.18	10097	3/8	381.31	11570	3/8	406.44	13147
1/2	331.44	8741	1/2	356.57	10119	1/2	381.70	11594	1/2	406.84	13172
5/8	331.83	8762	5/8	356.96	10141	5/8	382.10	11618	5/8	407.23	13198
3/4	332.22	8783	3/4	357.36	10163	3/4	382.49	11642	3/4	407.62	13223
7/8	332.62	8804	7/8	357.75	10185	7/8	382.88	11666	7/8	408.02	13248
106.	333.01	8825	114.	358.14	10207	122.	383.28	11690	130.	408.41	13273
1/8	333.40	8845	1/8	358.54	10230	1/8	383.67	11714	1/8	408.80	13299
1/4	333.80	8866	1/4	358.93	10252	1/4	384.06	11738	1/4	409.19	13324
3/8	334.19	8887	3/8	359.32	10275	3/8	384.45	11762	3/8	409.59	13350
1/2	334.58	8908	1/2	359.71	10297	1/2	384.85	11786	1/2	409.98	13375
5/8	334.97	8929	5/8	360.11	10320	5/8	385.24	11810	5/8	410.37	13401
3/4	335.37	8950	3/4	360.50	10342	3/4	385.63	11834	3/4	410.76	13426
7/8	335.76	8971	7/8	360.89	10365	7/8	386.02	11858	7/8	411.16	13452

Table B.1 Circumferences and areas of circles (cont.)

Dia.	Circum.	Area	Dia.	Circum.	Area	Dia.	Circum.	Area	Dia.	Circum.	Area
131.	411.55	13478	139.	436.68	15175	147.	461.82	16972	155.	486.95	18869
1/8	411.94	13504	1/8	437.08	15203	1/8	462.21	17000	1/8	487.34	18900
1/4	412.34	13529	1/4	437.47	15230	1/4	462.60	17029	1/4	487.73	18930
3/8	412.73	13555	3/8	437.86	15258	3/8	462.99	17058	3/8	488.13	18961
1/2	413.12	13581	1/2	438.25	15285	1/2	463.39	17087	1/2	488.52	18991
5/8	413.51	13607	5/8	438.65	15313	5/8	463.78	17116	5/8	488.91	19022
3/4	413.91	13633	3/4	439.04	15340	3/4	464.17	17145	3/4	489.30	19052
7/8	414.30	13659	7/8	439.43	15367	7/8	464.56	17174	7/8	489.70	19083
132.	414.69	13685	140.	439.82	15394	148.	464.96	17203	156.	490.09	19113
1/8	415.08	13711	1/8	440.22	15422	1/8	465.35	17232	1/8	490.48	19144
1/4	415.48	13737	1/4	440.61	15449	1/4	465.74	17262	1/4	490.88	19174
3/8	415.87	13763	3/8	441.00	15477	3/8	466.14	17291	3/8	491.27	19205
1/2	416.26	13789	1/2	441.40	15504	1/2	466.53	17321	1/2	491.66	19235
5/8	416.66	13815	5/8	441.79	15532	5/8	466.92	17350	5/8	492.05	19266
3/4	417.05	13841	3/4	442.18	15559	3/4	467.31	17379	3/4	492.45	19297
7/8	417.44	13867	7/8	442.57	15587	7/8	467.71	17408	7/8	492.84	19328
133.	417.83	13893	141.	442.97	15615	149.	468.10	17437	157.	493.23	19359
1/8	418.23	13919	1/8	443.36	15642	1/8	468.49	17466	1/8	493.62	19390
1/4	418.62	13946	1/4	443.75	15670	1/4	468.88	17496	1/4	494.02	19421
3/8	419.01	13972	3/8	444.14	15697	3/8	469.28	17525	3/8	494.41	19452
1/2	419.40	13999	1/2	444.54	15725	1/2	469.67	17555	1/2	494.80	19483
5/8	419.80	14025	5/8	444.93	15753	5/8	470.06	17584	5/8	495.20	19514
3/4	420.19	14051	3/4	445.32	15781	3/4	470.46	17614	3/4	495.59	19545
7/8	420.58	14077	7/8	445.72	15809	7/8	470.85	17643	7/8	495.98	19576
134.	420.97	14103	142.	446.11	15837	150.	471.24	17672	158.	496.37	19607
1/8	421.37	14130	1/8	446.50	15865	1/8	471.63	17702	1/8	496.77	19638
1/4	421.76	14156	1/4	446.89	15893	1/4	472.03	17731	1/4	497.16	19669
3/8	422.15	14183	3/8	447.29	15921	3/8	472.42	17761	3/8	497.55	19701
1/2	422.55	14209	1/2	447.68	15949	1/2	472.81	17790	1/2	497.94	19732
5/8	422.94	14236	5/8	448.07	15977	5/8	473.20	17820	5/8	498.34	19763
3/4	423.33	14262	3/4	448.46	16005	3/4	473.60	17849	3/4	498.73	19794
7/8	423.72	14288	7/8	448.86	16033	7/8	473.99	17879	7/8	499.12	19825

Table B.1 Circumferences and areas of circles (cont.)

Diam.	Circumf.	Area	Diam.	Circumf.	Area	Diam.	Circumf.	Area	Diam.	Circumf.	Area
135.	424.12	14314	143.	449.25	16061	151.	474.38	17908	159.	499.51	19856
1/8	424.51	14341	1/8	449.64	16089	1/8	474.77	17938	1/8	499.91	19887
1/4	424.90	14367	1/4	450.03	16117	1/4	475.17	17967	1/4	500.30	19919
3/8	425.29	14394	3/8	450.43	16145	3/8	475.56	17997	3/8	500.69	19950
1/2	425.69	14420	1/2	450.82	16173	1/2	475.95	18026	1/2	501.09	19982
5/8	426.08	14447	5/8	451.21	16201	5/8	476.35	18056	5/8	501.48	20013
3/4	426.47	14473	3/4	451.61	16229	3/4	476.74	18086	3/4	501.87	20044
7/8	426.87	14500	7/8	452.00	16258	7/8	477.13	18116	7/8	502.26	20075
136.	427.26	14527	144.	452.39	16286	152.	477.52	18146	160.	502.66	20106
1/8	427.65	14553	1/8	452.78	16314	1/8	477.92	18175	1/8	503.05	20138
1/4	428.04	14580	1/4	453.18	16342	1/4	478.31	18205	1/4	503.44	20169
3/8	428.44	14607	3/8	453.57	16371	3/8	478.70	18235	3/8	503.83	20201
1/2	428.83	14633	1/2	453.96	16399	1/2	479.09	18265	1/2	504.23	20232
5/8	429.22	14660	5/8	454.35	16428	5/8	479.49	18295	5/8	504.62	20264
3/4	429.61	14687	3/4	454.75	16456	3/4	479.88	18325	3/4	505.01	20295
7/8	430.01	14714	7/8	455.14	16485	7/8	480.27	18355	7/8	505.41	20327
137.	430.40	14741	145.	455.53	16513	153.	480.67	18385	161.	505.80	20358
1/8	430.79	14768	1/8	455.93	16542	1/8	481.06	18415	1/8	506.19	20390
1/4	431.19	14795	1/4	456.32	16570	1/4	481.45	18446	1/4	506.58	20421
3/8	431.58	14822	3/8	456.71	16599	3/8	481.84	18476	3/8	506.98	20453
1/2	431.97	14849	1/2	457.10	16627	1/2	482.24	18507	1/2	507.37	20484
5/8	432.36	14876	5/8	457.50	16656	5/8	482.63	18537	5/8	507.76	20516
3/4	432.76	14903	3/4	457.89	16684	3/4	483.02	18567	3/4	508.15	20548
7/8	433.15	14930	7/8	458.28	16713	7/8	483.41	18597	7/8	508.55	20580
138	433.54	14957	146.	458.67	16742	154.	483.81	18627	162.	508.94	20612
1/8	433.93	14984	1/8	459.07	16770	1/8	484.20	18658	1/8	509.33	20644
1/4	434.33	15012	1/4	459.46	16799	1/4	484.59	18688	1/4	509.73	20675
3/8	434.72	15039	3/8	459.85	16827	3/8	484.99	18719	3/8	510.12	20707
1/2	435.11	15067	1/2	460.24	16856	1/2	485.38	18749	1/2	510.51	20739
5/8	435.50	15094	5/8	460.64	16885	5/8	485.77	18779	5/8	510.90	20771
3/4	435.90	15121	3/4	461.03	16914	3/4	486.16	18809	3/4	511.30	20803
7/8	436.29	15148	7/8	461.42	16943	7/8	486.56	18839	7/8	511.69	20835

Decimal Equivalents and Theoretical Weights of Steel Plates

Table C.1

Plate thickness			Weight, psf	Plate thickness			Weight, psf
Fractions of an inch		Decimal equivalent		Fractions of an inch		Decimal equivalent	
	1/64	0.01562	0.637		33/64	0.51562	21.037
1/32		0.03125	1.274	17/32		0.53125	21.675
	3/64	0.04687	1.911		35/64	0.54687	22.312
1/16		0.0625	2.55	9/16		0.5625	22.950
	5/64	0.07812	3.1875		37/64	0.57812	23.587
3/32		0.09375	3.825	19/32		0.59375	24.225
	7/64	0.10937	4.4625		39/64	0.60937	24.862
1/8		0.125	5.10	5/8		0.625	25.50
	9/64	0.14062	5.7375		41/64	0.64062	26.137
5/32		0.15625	6.375	21/32		0.65625	26.775
	11/64	0.17187	7.0125		43/64	0.67187	27.412
3/16		0.1875	7.65	11/16		0.6875	28.050
	13/64	0.20312	8.2875		45/64	0.70312	28.687
7/32		0.21875	8.925	23/32		0.71875	29.325
	15/64	0.23437	9.5625		47/64	0.73437	29.962
1/4		0.25	10.20	3/4		0.75	30.60
	17/64	0.26562	10.837		49/64	0.76562	31.237
9/32		0.28125	11.475	25/32		0.78125	31.875
	19/64	0.29687	12.112		51/64	0.79687	32.512
5/16		0.3125	12.75	13/16		0.8125	33.150
	21/64	0.32812	13.387		53/64	0.82812	33.787
11/32		0.34375	14.025	27/32		0.84375	34.425
	23/64	0.35937	14.662		55/64	0.85937	35.062
3/8		0.375	15.3	7/8		0.875	35.70
	25/64	0.39062	15.937		57/64	0.89062	36.337
13/32		0.40625	16.575	29/32		0.90625	36.975
	27/64	0.42187	17.212		59/64	0.92187	37.612
7/16		0.4375	17.85	15/16		0.9375	38.250
	29/64	0.45312	18.487		61/64	0.95312	38.887
15/32		0.46875	19.125	31/32		0.96875	39.525
	31/64	0.48437	19.762		63/64	0.98437	40.162
1/2		0.50	20.40	1		1.00	40.80

Table D.1 Pipe Wall Thicknesses

All wall thicknesses conform with ANSI B36.10 excepting 5S and 10S for stainless steel pipe, which conforms with ANSI B36.19. The thicknesses shown represent the nominal or average pipe wall dimensions and include an allowance for mill tolerance of 12.5 percent. Where minimum thickness is required,

$$t_n \times 0.875 = t_m$$

Nominal pipe size	Outside diameter, in.	5S	10S	20	30	40	STD.	60	80	Extra heavy	100	120	140	160	Double extra heavy
1/2	0.840	0.065	0.083			0.109	0.109		0.147	0.147				0.187	0.294
3/4	1.050	0.065	0.083			0.113	0.113		0.154	0.154				0.218	0.308
1	1.315	0.065	0.109			0.133	0.133		0.179	0.179				0.250	0.358
1 1/4	1.66	0.065	0.109			0.140	0.140		0.191	0.191				0.250	0.382
1 1/2	1.90	0.065	0.109			0.145	0.145		0.200	0.200				0.281	0.400
2	2.375	0.065	0.109			0.154	0.154		0.218	0.218				0.343	0.436
2 1/2	2.875	0.083	0.120			0.203	0.203		0.276	0.276				0.375	0.552
3	3.5	0.083	0.120			0.216	0.216		0.300	0.300				0.437	0.600
3 1/2	4.0	0.083	0.120			0.226	0.226		0.318	0.318					0.636
4	4.5	0.083	0.120			0.237	0.237		0.337	0.337		0.437		0.531	0.674
4 1/2	5.0	0.109	0.134				0.247			0.355					0.710
5	5.563	0.109	0.134			0.258	0.258		0.375	0.375		0.500		0.625	0.750
6	6.625					0.280	0.280		0.432	0.432		0.562		0.718	0.864
7	7.675						0.301								0.875
8	8.625	0.109	0.148	0.250	0.277	0.322	0.322	0.406	0.500	0.500	0.593	0.718	0.812	0.906	0.875
9	9.625						0.342								
10	10.75	0.134	0.165	0.250	0.307	0.365	0.365	0.500	0.593	0.500	0.718	0.843	1.000	1.125	
11	11.75						0.375			0.500					
12	12.75	0.165	0.180	0.250	0.330	0.406	0.375	0.562	0.687	0.500	0.843	1.000	1.125	1.312	
14 O.D.	14.0			0.312	0.375	0.437	0.375	0.593	0.750	0.500	0.937	1.093	1.250	1.406	
16 O.D.	16.0			0.312	0.375	0.500	0.375	0.656	0.843	0.500	1.031	1.218	1.437	1.593	
18 O.D.	18.0			0.312	0.437	0.562	0.375	0.750	0.937	0.500	1.156	1.375	1.562	1.781	
20 O.D.	20.0			0.375	0.500	0.593	0.375	0.812	1.031	0.500	1.280	1.500	1.750	1.968	
24 O.D.	24.0			0.375	0.562	0.687	0.375	0.968	1.218	0.500	1.531	1.812	2.062	2.343	
30 O.D.	30.0			0.500	0.625										

Nominal pipe wall thicknesses, in.

Dry Saturated Steam Temperatures

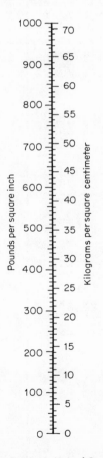

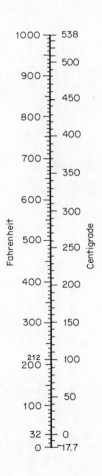

Fig. E.1 United States and metric conversion table.

Table E.1 Dry Saturated Steam Temperatures

Temperature, deg F	Absolute pressure, psi	Enthalpy			Specific volume			Entropy		
		Saturated liquid	Evaporation	Saturated vapor	Saturated liquid	Evaporation	Saturated vapor	Saturated liquid	Evaporation	Saturated vapor
32	0.08854	0.01602	3306	3306	0.00	1075.8	1075.8	0.0000	2.1877	2.1877
35	0.09995	0.01602	2947	2947	3.02	1074.1	1077.1	0.0061	2.1709	2.1770
40	0.12170	0.01602	2444	2444	8.05	1071.3	1079.3	0.0162	2.1435	2.1597
45	0.14752	0.01602	2036.4	2036.4	13.06	1068.4	1081.5	0.0262	2.1167	2.1429
50	0.17811	0.01603	1703.2	1703.2	18.07	1065.6	1083.7	0.0361	2.0903	2.1264
60	0.2563	0.01604	1206.6	1206.7	28.06	1059.9	1088.0	0.0555	2.0393	2.0948
70	0.3631	0.01606	867.8	867.9	38.04	1054.3	1092.3	0.0745	1.9902	2.0647
80	0.5069	0.01608	633.1	633.1	48.02	1048.6	1096.6	0.0932	1.9428	2.0360
90	0.6982	0.01610	468.0	468.0	57.99	1042.9	1100.9	0.1115	1.8972	2.0087
100	0.9492	0.01613	350.3	350.4	67.97	1037.2	1105.2	0.1295	1.8531	1.9826
110	1.2748	0.01617	265.3	265.4	77.94	1031.6	1109.5	0.1471	1.8106	1.9577
120	1.6924	0.01620	203.25	203.27	87.92	1025.8	1113.7	0.1645	1.7694	1.9339
130	2.2225	0.01625	157.32	157.34	97.90	1020.0	1117.9	0.1816	1.7296	1.9112
140	2.8886	0.01629	122.99	123.01	107.89	1014.1	1122.0	0.1984	1.6910	1.8894
150	3.718	0.01634	97.06	97.07	117.89	1008.2	1126.1	0.2149	1.6537	1.8685
160	4.741	0.01639	77.27	77.29	127.89	1002.3	1130.2	0.2311	1.6174	1.8485
170	5.992	0.01645	62.04	62.06	137.90	996.3	1134.2	0.2472	1.5822	1.8293
180	7.510	0.01651	50.21	50.23	147.92	990.2	1138.1	0.2630	1.5480	1.8109
190	9.339	0.01657	40.94	40.96	157.95	984.1	1142.0	0.2785	1.5147	1.7932
200	11.526	0.01663	33.62	33.64	167.99	977.9	1145.9	0.2938	1.4824	1.7762
210	14.123	0.01670	27.80	27.82	178.05	971.6	1149.7	0.3090	1.4508	1.7598
212	14.696	0.01672	26.78	26.80	180.07	970.3	1150.4	0.3120	1.4446	1.7566
220	17.186	0.01677	23.13	23.15	188.13	965.2	1153.4	0.3239	1.4201	1.7440
230	20.780	0.01684	19.365	19.382	198.23	958.8	1157.0	0.3387	1.3901	1.7288
240	24.969	0.01692	16.306	16.323	208.34	952.2	1160.5	0.3531	1.3609	1.7140
250	29.825	0.01700	13.804	13.821	216.48	945.5	1164.0	0.3675	1.3323	1.6998
260	35.429	0.01709	11.746	11.763	228.64	938.7	1167.3	0.3817	1.3043	1.6860
270	41.858	0.01717	10.044	10.061	238.84	931.8	1170.6	0.3958	1.2769	1.6727
280	49.203	0.01726	8.628	8.645	249.06	924.7	1173.8	0.4096	1.2501	1.6597
290	57.556	0.01735	7.444	7.461	259.31	917.5	1176.8	0.4234	1.2238	1.6472

Table E.1 (cont.)

Temperature, deg F	Absolute pressure, psi	Specific volume			Entropy			Enthalpy		
		Saturated liquid	Evaporation	Saturated vapor	Saturated liquid	Evaporation	Saturated vapor	Saturated liquid	Evaporation	Saturated vapor
300	67.013	0.01745	6.449	6.466	269.59	910.1	1179.7	0.4369	1.1980	1.6350
310	77.68	0.01755	5.609	5.626	279.92	902.6	1182.5	0.4504	1.1727	1.6231
320	89.66	0.01765	4.896	4.914	290.28	894.9	1185.2	0.4637	1.1478	1.6115
330	103.06	0.01776	4.289	4.307	300.68	887.0	1187.7	0.4769	1.1233	1.6002
340	118.01	0.01787	3.770	3.788	311.13	879.0	1190.1	0.4900	1.0992	1.5891
350	134.63	0.01799	3.324	3.342	321.63	870.7	1192.3	0.5029	1.0754	1.5783
360	153.04	0.01811	2.939	2.957	332.18	862.2	1194.4	0.5158	1.0519	1.5677
370	173.37	0.01823	2.606	2.625	342.79	853.5	1196.3	0.5286	1.0287	1.5573
380	195.77	0.01836	2.317	2.335	353.45	844.6	1198.1	0.5413	1.0059	1.5471
390	220.37	0.01850	2.0651	2.0836	364.17	835.4	1199.6	0.5539	0.9832	1.5371
400	247.31	0.01864	1.8447	1.8633	374.97	826.0	1201.0	0.5664	0.9608	1.5272
410	276.75	0.01878	1.6512	1.6700	385.83	816.3	1202.1	0.5788	0.9386	1.5174
420	308.83	0.01894	1.4811	1.5000	396.77	806.3	1203.1	0.5912	0.9166	1.5078
430	343.72	0.01910	1.3308	1.3499	407.79	796.0	1203.8	0.6035	0.8947	1.4982
440	381.59	0.01926	1.1979	1.2171	418.90	785.4	1204.3	0.6158	0.8730	1.4887
450	422.6	0.0194	1.0799	1.0993	430.1	774.5	1204.6	0.6280	0.8513	1.4793
460	466.9	0.0196	0.9748	0.9944	441.4	763.2	1204.6	0.6402	0.8298	1.4700
470	514.7	0.0198	0.8811	0.9009	452.8	751.5	1204.3	0.6523	0.8083	1.4606
480	566.1	0.0200	0.7972	0.8172	464.4	739.4	1203.7	0.6645	0.7868	1.4513
490	621.4	0.0202	0.7221	0.7423	476.0	726.8	1202.8	0.6766	0.7653	1.4419
500	680.8	0.0204	0.6545	0.6749	487.8	713.9	1201.7	0.6887	0.7438	1.4325
520	812.4	0.0209	0.5385	0.5594	511.9	686.4	1198.2	0.7130	0.7006	1.4136
540	962.5	0.0215	0.4434	0.4649	536.6	656.6	1193.2	0.7374	0.6568	1.3942
560	1133.1	0.0221	0.3647	0.3868	562.2	624.2	1186.4	0.7621	0.6121	1.3742
580	1325.8	0.0228	0.2989	0.3217	588.9	588.4	1177.3	0.7872	0.5659	1.3532
600	1542.9	0.0236	0.2432	0.2668	617.0	548.5	1165.5	0.8131	0.5176	1.3307
620	1786.6	0.0247	0.1955	0.2201	646.7	503.6	1150.3	0.8398	0.4664	1.3062
640	2059.7	0.0260	0.1538	0.1798	678.6	452.0	1130.5	0.8679	0.4110	1.2789
660	2365.4	0.0278	0.1165	0.1442	714.2	390.2	1104.4	0.8987	0.3485	1.2472
680	2708.1	0.0305	0.0810	0.1115	757.3	309.9	1067.2	0.9351	0.2719	1.2071
700	3093.7	0.0369	0.0392	0.0761	823.3	172.1	995.4	0.9905	0.1484	1.1389
705.4	3206.2	0.0503	0	0.0503	902.7	0	902.7	1.0580	0	1.0580

Abridged from "Thermodynamic Properties of Steam" by Joseph H. Keenan and Frederick G. Keyes. John Wiley & Sons. New York. 1937

Table F.1 Corrosion resistance data

These data, reprinted by courtesy of Star Tank and Filter Corporation, represent results of laboratory tests and are indicative only of the conditions under which the tests were run. They may therefore be considered as recommendations but not as guarantees.   As actual operating conditions can rarely be duplicated in the laboratory, it is desirable to make further tests under actual service.   The letters A, B, C, D, and E represent the approximate corrosion penetration per month as follows:

A—fully resistant (less than 0.00035 in)
B—satisfactorily resistant (0.00035 to 0.0035 in)
C—fairly resistant (0.0035 to 0.010 in)
D—slightly resistant (0.010 to 0.035 in)
E—not resistant (over 0.035 in)

Substance and condition	Temperature, deg F	Type 304–S/S	Type 316–S/S	Other alloys	Carbon steel
Acetic acid					
5 per cent agitated	70	A	A		E
aerated	70	A	A		E
10 per cent agitated	70	A	A		E
aerated	70	A	A		E
20 per cent agitated	70	A	A		E
aerated	70	A	A		E
50 per cent	70	A	A		E
	boiling	C	B		E
80 per cent	70	A	A		E
	boiling	D	B		E
100 per cent	70	A	A		E
	boiling	C	B		E
100 per cent (150-lb pressure)	400	E	C		E
Acetic anhydride	boiling	A	A		E
	70	A	A		E
Acetic vapors					
100 per cent	hot	E	C		E
30 per cent	hot	C	B		E
Acetone	boiling	A	A		A
	70	A	A		A
Acetylene					
concentrated	70	A			
commercially pure	70				
Alcohol, ethyl	70	A	A		A
	boiling	A	A		A
Alcohol, methyl	70	A	A		A
	150	C*	B		E
Alum, chrome (5 per cent)	70	A	A		
Aluminum	molten	E	E		
Aluminum acetate (saturated)		A	A		

Table F.1   Corrosion resistance data (cont.)

Substance and condition	Temperature, deg F	Type 304–S/S	Type 316–S/S	Other alloys	Carbon steel
Aluminum chloride	70	D	C	Hast. B (A)	E
Aluminum fluoride	70	D	C		
Aluminum hydroxide					
(saturated)		A	A		
Aluminum potassium					
sulphate					
2 per cent (Alum.)	70	A	A		E
10 per cent	70	A	A		E
	boiling	B	A		E
saturated	boiling	C	B	Carp. 20 (A)	E
Aluminum sulphate					
10 per cent	70	A	A		E
	boiling	B	A		E
saturated	70	A	A		E
	boiling	B	A		E
Ammonia					
all concentrations	70	A	A		
gas	hot	D			
Ammonia bicarbonate	70	A	A		
	hot	A	A		
Ammonia carbonate					
1 and 5 per cent still	70	A	A		
aerated	70	A	A		
agitated	70	A	A		
Ammonia liquor	70	A	A		
	boiling	A	A		
Ammonium chloride					
1 per cent still	70	A	A		E
aerated	70	A	A		E
agitated	70	A	A		E
10 per cent solution	boiling	A*	A*		E
28 per cent solution	boiling	B*	A*		E
50 per cent solution	boiling	B*	A*		E
Ammonium nitrate					
All concentrations (agi-					
tated and aerated)	70	A	A		B
saturated	boiling	A	A		
Ammonium oxalate					
(5 per cent)	70	A	A		E
Ammonium persulphate					
(5 per cent)	70	A	A		E
Ammonium phosphate					
(5 per cent)	70	A	A		B

* Subject to pitting at air line or when allowed to dry.

Table F.1   Corrosion resistance data (cont.)

Substance and condition	Temperature, deg F	Type 304–S/S	Type 316–S/S	Other alloys	Carbon steel
Ammonium sulphate					
1 and 5 per cent					
agitated	70	A	A		E
aerated	70	A	A		E
10 per cent	boiling	B*	A*		E
saturated	boiling	B*	A*		E
Ammonium sulphite	cold	A	A		E
	boiling	A	A		E
Amyl acetate (concentrated solution)	70	A			A
Amyl chloride		A			A
Aniline					
3 per cent	70	A	A		A
concentrated crude	70	A	A		A
Aniline hydrochloride	70	E	D	Carp. 20 (B)	E
Antimony trichloride	70	E	D	Hast. B (B)	E
Barium carbonate	70	A	A		
Barium chloride					
5 per cent	70	A	A		E
saturated	70	A	A		E
aqueous solution	hot	B*	A*		E
Barium nitrate (aqueous solution)	hot	A	A		A
Barium sulphate (barytes-blanc fixe)	70	A	A		
Barium sulphide					
saturated solution	70	A			E
solution	70		A		E
Beer		A	A		E
Benzene	70	A	A		A
Benzoic acid	70	A	A		E
Benzol	hot	A	A		E
Blood (meat juices)	cold	A*	A		E
Borax 5 per cent	hot	A	A		A
Boric acid					
5 per cent solution	70	A	A		E
	hot				E
	boiling		A*		E
saturated solution	70	A*	A*		E
	boiling		A*		E
Bromine water	70	E	D	Hastelloy C (C)	E
Buttermilk	70	A	A		E
Butyl acetate			A		B

* Subject to pitting at air line or when allowed to dry.

Table F.1   Corrosion resistance data (cont.)

Substance and condition	Temperature, deg F	Type 304–S/S	Type 316–S/S	Other alloys	Carbon steel
Butyric acid					
5 per cent	70	A	A		E
	150	A	A		E
aqueous solution					
(sp. gr. 0.964)	boiling	A	A		E
Calcium carbonate	70	A	A		
Calcium chlorate					
dilute solution	70	A	A		
dilute solution	hot	A	A		
Calcium chloride					
dilute solution	70	B**	A*		E
concentrated solution	70	B**	A*		E
Calcium hydroxide					
10 per cent	boiling	A	A		E
20 per cent	boiling	A	A		E
50 per cent	boiling	C	B	Carpenter 20 (B)	E
Calcium hypochlorite					
(2 per cent)	70	B*	A*		E
Calcium sulphate					
(saturated)	70	A	A		
Carbolic acid					
C.P.	boiling	A	A		E
Crude	boiling	A	A		E
C.P.	70	A	A		E
Carbon bisulphide	70	A	A		A
Carbon monoxide gas	870 (deg C)	A	A		
	760 (deg C)	A	A		
Carbon tetrachloride					
pure	70	A	A		A
aqueous solution					
5–10 per cent	70	C*			
Carbonated water		A	A		E
Carbonic acid (saturated					
solution)	70	A			B
Chloracetic acid	70	D	C	Hastelloy (B)	E
Chlorbenzol concentrated					
pure	70	A	A		A
Chloric acid	70	E	D	Carpenter 20 (B)	E
Chlorinated water					
(saturated)	70	C*	B*	Carpenter 20 (B)	E
Chlorine gas					
dry	70	C	B		B
moist	70	D	C	Hastelloy C (B)	E
	100 (deg C)	E	D		

** Keep solutions alkaline.

Table F.1   Corrosion resistance data (cont.)

Substance and condition	Temperature, deg F	Type 304–S/S	Type 316–S/S	Other alloys	Carbon steel
Chloroform	70	A	A		A
Chromic acid					
5 per cent	70	A	A		E
10 per cent C.P.	boiling	C	B		E
50 per cent commercial (cont. SO₃)	boiling	D*	C		E
Chromium alum (see Alum, chrome)					
Chromium plating bath	70	A	A		E
Cider	70	A	A		
Citric acid					
5 per cent still	70	A	A		E
	150	A	A		E
15 per cent	70	A	A		E
	boiling	B	A		E
concentrated	boiling	C	B		E
Coca cola syrup (pure)	70	A	A		E
Coffee	boiling	A	A		E
Copper acetate (saturated solution)	70	A	A		
Copperas (see Ferrous sulphate)					
Copper carbonate (sat. sol. in 50 per cent NH₃OH)		A	A		
Copper chloride					
1 per cent agitated	70	B*	A*		E
aerated	70	B*	A*		E
5 per cent agitated	70	C*	B*		E
aerated	70	E*	D*	Hastelloy C (B)	E
Copper cyanide (saturated solution)	boiling	A	A		
Copper nitrate					
1 per cent still	70	A	A		E
agitated	70	A	A		E
aerated	70	A	A		E
5 per cent still	70	A	A		E
agitated	70	A	A		E
aerated	70	A	A		E
50 per cent aqueous solution	hot	A	A		E
Copper sulphate					
5 per cent agitated or still	70	A	A		E
aerated	70	A	A		E

Table F.1   Corrosion resistance data (cont.)

Substance and condition	Temperature, deg F	Type 304–S/S	Type 316–S/S	Other alloys	Carbon steel
Copper sulphate (cont.)					
saturated solution	boiling	A	A		E
Creosote (coal tar)	hot	A	A		A
Creosote oil	hot	A	A		
Cyanogen gas	70	A	A		
Denitrochlorbenzol					
(melted and solidified)	70	A	A		E
Developing solutions	70	A	A		E
Dyewood liquor	70	A†	A		E
Epsom salt	hot and cold	A	A		
Ether	70	A	A		B
Ethyl acetate (concen-					
trated solution)	70	A			A
Ethyl chloride	70	A	A		
Ethylene chloride	70	A	A		
Ethylene glycol (con-					
centrated)	70	A			A
Ferric chloride					
1 per cent still solution	70	B*††	A*	Hastelloy C (A)	E
	boiling	D*††	C*		
5 per cent still solution	70	D*†	C*	Hastelloy C (B)	E
agitated solution	70	C*†	C*	Hastelloy C (B)	E
aerated solution	70	C*†	C*	Hastelloy C (B)	E
Ferric hydroxide (hy-					
drated iron oxide)	70	A	A		
Ferric nitrate					
1 per cent still, agi-					
tated, or aerated	70	A	A		E
5 per cent still, agi-					
tated, or aerated	70	A	A		E
Ferric sulphate (1 and 5					
per cent still, aer-					
ated, or agitated)	70	A*	A		E
Ferrous chloride (satu-					
rated solution)	70		A		E
Ferrous sulphate (dilute					
solution)	70	A	A		E
Fluorine	70	E	E		E
Formaldehyde (40 per					
cent solution)		A*	A*		B

* Subject to pitting at air line or when allowed to dry.
† May attack when sulphuric acid is present.
†† May attack when hydrochloric acid is present.

Table F.1   Corrosion resistance data (cont.)

Substance and condition	Temperature, deg F	Type 304–S/S	Type 316–S/S	Other alloys	Carbon steel
Formic acid					
5 per cent still	70	B	A		E
	150	B	A		E
Fruit juices	70	A	A		E
Fuel oil	hot	A	A		A
containing sulphuric		C	B	Carpenter 20 (A)	E
Furfural	70	A	A		A
Gallic acid (5 per cent					
solution)	70	A	A		E
	150	A	A		E
Gasoline	70	A	A		A
Gelatin			A		E
Glue					
dry	70	A	A		A
acid solution	70	B*	A		B
	140	B*	A		E
Glycerine	70	A	A		A
Hydrochloric acid					
all concentrations	70	E	E	Hastelloy	E
Hydrocyanic acid		A	A		E
Hydrofluosilicic acid	70	E	D	Hastelloy	E
Hydrogen peroxide	70	A†	A		
	boiling	B†	A		
Hydrogen sulphide					
dry		A	A		B
wet		B†	A†		E
Ink		B†	A		E
Iodine		E	D	Hastelloy	E
Iodoform		A	A		E
Kerosene	70	A	A		A
Ketchup	70	A*	A		E
Lactic acid					
5 per cent	70	A	A		E
	150	B	A		E
10 per cent	boiling	D	B	Carpenter 20	E
	150	C	B	Carpenter 20	E
Lard	70	A	A		A
Lead	molten	B	B		
Linseed oil	70	A	A		A

* Subject to pitting at air line or when allowed to dry.

Table F.1   Corrosion resistance data (cont.)

Substance and condition	Temperature, deg F	Type 304–S/S	Type 316–S/S	Other alloys	Carbon steel
Magnesium chloride (1 and 5 per cent still)	70	A*	A		E
	hot	C*	B*	Carpenter 20	E
Magnesium sulphate	cold and hot	A	A		
Malic acid	cold and hot	B	A		E
Mayonnaise	70	A*	A		E
Mercuric chloride (dilute solutions)		E*	D*	Hastelloy	E
Mercury		A	A		
Methanol (methyl alcohol)		A	A		A
Milk (fresh or sour)	hot and cold	A	A		E
Mixed acids					
53 per cent H_2SO_4	cold	A	A		E
45 per cent HNO_3	cold	A	A		E
Molasses		A	A		A
Muriatic acid	70	E	E	Hastelloy	E
Mustard	70	A*	A*		E
Naptha	70	A	A		A
Naptha (crude)	70	A	A		A
Nickel chloride solution	70	A*	A*		E
Nickel sulphate	hot and cold	A	A		E
Niter cake	fused	B	A		E
Nitric acid					
5 per cent solution	70	A	A		E
20 per cent solution	70	A	A		E
50 per cent solution	70	A	A		E
	boiling	A	A		E
65 per cent solution	boiling	B	B	Duriron	E
concentrated	70	A	A		E
	boiling	D	D	Duriron	E
Nitrous acid					
5 per cent solution	70	A	A		E
Oils (crude)	hot and cold	A†	A†		B
Oils (vegetable, mineral)	hot and cold	A†	A		A
Oleic acid	70	A*	A		E
Oxalic acid					
5 per cent	hot and cold	A	A		E
10 per cent	70	A	A		E
	boiling	D	C	Carpenter 20	E
Paraffine	hot and cold	A	A		A
Petroleum ether		A	A		

† May attack when sulphuric acid is present.

Table F.1   Corrosion resistance data (cont.)

Substance and condition	Temperature, deg F	Type 304–S/S	Type 316–S/S	Other alloys	Carbon steel
Phenol		A	A		E
Phosphoric acid					
1 per cent	70	A††	A††		E
5 per cent still, agitated, or aerated	70	A	A		E
10 per cent still,	70	C	A		E
agitated, or aerated	70	C	B	Carpenter 20	E
Picric acid	70	A	A		E
Potassium bichromate	70	A	A		E
Potassium bromide	70	B*	A*		E
Potassium carbonate (1 per cent still, agitated, or aerated)	70	A	A		A
Potassium carbonate	hot	A	A		A
Potassium chlorate		A	A		A
Potassium chloride					
1 per cent still	70	A*	A*		E
agitated or aerated	70	A	A		E
5 per cent still	70	A*	A*		E
agitated or aerated	70	A	A		E
	boiling	A	A		E
Potassium ferricyanide (5 per cent)	70	A	A		E
Potassium ferrocyanide (5 per cent)	70	A	A		E
Potassium hydroxide					
5 per cent still, agitated, or aerated	70	A	A		B
27 per cent	boiling	A	A		E
50 per cent	boiling	B	A		E
Potassium nitrate (1 and 5 per cent still, agitated, or aerated)	70	A	A		
Potassium nitrate	hot	A	A		
Potassium oxalate		A	A		
Potassium permanganate (5 per cent)	70	A	A		A
Potassium sulphate					
1 and 5 per cent still, agitated, or aerated	70	A	A		E
5 per cent aerated	hot	A	A		E
Potassium sulphide (salt)		A	A		E
Pyrogallic acid		A	A		A

* Subject to pitting at air line or when allowed to dry.
†† May attack when hydrochloric acid is present.

Table F.1   Corrosion resistance data (cont.)

Substance and condition	Temperature, deg F	Type 304–S/S	Type 316–S/S	Other alloys	Carbon steel
Quinine bisulphate (dry)		B	A		E
Quinine sulphate (dry)		A	A		E
Rosin	molten	A	A		E
Sea water		A*	A*	Monel	E
Sewage		A†	A†		
Silver bromide		B*	A*		E
Silver chloride		E	E	Carpenter 20	E
Silver nitrate		A	A		
Soap	70	A	A		A
Sodium acetate (moist)		A*	A		E
Sodium bicarbonate					
all concentrations	70	A	A		A
5 per cent still	150	A	A		A
Sodium bisulphate					
solution	70	A	A		E
saturated solution	70	E			E
$2_g + 1_g H_2SO_4$ per liter	68		A		E
	212		B		E
Sodium carbonate					
5 per cent	70	A	A		A
	150	A	A		A
Sodium chloride					
5 per cent still	70	A*	A		E
	150	A*	A		E
20 per cent aerated	70	A*	A		E
saturated	70	A*	A		E
saturated	boiling	B*	A		E
Sodium fluoride (5 per cent solution)		B*	A*		E
Sodium hydroxide		A	A		E
Sodium hypochlorite (5 per cent still)		B*	A*		E
Sodium hyposulphite	70	A†	A		E
Sodium nitrate	fused	C	B		
Sodium sulphate (5 per cent still)	70	A	A		A
Sodium sulphate (all concentrations)	70	A	A		A
Sodium sulphide (saturated)		B*	A		E
Sodium sulphite					
5 per cent	70	A	A		E
10 per cent	150	A	A		E

† May attack when sulphuric acid is present.

Table F.1 Corrosion resistance data (cont.)

Substance and condition	Temperature, deg F	Type 304–S/S	Type 316–S/S	Other alloys	Carbon steel
Sodium thiosulphate					
saturated solution	70	A	A†		E
acid fixing bath (hypo)	70		A		E
25 per cent solution	70		A†		E
	boiling		A†		E
Stannic chloride (sp. gr., 1.21)	boiling	E	E	Hastelloy	E
Stannic chloride solution	70	D	C	Hastelloy	E
Stannous chloride (saturated)		C	A	Carpenter 20	E
Stearic acid		A	A		E
Sugar juice		A	A		E
Sulphur					
dry	molten	A	A		E
wet		B*	A*		E
Sulphur chloride		E	D		E
Sulphur dioxide gas					
moist	70	B	A		E
gas	575	A	A		
Sulphuric acid					
5 per cent	70	C	B	Carpenter 20	E
	boiling	E	C	Carpenter 20	E
10 per cent	70	C	B	Carpenter 20	E
	boiling	E	D	Carpenter 20	E
50 per cent	70	D	C	Carpenter 20	E
	boiling	E	D	Duriron	E
concentrated	70	A	A		A
	boiling	D	D	Duriron	E
	300	E	E		E
fuming	70	C	B	Carpenter 20	E
Sulphurous acid					
saturated	70	C	B	Carpenter 20	E
saturated (60-lb pressure)	250	C	B	Carpenter 20	E
saturated (70 to 125-lb pressure)	310	C	B	Carpenter 20	E
saturated (150-lb pressure)	375	C	B	Carpenter 20	E
Sulphurous spray	70	D*	D*		E
Tannic acid	70	A	A		E
	150	B	A		E
Tartaric acid	70	A	A		E
	150	B	A		E

Table F.1   Corrosion resistance data (cont.)

Substance and condition	Temperature, deg F	Type 304–S/S	Type 316–S/S	Other alloys	Carbon steel
Tin	molten	C	C		E
Trichloracetic acid	70	E	E		
Trichlorethylene	70	A*			A
Varnish	70	A	A		E
	hot	A	A		E
Vegetable juices		A	A		E
Vinegar fumes		B	A		E
Vinegar					
still or aerated	70	A	A		E
agitated		A	A		E
Zinc	molten	E	E		
Zinc chloride (5 per cent					
still)	70	A*	A*		E
	boiling	B*	B*		E
Zinc sulphate					
5 per cent	70	A	A		E
saturated	70	A	A		E
25 per cent	boiling	A	A		E

* Subject to pitting at air line or when allowed to dry.
† May attack when sulphuric acid is present.

Metric (SI) Units, English and Metric Conversions

Metric notation and terminology employs a base term, preceded by a prefix, which indicates a multiple of that base term. For example, mass is measured in grams. To indicte 1000 grams, the prefix "kilo" is placed in front of gram, and thus, we obtain kilogram. The multiples of the base are always expressed as powers of ten, as follows:

micro	$1/1,000,000$
milli	$1/1,000$
centi	$1/100$
deci	$1/10$
deca	10
hecta	100
kilo	1,000
mega	1,000,000

An example of pressure vessel thickness calculations is demonstrated for both English and SI metric units as follows:

English Units	Metric Units
Pressure $= P = 250$ psi	1723.7 kPa
Diameter $= D = 24$ in	609.6 mm
Radius $= R = 12$ in	304.8 mm
Allowable stress $= S = 13,800$ psi	95,148.2 kPa
Joint efficiency $= E = 1.0$	1.0
Crown radius $= L = 24$ in	609.6 mm

TABLE G.1 **Sample calculations**

English	Metric

Shell thickness

$$T = \frac{P(R)}{SE - 0.6\,P}$$

$$= \frac{250(12)}{13,800(1) - 0.6(250)}$$

$$= 0.220 \text{ in}$$

$$T = \frac{P(R)}{SE - 0.6\,P}$$

$$= \frac{1723.7(304.8)}{95,148.2(1) - 0.6(1723.7)}$$

$$= 5.582 \text{ mm}$$

Ellipsoidal head thickness

$$T = \frac{P(D)}{2SE - 0.2\,P}$$

$$= \frac{250(24)}{2(13,800)(1) - 0.2(250)}$$

$$= 0.218 \text{ in}$$

$$T = \frac{P(D)}{2SE - 0.2\,P}$$

$$= \frac{1723.7(609.6)}{2(95,148.2)(1) - 0.2(1723.7)}$$

$$= 5.532 \text{ mm}$$

Torispherical head thickness

$$T = \frac{.885\,PL}{SE - 0.1\,P}$$

$$= \frac{.885(250)(24)}{13,800(1) - 0.1(250)}$$

$$= 0.385 \text{ in}$$

$$T = \frac{.885\,PL}{SE - 0.1\,P}$$

$$= \frac{.885(1723.7)(609.6)}{95,148(1) - 0.1(1723.7)}$$

$$= 9.791 \text{ mm}$$

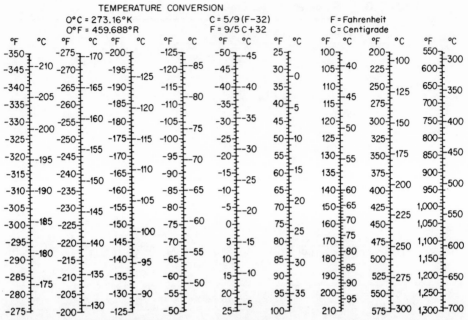

TEMPERATURE CONVERSION

0°C = 273.16°K   C = 5/9 (F−32)   F = Fahrenheit
0°F = 459.688°R   F = 9/5 C+32   C = Centigrade

Fig. G.1 Temperature conversion. F = Fahrenheit and C = Celsius. $C = \frac{5}{9}(F - 32)$; $F = \frac{9}{5}C + 32$; $0°C = 273.16°K$; $0°F = 459.688°R$.

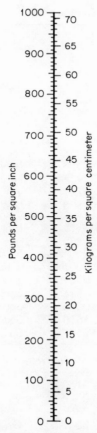

Fig. G.2 English and metric conversion table, pounds per square inch and kilograms per square centimeter.

TABLE G.2 **English and Metric Conversions**

Multiply	By	To obtain
	Length	
Inches	25.4	Millimeters
Inches	0.0254	Meters
Inches	2.540	Centimeters
Feet	304.8	Millimeters
Feet	30.48	Centimeters
Feet	0.3048	Meters
Millimeters	0.03937	Inches
Centimeters	0.3937	Inches
Meters	39.37	Inches
Meters	3.2808	Feet
	Area	
Square inches	645.2	Square millimeters
Square inches	0.0006452	Square meters
Square feet	0.09290	Square meters
Square inches	6.4516	Square centimeters
Square millimeters	0.00155	Square inches
Square centimeters	0.1550	Square inches
Square meters	10.764	Square feet
	Volume and volumetric flow rates	
Cubic Inches	0.01639	Liters
Cubic inches	16390.0	Cubic millimeters
Gallons	3.785	Liters
Liters	0.2642	Gallons
Gallons/minute	0.06308	Liters/second
Gallons	0.003785	Cubic meters
Gallons	3785000.0	Cubic millimeters
Cubic feet	0.02832	Cubic meters
Cubic feet	28.32	Liters
Liters	0.03531	Cubic feet
Liters	61.02	Cubic inches
Cubic meters	264.2	Gallons
Cubic millimeters	0.00006102	Cubic inches
Cubic meters	35.31	Cubic feet
Liters/second	15.8529	Gallons/minute
	Weight or mass	
Pounds	0.4536	Kilograms
Pounds	453.6	Grams
Kilograms	2.2046	Pounds
Grams	0.0022046	Pounds

Multiply	By	To obtain
Force		
Pound force	4.448	Newtons
Newtons	0.2248	Pound force
Bending (torque)		
Pound-foot	1.3558	Newton-meters
Pound-inch	0.11298	Newton-meters
Newton-meters	0.73757	Pound-foot
Newton-meters	8.8511	Pound-foot
Pressure, stress		
Pound/square inch (psi)	6894.757	Pascals
Kips/square inch	6894757.	Pascals
Pound/square inch	6.8948	Kilopascals
Kips/square inch	6.8948	Megapascals
Kilopascals	0.14504	Pounds/square inch
Kilogram/square centimeter	14.22	Pounds/square inch
Pounds/square inch	0.0703	Kilograms/square inch
Pound/square foot	47.8803	Pascals
Kilogram/square meter	9.8067	Pascals
Fracture toughness		
Thousand pounds-$\sqrt{\text{inch}}$	1.0988×10^6	Pascal-$\sqrt{\text{meter}}$

Bibliography

ASME *Boiler and Pressure Vessel Code*. The American Society of Mechanical Engineers, New York, 1980.

"American Society of Mechanical Engineers Boiler Code," *The Locomotive*, Hartford Steam Boiler Inspection and Insurance Co., Hartford, Conn., April 1915.

Eisenberg, G. M., "Elements of Joint Design," Transactions of the ASME–Series J, *Journal of Pressure Vessel Technology*, February 1975 and February 1977.

Garvin, W. L., "Testing, Maintenance, and Installation of Safety Valves and Safety Relief Valves," *National Board Bulletin*, Columbus, July 1971.

Gillissie, John G., *The Authorized Inspector and the Manufacturer's Quality Control System*, The National Board of Boiler and Pressure Vessel Inspectors, Columbus, 1975.

Greene, Arthur M., Jr., *History of the ASME Boiler Code*, The American Society of Mechanical Engineers, New York, 1955.

Harrison, S. F., "Inspector's Responsibility," *National Board Bulletin*, Columbus, October 1969.

Industrial Radiography, Gevaert, Teterboro, N.J., 1965.

Introductory Welding Metallurgy, The American Welding Society, Miami, 1968.

The National Board Inspection Code, The National Board of Boiler and Pressure Vessel Inspectors, Columbus, 1981.

"The National Board of Boiler and Pressure Vessel Inspectors Organizes," *Power*, New York, Feb. 15, 1921.

Panel discussion by Canadian Inspectors, *Proceedings of the Twenty-Third General Meeting of the National Board of Boiler and Pressure Vessel Inspectors*, Columbus, May 1954.

Recommended Practice No. SNT-TC-1A. American Society for Nondestructive Testing, Columbus, 1980.

"Set Screw Failure," *The Locomotive*, Hartford Steam Boiler Inspection and Insurance Co., Hartford, Conn., April 1959.

"Tabulation of the Boiler and Pressure Vessel Laws of the United States and Canada," *Data Sheet*, Uniform Boiler and Pressure Vessel Laws Society, Hartford, Conn., June 1982.

Welding Handbook, 7th ed., American Welding Society, Miami, 1982.

Index

ABOUT THE AUTHORS

Robert Chuse graduated from the Pennsylvania State
Nautical School and served as an engineering officer in the
Merchant Marine.

He is a qualified inspection specialist of nuclear power
plant components as well as an Authorized Inspector of
boilers and pressure vessels with over forty years experience
in quality assurance and quality control and design relating
to the *ASME Boiler and Pressure Vessel Code*.

He formerly served as consultant to the National Board of
Boiler and Pressure Vessel Inspectors and as a member of
the American Society of Mechanical Engineers survey teams.

During his forty-three year career, Mr. Chuse has
published more than forty-five articles on pressure vessel
design, fabrication, and quality control in engineering and
industrial journals. His articles and books are widely used by
pressure vessel fabricators, designers, and users.

He has given lectures on quality control, welding
procedure qualification, and pressure vessel design and
fabrication for the American Welding Society and utility and
manufacturing engineering departments.

Mr. Chuse is a member of the American Society of
Mechanical Engineers and the American Welding Society.

Stephen Eber graduated from Rensselaer Polytechnic
Institute in 1970. He obtained his M.S.M.E. Degree from
Manhattan College in 1980. Mr. Eber is a registered
professional engineer in the State of New York.

Mr. Eber's involvement with the ASME Codes began with
a position as an engineer at the American Society of
Mechanical Engineers, Division of Pressure Vessel
Technology. Following this, he joined the engineering staff
of Jacoby-Tarbox Corp., where his responsibilities included
establishing and implementing a quality assurance program
per Section III of the ASME Code. He was also responsible
for product development in conjunction with the
requirements of the Code.

Mr. Eber has also worked on the engineering staff of the
American Electric Power Service Corporation. He is
currently employed by Ebasco Services Inc. as a performance
engineer, where he utilizes the ASME Performance Test
Codes regularly in order to write specialized computer
software useful for power plant component testing and
monitoring.